1548473-1
7/23/03

Neurobiology of Aggression

Contemporary Neuroscience

Neurobiology of Aggression: Understanding and Preventing Violence, edited by **Mark P. Mattson**, 2003

Neuroinflammation: Mechanisms and Management, Second Edition, edited by **Paul L. Wood**, 2003

Neural Stem Cells for Brain and Spinal Cord Repair, edited by **Tanja Zigova, Evan Y. Snyder, and Paul R. Sanberg**, 2002

Neurotransmitter Transporters: Structure, Function, and Regulation, 2/e, edited by **Maarten E. A. Reith**, 2002

The Neuronal Environment: Brain Homeostasis in Health and Disease, edited by **Wolfgang Walz**, 2002

Pathogenesis of Neurodegenerative Disorders, edited by **Mark P. Mattson**, 2001

Stem Cells and CNS Development, edited by **Mahendra S. Rao**, 2001

Neurobiology of Spinal Cord Injury, edited by **Robert G. Kalb and Stephen M. Strittmatter**, 2000

Cerebral Signal Transduction: From First to Fourth Messengers, edited by **Maarten E. A. Reith**, 2000

Central Nervous System Diseases: Innovative Animal Models from Lab to Clinic, edited by **Dwaine F. Emerich, Reginald L. Dean, III, and Paul R. Sanberg**, 2000

Mitochondrial Inhibitors and Neurodegenerative Disorders, edited by **Paul R. Sanberg, Hitoo Nishino, and Cesario V. Borlongan**, 2000

Cerebral Ischemia: Molecular and Cellular Pathophysiology, edited by **Wolfgang Walz**, 1999

Cell Transplantation for Neurological Disorders, edited by **Thomas B. Freeman and Håkan Widner**, 1998

Gene Therapy for Neurological Disorders and Brain Tumors, edited by **E. Antonio Chiocca and Xandra O. Breakefield**, 1998

Highly Selective Neurotoxins: Basic and Clinical Applications, edited by **Richard M. Kostrzewa**, 1998

Neuroinflammation: Mechanisms and Management, edited by **Paul L. Wood**, 1998

Neuroprotective Signal Transduction, edited by **Mark P. Mattson**, 1998

Clinical Pharmacology of Cerebral Ischemia, edited by **Gert J. Ter Horst and Jakob Korf**, 1997

Molecular Mechanisms of Dementia, edited by **Wilma Wasco and Rudolph E. Tanzi**, 1997

Neurotransmitter Transporters: Structure, Function, and Regulation, edited by **Maarten E. A. Reith**, 1997

Motor Activity and Movement Disorders: Research Issues and Applications, edited by **Paul R. Sanberg, Klaus-Peter Ossenkopp, and Martin Kavaliers**, 1996

Neurotherapeutics: Emerging Strategies, edited by **Linda M. Pullan and Jitendra Patel**, 1996

Neuron–Glia Interrelations During Phylogeny: II. Plasticity and Regeneration, edited by **Antonia Vernadakis and Betty I. Roots**, 1995

Neuron–Glia Interrelations During Phylogeny: I. Phylogeny and Ontogeny of Glial Cells, edited by **Antonia Vernadakis and Betty I. Roots**, 1995

The Biology of Neuropeptide Y and Related Peptides, edited by **William F. Colmers and Claes Wahlestedt**, 1993

Psychoactive Drugs: Tolerance and Sensitization, edited by **A. J. Goudie and M. W. Emmett-Oglesby**, 1989

Experimental Psychopharmacology, edited by **Andrew J. Greenshaw and Colin T. Dourish**, 1987

Neurobiology of Aggression

Understanding and Preventing Violence

Edited by

Mark P. Mattson, PhD

National Institute on Aging, Baltimore, MD

Humana Press ✹ Totowa, New Jersey

© 2003 Humana Press Inc.
999 Riverview Drive, Suite 208
Totowa, New Jersey 07512

www.humanapress.com

All rights reserved.

No part of this book may be reproduced, stored in a retrieval system, or transmitted in any form or by any means, electronic, mechanical, photocopying, microfilming, recording, or otherwise without written permission from the Publisher.

All papers, comments, opinions, conclusions, or recommendations are those of the author(s), and do not necessarily reflect the views of the publisher.

For additional copies, pricing for bulk purchases, and/or information about other Humana titles, contact Humana at the above address or at any of the following numbers: Tel.: 973-256-1699; Fax: 973-256-8341; E-mail: humana@humanapr.com, or visit our Website: www.humanapress.com

This publication is printed on acid-free paper. ∞
ANSI Z39.48-1984 (American Standards Institute) Permanence of Paper for Printed Library Materials.

Brain illustration provided by Mark P. Mattson.

Cover design by Patricia F. Cleary.

Photocopy Authorization Policy:
Authorization to photocopy items for internal or personal use, or the internal or personal use of specific clients, is granted by Humana Press Inc., provided that the base fee of US $20 is paid directly to the Copyright Clearance Center at 222 Rosewood Drive, Danvers, MA 01923. For those organizations that have been granted a photocopy license from the CCC, a separate system of payment has been arranged and is acceptable to Humana Press Inc. The fee code for users of the Transactional Reporting Service is: [1-58829-188-X/03 $20].

Printed in the United States of America. 10 9 8 7 6 5 4 3 2 1

Library of Congress Cataloging-in-Publication Data

Neurobiology of aggression: understanding and preventing violence / edited by Mark P. Mattson.
        p. cm.
  Includes bibliographical references and index.
  ISBN 1-58829-188-X (hb); 1-59259-382-8 (e-book)
  1. Aggressiveness—Physiological aspects. 2. Aggressiveness—Social aspects. 3. Aggressiveness—Treatment. 4. Violence—Prevention. 5. Violence—Physiological aspects.
  I. Mattson, Mark Paul.
  RC569.5.A342 N486 2003
  616.85'–dc21
                                                                                       2002032942

Preface

Aggression is a highly conserved behavioral adaptation that evolved to help organisms compete for limited resources and thereby ensure their survival. However, in modern societies where resources such as food, shelter, etc. are not limiting, aggression has become a major cultural problem worldwide presumably because of its deep seeded roots in the neuronal circuits and neurochemical pathways of the human brain. In *Neurobiology of Aggression: Understanding and Preventing Violence*, leading experts in the fields of the neurobiology, neurochemistry, genetics, and behavioral and cultural aspects of aggression and violence provide a comprehensive collection of review articles on one of the most important cross-disciplinary issues of our time. Rather than summarize the topics covered by each author in each chapter, I present a schematic diagram to guide the reader in thinking about different aspects of aggressive and violent behavior from its neurobiological roots to environmental factors that can either promote or prevent aggression to visions of some of the most horrific acts of violence of our times, and then towards the development of strategies to reduce aggressive behavior and prevent violence.

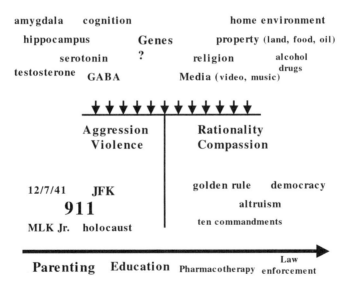

It is hoped that *Neurobiology of Aggression: Understanding and Preventing Violence* will foster further research aimed at understanding the environmental genetic and neurochemical roots of aggression and how such information can be used to move forward towards the goal of eliminating violence.

Mark P. Mattson, PhD

Contents

Preface .... *v*
Contributors .... *ix*

1 Cortical and Limbic Neural Circuits Mediating Aggressive Behavior
   Thomas R. Gregg 1

2 Emotion Regulation: *An Affective Neuroscience Approach*
   R. James R. Blair and Dennis S. Charney 21

3 The Serotonergic Dimension of Aggression and Violence
   Klaus Peter Lesch 33

4 The Neurochemical Genetics of Serotonin in Aggression,
   Impulsivity, and Suicide
   Mark D. Underwood and J. John Mann 65

5 Behavioral and Neuropharmacological Differentiation
   of Offensive and Defensive Aggression in Experimental
   and Seminaturalistic Models
   Philip M. Wall, D. Caroline Blanchard, and Robert J. Blanchard 73

6 Neuroendocrine Stress Responses and Aggression
   Jozsef Haller and Menno R. Kruk 93

7 Y Chromosome and Antisocial Behavior
   Pierre L. Roubertoux and Michèle Carlier 119

8 Aggression in Psychiatric Disorders
   Shari R. Kohn and Gregory M. Asnis 135

9 Aggression in Brain Injury, Aging, and Neurodegenerative
   Disorders
   Mark P. Mattson 151

10 Environmental Factors and Aggression in Nonhuman Primates
    Michael Lawrence Wilson 167

11 Aggression, Biology, and Context: *Dejà-Vù All Over Again?*
    Rebecca M. Young and Evan Balaban 191

12 The Family Environment in Early Life and Aggressive Behavior
    in Adolescents and Young Adults
    Sven Barnow and Harald-J. Freyberger 213

13 Television and Movies, Rock Music and Music Videos,
    and Computer and Video Games: *Understanding and
    Preventing Learned Violence in the Information Age*
    Susan Villani and Nandita Joshi 231

14 Social Drinking and Aggression
    Kathryn Graham 253

15 Cognitive-Behavioral Intervention for Childhood Aggressions
    Cynthia Hudley 275

16 Pharmacological Intervention in Aggression
    Debra V. McQuade, R. Joffree Barrnett, and Bryan H. King 289

    Index 315

Contributors

GREGORY M. ASNIS, MD • *Department of Psychiatry, Albert Einstein College of Medicine, Montefiore Medical Center, Bronx, NY*

EVAN BALABAN • *Neurosciences Program, CUNY CSI, Staten Island, NY*

R. JOFFREE BARRNETT • *Dartmouth Medical School, Hanover, NH*

SVEN BARNOW, PhD • *Department of Psychiatry and Psychotherapy of the Ernst-Moritz-Arndt University Greifswald, Medical Centre of Stralsund, Germany*

R. JAMES R. BLAIR • *Mood and Anxiety Disorders Program, National Institute of Mental Health, Bethesda, MD*

D. CAROLINE BLANCHARD, PhD • *Pacific Biomedical Research Center, and Division of Neuroscience, John A. Burns School of Medicine, University of Hawaii, Honolulu, HI*

ROBERT J. BLANCHARD, PhD • *Department of Psychology and Pacific Biomedical Research Center, University of Hawaii, Honolulu, HI*

MICHÈLE CARLIER • *Centre de recherche PsyCLÉ (Connaissance, Langage, Émotion), Université de Provence, France*

DENNIS S. CHARNEY • *Mood and Anxiety Disorders Program, National Institute of Mental Health, Bethesda, MD*

HARALD-J. FREYBERGER • *Department of Psychiatry and Psychotherapy of the Ernst-Moritz-Arndt University Greifswald, Medical Centre of Stralsund, Germany*

KATHRYN GRAHAM, PhD • *Senior Scientist and Head, Social Factors and Prevention Initiatives, Social, Prevention and Health Policy Research Department, Centre for Addiction and Mental Health, London, Ontario, Canada*

THOMAS R. GREGG, MA • *Department of Neurosciences, University of Medicine and Dentistry, Newark, NJ*

JOZSEF HALLER, PhD • *Institute of Experimental Medicine, Budapest, Hungary*

CYNTHIA HUDLEY • *Rossier School of Education, University of Southern California, Los Angeles, CA*

NANDITA JOSHI, MBBS • *Department of Child Psychiatry, Johns Hopkins University, Baltimore, MD*

BRYAN H. KING • *Dartmouth Medical School, Hanover, NH*

SHARI R. KOHN, PhD • *Department of Psychiatry, Albert Einstein College of Medicine, Montefiore Medical Center, Bronx, NY*

MENNO R. KRUK • *Department of Medical Pharmacology, Leiden University, The Netherlands*

KLAUS PETER LESCH, MD • *Department of Psychiatry and Psychotherapy, University of Würzburg, Würzberg, Germany*

J. JOHN MANN, MD • *Department of Neuroscience, New York State Psychiatric Institute, and Department of Psychiatry, Columbia University College of Physicians and Surgeons, New York, NY*

MARK P. MATTSON, PhD • *Laboratory of Neurosciences, Gerontology Research Center, National Institute on Aging, Baltimore, MD*
DEBRA V. MCQUADE • *Dartmouth Medical School, Hanover, NH*
PIERRE L. ROUBERTOUX • *Institut de Neurosciences Physiologiques et Cognitives, INPC.CNRS, Marseille, and Université d'Orléans, Orléans, France*
MARK D. UNDERWOOD, PhD • *Department of Neuroscience, New York State Psychiatric Institute, and Department of Psychiatry, Columbia University College of Physicians and Surgeons, New York, NY*
SUSAN VILLANI, MD • *Medical Director, Kennedy Krieger School Programs, Kennedy Krieger Institute, and Assistant Professor of Psychiatry, Johns Hopkins School of Medicine, Baltimore, MD*
PHILIP M. WALL, PhD • *Department of Psychology, University of Hawaii, Honolulu, HI*
MICHAEL LAWRENCE WILSON, PhD • *The Jane Goodall Institute's Center for Primate Studies, Department of Ecology, Evolution and Behavior, University of Minnesota, St. Paul, MN*
REBECCA M. YOUNG • *Barnard College, and NDRI, Inc., New York, NY*

1
Cortical and Limbic Neural Circuits Mediating Aggressive Behavior

Thomas R. Gregg

INTRODUCTION

When a cat is cornered by an attacker, and escape is impossible, the cat displays an unambiguous set of behaviors: the fur stands on end, the pupils dilate, the back arches, the claws are unsheathed, the ears are laid back, the heart rate increases, and the cat may yowl, growl, or bare its teeth and hiss. Finally, if the aggressor persists in spite of these warning signs, the cat will strike to defend itself and fight furiously. This intense defensive–aggression response has been called "affective defense," "defensive rage behavior," or simply "rage," and a similar response occurs in other species. Many brain areas can influence defensive rage (Goddard, 1964; Kaada, 1967; Clemente and Chase, 1973; Ursin, 1981; Albert et al., 1984, 1993; Siegel and Brutus, 1990; Gregg and Siegel, 2001). The available evidence suggests that in humans and animals, similar brain areas are involved (reviewed in Perachio and Alexander, 1975; Lipp and Hunsperger, 1978; Trieman, 1991; Albert et al., 1993; Gregg and Siegel, 2001). This chapter will focus on seven of these areas: the periaqueductal gray of the midbrain (PAG), hypothalamus, septal nuclei, amygdala, prefrontal cortex, bed nucleus of the stria terminalis (BNST), and nucleus accumbens (Fig. 1).

THE PERIAQUEDUCTAL GRAY OF THE MIDBRAIN

Although a few contrary results have been found, most data suggest that the dorsal, rostral PAG is the organizing center for the orchestrated expression of all the behavioral components of the defensive rage response (Masserman, 1941; Bailey and Davis, 1942; Davis and Bailey, 1944; Hunsperger, 1956; Glusman et al., 1962; Fenandez De Molina and Hunsperger, 1962; Skultety, 1963; Woods, 1964; Kaada, 1967; Flynn, 1967; Ellison, 1967; Ellison and Flynn, 1968; Adams, 1968; Blanchard et al., 1981; Schubert et al., 1996; Gregg and Siegel, 2001.) The contrary results were reported by Bazett and Penfield (1922), Keller (1932); Bard and Mountcastle (1948), and Kaada (1967) (*see also* the reviews in Rothfield and Harman, 1954; and Siegel et al., 1999).

Anatomy and Behavior

The PAG is a long, roughly cylindrical structure that surrounds the cerebral aqueduct. In cats, activation of this structure by electrical or chemical stimulation produces a defensive rage response (Siegel et al., 1999; Gregg and Siegel, 2001). The PAG has also been

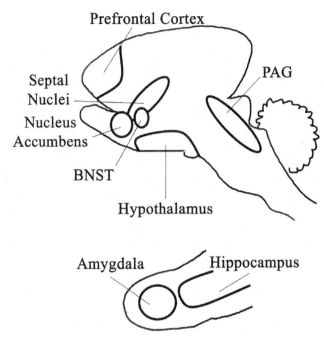

Fig. 1. Diagram of mammalian brain indicates relative positions of nuclei important for defensive rage. Lower part of diagram indicates structures located in temporal lobe. Abbreviations: BNST, bed nucleus of the stria terminalis; PAG, periaqueductal gray of the midbrain.

implicated in pain processing and antinociception, vocalization, lordosis (i.e., the copulatory behavior of female rats), autonomic regulation, cardiovascular regulation, fear, aversion, and flight–escape behavior (reviewed in Behbehani, 1995; Bandler et al., 2000; Dielenberg et al., 2001; Omori et al., 2001). The ventral PAG may play some role in predatory and fear-related behavior.

The dorsal PAG receives projections from the hypothalamus, the central and basal nuclei of the amygdala, the prefrontal cortex, and the septal nuclei, among other areas. The ventral PAG receives projections from the central nucleus of the amygdala. It should be noted that the area referred to by some authors as the "lateral" PAG corresponds to the dorsal PAG.

Descending Projections from the PAG
Are Responsible for the Expression of Defensive Rage

Neurons in the PAG project to brainstem areas that are responsible for the different components of the defensive rage response (Fig. 2). I will refer to these brainstem areas as "effector" areas. PAG neurons project to locus ceruleus (responsible for sympathetic arousal), the central tegmental fields, the motor nucleus of the trigeminal nerve [responsible for jaw-opening, a component of hissing (Siegel and Brutus, 1990)], the mesencephalic locomotor region (Brudzynski and Wang, 1996), motor neurons in the spinal cord (Vanderhorst et al., 2000, Mouton and Holstege, 2000), and neurons responsible for vocalization, exhalation (reviewed in Vanderhorst et al., 2000), and tongue movements (hypoglossal nucleus; Jurgens, 1994), which may be necessary for the expression of the

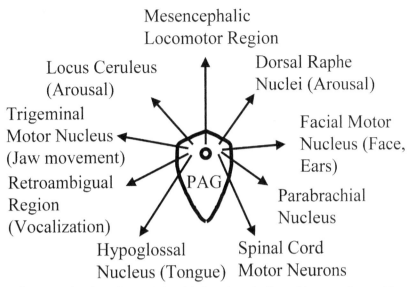

Fig. 2. Efferent projections from PAG, with functions indicated in parentheses. These projections originate in the dorsal, rostral PAG.

hiss. PAG neurons also project, via a disynaptic projection, to neurons responsible for ear movements (possibly for the laying-back of the ears in rage; see below). A descending projection may also be the basis for the pawstrike component of rage (Edwards and Flynn, 1972; Mouton and Holstege, 2000; Li and Mitchell, 2000). The ascending projections from the PAG are of unknown function.

PAG and Vocalization

PAG activation causes animals to vocalize. The pathway responsible for this effect descends from the PAG to the nucleus ambiguus and retroambigual region (RAb), giving off fibers to the pontine reticular formation, parabrachial nucleus (PBN), and trigeminal motor nucleus as it descends (Berntson 1972; Jurgens, 1998; *see also* Ennis et al., 1997; Luthe et al., 2000). RAb projects to spinal motor neurons, which are likely involved in vocalization (Jurgens, 1998), to the dorsomedial facial subnucleus, which is involved in ear movements (Holstege, 1989), and to muscles probably involved in lordosis (Vanderhorst et al., 2000). It is interesting to note that even fish make noises, using the swimbladder, during aggressive behavior (Amorim and Hawkins, 2000).

Because PAG stimulation elicits species-specific vocalizations in many species (Holstege, 1989), and laughter is likely to be a species-specific vocalization (Jurgens, 1998), which can be evoked by stimulation of the PAG in humans (reviewed in Jurgens, 1994), it may be speculated that damage to the PAG in amyotrophic lateral sclerosis in humans (Kobayashi et al., 1999) causes the pathological laughter that is sometimes expressed in this disorder. It may also be speculated that marijuana use influences laughter, aggression, and other types of emotional expression in humans by a site of action in the PAG or limbic areas (*see* Lester and Dreher, 1989; Cherek and Dougherty, 1995; Arnold et al., 2001; Gregg and Siegel, 2001; and Budney et al., 2001 for some suggestive data).

Localization of Emotional-Expression Functions within the PAG

An influential theory states that an immediate threat, such as an attack by a predator, will activate the dorsal PAG, resulting in escape and defensive aggression, as well as nonopioid analgesia, while less immediate threats activate the amygdala, which in turn commands the ventral PAG to produce alert freezing behavior and opioid analgesia (Fanselow, 1991). Fanselow also proposed the existence of inhibitory projections from dorsal PAG to ventral PAG that suppress freezing when a vigorous defense is required. This idea has recently gained support based on experimental data (Walker et al., 1997; Bandler et al., 2000).

Conclusion

The PAG is crucially important for the expression of defensive rage behavior. This behavior occurs when the PAG sends commands, through descending projections, to effector regions in the brainstem, and the effector regions send commands to the muscles and glands, which produce the components of defensive rage (e.g., pupillary dilation, increased heart rate, vocalization, pawstrike). Activation of the medial hypothalamus causes defensive rage via its projections to the PAG.

Medial Hypothalamus

The hypothalamus is involved in reproduction, ingestion, homeostasis, and sleep (Swanson, 1987) and is second in importance only to the PAG for the expression of defensive rage. The medial hypothalamus receives projections from widespread areas of the limbic system and brainstem and projects to the PAG, among other areas (Swanson, 1987). Electrical stimulation of many sites throughout the rostrocaudal extent of the medial hypothalamus gives rise to rage behavior. It can be estimated that electrodes eliciting rage have been located in the ventromedial hypothalamus (VMH), medial preoptic area or medial preoptic nucleus, dorsomedial hypothalamic nucleus, and anterior hypothalamic nucleus, as well as the paraventricular nucleus (Siegel et al., 1999). Unfortunately, nobody has systematically studied the biochemical and morphological characteristics of the neurons responsible for defensive rage.

Anterior Hypothalamic Area

The importance of the anterior hypothalamic area (AHA) in the cat was shown by Fuchs et al., in 1985. In the first stage of this experiment, it was confirmed that electrical stimulation of neurons in the VMH produced defensive rage. In the second stage, it was discovered that these neurons had few, if any, projections to the PAG. This was unexpected, because it had been assumed that stimulation of the VMH gives rise to aggression by means of direct projections to the PAG. Fuchs et al. (1985) then showed that the neurons located at VMH rage sites projected to AHA and that neurons located at AHA rage sites projected to the PAG. This finding gave rise to the hypothesis that stimulation of the VMH produces defensive rage via a two-limbed projection, which first ascends to the AHA and then, from AHA, descends to the PAG. Fuchs et al. (1985) provided evidence supporting this hypothesis, by showing that AHA lesions blocked rage elicited by VMH stimulation (reviewed by Siegel and Brutus, 1990). It is not clear which subsets of neurons within the AHA and VMH are involved.

VMH

VMH is an unusual structure. Paradoxically, aggression can be produced either by electrical stimulation of VMH or by destruction or temporary inactivation of this structure (Kaada, 1967; Albert et al., 1985). There has not yet been a satisfactory explanation for this finding (Glusman et al., 1962; Kaada, 1967; Nishiyama et al., 2001). The effect of VMH inactivation is apparently mediated through the PAG, since a PAG lesion abolishes the VMH lesion-induced aggression for 4–6 wk (Glusman et al., 1962). An unanswered question is whether an AHA lesion would abolish it, too. Lu et al. (1992), in an immunohistochemical study, found many glutamate-containing cells in the cat hypothalamus, but only rarely were they found in VMH. Adding to the puzzle, stria terminalis activation inhibits VMH neurons (McBride and Sutin, 1975), but activation of projections from the amygdala to VMH facilitates aggression (Stoddard-Apter and MacDonnell, 1980; see below for a discussion of the relationship of the stria terminalis to the amygdala). Projections from VMH to PAG play a role in lordosis in the rat (Chung et al., 1990; Tsukahara and Yamanouchi, 2001).

Other Nuclei in the Medial Hypothalamus

The dorsomedial nucleus and premammillary nuclei may be important in aggression (Olivier et al., 1983; Bernardis and Bellinger, 1987; Yokosuka and Hayashi, 1996; Comoli et al., 2000; Dielenberg et al., 2001).

NEUROTRANSMITTERS WITHIN THE HYPOTHALAMUS-PAG CIRCUIT

Hypotheses stating that certain neurotransmitters have a unitary function in the brain should be examined skeptically. One example of a neurotransmitter proposed to have a function in aggression is nitric oxide, but, recently, this has been called into question (Le Roy et al., 2000). The roles of several neurotransmitters in the neural circuits for aggression have been reviewed elsewhere (Siegel et al., 1999; Gregg and Siegel, 2001).

Neurokinin (NK)-1 receptors in the hypothalamus and PAG appear to exert a facilitatory influence on aggression (Gregg and Siegel, 2001). In cats, injections of an NK-1 receptor agonist into these areas also produce a behavioral syndrome that appears similar to anxiety (observations of the author). Oddly enough, substance P injection into the dorsal PAG also facilitates lordosis (Behbehani, 1995). Defensive responses can be elicited in the cat by injecting the cholinergic agonist carbachol into the hypothalamus (reviewed by Siegel et al., 1999). Recently, Brudzynski et al. (1998) have postulated that cholinergic projections from the laterodorsal tegmental nucleus of the midbrain to the PAG, hypothalamus, and septum facilitate defensive responses. Serotonin in the dorsal raphe nuclei may be involved in aggression (Mos et al., 1993; Koprowska and Romaniuk, 1997).

LIMBIC AND CORTICAL STRUCTURES MODULATE THE INTENSITY OF AGGRESSION

Limbic and cortical areas are also involved in defensive rage in cats, but their projections to the brainstem effector regions of emotional expression are apparently not strong enough to produce rage behavior independently under normal circumstances (reviewed by Siegel and Brutus, 1990). Instead, these areas modify the propensity of the hypothala-

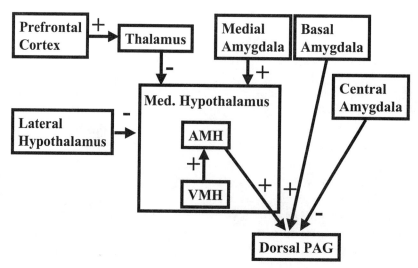

Fig. 3. Summary of best-characterized functional anatomical connections relevant for defensive rage behavior. Abbreviations: +, facilitates defensive rage; –, suppresses defensive rage; AMH, anterior medial hypothalamus; VMH, ventromedial hypothalamus; PAG, periaqueductal gray of the midbrain. Arrows symbolize axonal projections, and boxes symbolize brain nuclei.

mus and PAG to produce aggression (Gregg and Siegel, 2001). Such areas are termed *modulatory* structures (Fig. 3).

Modulatory Structures: Septal Nuclei

Anatomy and Physiology

The septal nuclei are a pair of midline structures in the cerebrum, just dorsal and rostral to the anterior hypothalamus. The major connections of the septal nuclei are with the hippocampus (via the large, compact fiber bundle, the fornix), hypothalamus (via the medial forebrain bundle) (Risold and Swanson, 1997), brainstem, thalamus, and cerebral cortex. There are probably GABAergic inhibitory projections from the septum to the anterior hypothalamus (Risold and Swanson, 1997). The septal area, along with the nearby diagonal band of Broca and nucleus basalis of Meynert–substantia innominata, contains cholinergic cells that project to widespread areas in the forebrain. The septal nuclei are just dorsal, medial, and caudal to the nucleus accumbens and ventral pallidum, which are structures that support "rewarding" electrical self-stimulation behavior. The medial septal nucleus plays an important role in modulating the activity of hippocampal neurons.

The Septal Nuclei are Involved in Several Behavioral Functions

In addition to aggression, the septal nuclei appear to influence water intake (Fried, 1972; Swanson and Lind, 1986; Camargo and Saad, 2001; *but see* Thorne, 1993), lordosis (through projections to PAG; Tsukahara and Yamanouchi, 2001), and maternal behavior (Flannelly et al., 1986). It is interesting that the limbic brain areas that modulate maternal behaviors also modulate defensive aggression (Sheehan et al., 2001).

Lesion Studies

Since 1940, numerous studies have found that lesions of the septal nuclei cause an increase in aggression (Blanchard et al., 1979; Albert and Walsh, 1984; Flannelly et al., 1986; Maeda and Maki, 1987; *but see* Grossman, 1976). However, this increase is usually short-lived, can be influenced by experience and possibly other factors, and is not found in all species (Brady and Nauta, 1954; Phillips and Lieblich, 1972; Poplawsky and Johnson, 1973; Grossman, 1976, 1978; Goodlett et al., 1982; Albert et al., 1985; *see also* Raisman, 1966. One group of researchers concluded that this increase is due to an increase in water intake (Flynn et al., 1979; *but see* Thorne, 1993). Several studies have found that septal lesions reduce dominance behaviors and aggression in some experimental situations (Grossman, 1976, 1978; Blanchard et al., 1977; Flannelly et al., 1986; *but see* Poplawsky and Johnson, 1973; Lau and Miczek, 1977; and Gage et al., 1978.

NEIGHBORING STRUCTURES

Based on these studies, it was concluded that, in rats and cats, lesions involving the septal area enhance aggression, at least temporarily. However, because these lesions were often large, several authors put forth the hypothesis that "septal rage" was due not to lesions of the septal nuclei, but to damage to nearby structures such as the BNST or nucleus accumbens. Experiments undertaken to test this hypothesis demonstrated that lesions of the septal nuclei do indeed increase aggression, but lesions of the nearby BNST and nucleus accumbens also increased aggression (reviewed in Wallace and Thorne, 1978; Albert et al., 1985).

FIBERS OF PASSAGE

Another question, the question of fibers of passage, is more difficult to resolve. Many axons travel from the hippocampus, through the fornix, then through the septal area, to areas such as the nucleus accumbens and hypothalamus (Powell and Hines, 1974; *see* review in Bannerman et al., 2001). Placing lesions in the septal area would inactivate not only the cell bodies of the septal nuclei, but also these fibers of passage; therefore, it is possible that the ensuing rage is actually caused by inactivation of the fibers of passage. However, fornix lesions do not cause rage in rats (Paxinos, 1975), which implies that the inactivation of these fibers of passage is probably not the cause of septal rage. In order to confirm this conclusion, excitotoxic lesions limited to the septal nuclei would have to be made and aggression observed. This has not yet been done (*but see* Munoz and Grossman, 1980; and Johansson et al., 1999).

It can be tentatively concluded that destruction or inactivation of the septal area temporarily enhances some types of aggression in rats and possibly in cats. However, this effect may not occur in other species.

Stimulation Studies

Contrary to expectation, electrical stimulation of some areas in the septal nuclei *enhances* aggression, and stimulation of other areas inhibits it in cats (Stoddard-Apter and MacDonnell, 1980; Siegel and Brutus, 1990). Such results are difficult to interpret, because electrical stimulation of the septal area would excite fibers of passage as well as septal efferents. Also difficult to interpret is the single report from 1963 that cholinergic stimulation of the septal nuclei induced rage in cats (reviewed in Siegel et al., 1999).

Human Relevance

In a few cases in human patients, septal-area tumors produced irritability and enhanced the startle response, temper outbursts, and aggression symptoms closely resembling those of septal-lesioned rats (reviewed in Albert et al., 1985; *see also* Gage et al., 1978; and Melia et al., 1992).

Conclusion

It may be concluded that septal lesions temporarily enhance some types of aggression in at least some species, but the mechanism of this effect is not clear.

Modulatory Structures: Amygdala

Anatomy

The amygdala is located anterior to the hippocampus in the temporal lobe. Three nuclei (the lateral, basal, and accessory basal nuclei) are located in the center of the amygdaloid complex, and several others are located on the edges (including the central nucleus of the amygdala, the medial nucleus, the cortical nuclei, and the anterior amygdaloid area). The basal nucleus is sometimes called the basolateral nucleus, and the accessory basal nucleus is sometimes called the basomedial nucleus (Pitkanen, 2000). Recently, it has been argued that the amygdala is a collection of several structures of different embryonic origins (Aggleton and Saunders, 2000).

The flow of information in the amygdala is from the lateral nucleus to the central, medial, basomedial, and basolateral nuclei. The lateral nucleus is the major input nucleus of the amygdaloid complex. It receives most of its input from cerebral cortex and thalamus and projects to several subnuclei, including the basal and central nuclei. The basal nucleus also projects to the central nucleus. The central nucleus projects to brainstem areas, including the PAG (Behbehani, 1995; Ledoux, 2000). It is thought that the central nucleus "executes the responses evoked by the other amygdaloid nuclei that innervate the central nucleus" (Jolkkonen and Pitkänen, 1998). The cortical amygdala receives projections from the retina in some species of rodents (Cooper et al., 1993).

The stria terminalis and ventral "amygdalofugal" pathway are the two major connection pathways of the amygdala (Richardson, 1973). The ventral amygdalofugal pathway is diffuse; it runs from the amygdala, in a medial direction, coursing over the optic tract, into the hypothalamic region. The stria terminalis, in contrast, is a compact bundle of fibers. Similar to the fornix, it travels posteriorly from the amygdala, then arches dorsally, then travels anterior and medially to the area of the anterior commissure. There it splits, like the fornix, into two components, one anterior (precommissural) and one posterior to the anterior commissure (postcommissural). This split occurs near the septal nuclei, dorsal to the hypothalamus. The precommissural part projects to the BNST, which is a collection of cells located ventrolateral to the septal nuclei, while the postcommissural part distributes to hypothalamic areas, such as VMH (Millhouse, 1973) and the anterior hypothalamus. Both the stria terminalis and ventral amygdalofugal pathway are afferent as well as efferent. The nuclei contributing to the stria terminalis include the medial, basomedial, and basolateral nuclei, and several nuclei contribute to the ventral pathway, including the central nucleus.

Lesion Studies

Forty-seven papers have reported that amygdalar lesions cause taming effects in diverse species (reviewed in Goddard, 1964; Clemente and Chase, 1973; and Albert and Walsh, 1984). The taming effect can be potent, as evidenced by the remarkable comment by Kling and Hutt (1958) that amygdala-lesioned cats "could be swung around in the air by the tail with impunity, and purred and showed playful behavior when petted immediately afterward." However, this anecdote is difficult to reconcile with the reports that decerebrate cats and cats with hypothalamic lesions express aggression in response to nociceptive (painful) stimuli.

A few studies have reported that large amygdalar lesions enhanced aggressivity (Bard and Mountcastle, 1948; *see* the reviews by Ursin, 1965, 1981; and Sweet et al., 1969). It is likely that most of these anomalous results were due to inadvertent damage to the hippocampus during surgery, resulting in seizure activity, which in turn caused aggression (Bard and Mountcastle, 1948; Ursin, 1965). In addition, other studies have shown that small lesions restricted to particular nuclei within the amygdala, especially the central nucleus, enhance aggression (Ursin, 1965; Miczek et al., 1974; Zagrodzka et al., 1998).

Differential Roles of the Central and Medial Nuclei

These results of research involving lesions of the central nucleus of the amygdala led to the hypothesis that the central nucleus acts to inhibit aggression. In support of this hypothesis, Siegel and coworkers found that stimulation of the central nucleus suppresses defensive rage in cats, and stimulation of the medial amygdala enhances aggression due to a substance P projection (reviewed by Gregg and Siegel, 2001). In addition, Vochteloo and Koolhaas (1987) found that a lesion of the medial amygdala reduced aggression in naïve, but not experienced rats.

These data are consistent with the idea that the central and medial nuclei play different roles in the regulation of defensive rage: the central nucleus inhibits it, and the medial nucleus enhances it. However, there is one paradoxical finding: stimulation of the central nucleus causes an *elevation* of heart rate (Iwata et al., 1987; Baklavadzhyan et al., 2000). This finding is unexpected: since stimulation of the central nucleus inhibits rage behavior, it would be expected that such stimulation would also decrease heart rate. The neuroanatomical basis for the reduction in heart rate seems to be a projection from the central nucleus to the PAG. Perhaps the paradoxical finding can be explained by the idea that the central and medial nuclei are differentially involved in fear and aggressivity, similar to the hypothesis of Fanselow (1991). Namely, perhaps activation of the medial amygdala gives rise to aggressivity ("anger"), accompanied by an increase in heart rate (Kovach et al., 2001) and a decrease of freezing behavior, while central nucleus activation gives rise to fear, accompanied by an increase in heart rate and freezing behavior, and suppression of aggressive behavior. The idea that the central nucleus is involved in fear is supported by much evidence (Walker et al., 1997; Ledoux, 2000).

One caveat is as follows. Davis and Whalen (2001) reported that some fibers pass from the basolateral nucleus of the amygdala through the central nucleus to the BNST. Thus, lesions or stimulation of the central nucleus may exert unintended effects on BNST activity.

Stimulation Studies

In early studies, stimulation of the amygdala with high current levels caused rage-like responses. However, these high current levels produced seizure-like motor activity (e.g., Gastaut et al., 1952). Thus, it is likely that these results are not generalizable to nonseizing animals, because they were due to seizures in the amygdala causing abnormally intense activation in the hypothalamus. More refined studies have found that subseizure levels of stimulation do not produce rage, but modulate it (Siegel and Brutus, 1990). At least one study found that injection of acetylcholine directly into the amygdala can elicit defensive rage, but this, too, may have been due to seizure activity (Siegel et al., 1999; Gregg and Siegel, 2001). Electrical stimulation of the stria terminalis as it exits the amygdala can give rise to rage (A. Siegel, personal communication).

Other Functions of the Amygdala

In addition to fear and aggression, the different subnuclei within the amygdala seem to be involved in olfactory processing, appetitive conditioning, conditioned taste aversion, memory (Aggleton and Saunders, 2000), and feeding behavior (Rollins et al., 2001). The amygdala may be an important relay nucleus by which olfactory information influences maternal behavior (Sheehan et al., 2001), and, in fact, the amygdala probably evolved as an olfactory structure. It should be noted in passing that some effects of septal lesions can be blocked by amygdaloid lesions (Blanchard et al., 1979; Melia et al., 1992).

Amygdala and Personality

One series of experiments has strongly suggested that cats with defensive "personalities" have a stronger connection between the amygdala and hypothalamus than do other cats (Adamec, 1991).

Human Behavior

It has been proposed that the amygdala is involved in cognitive and emotional symptoms of autism (Sweeten et al., 2002) and Alzheimer's disease (see below). The idea that abnormalities in amygdalar function can lead to violence is supported by Sweet et al. (1969) (*see* also Mark and Sweet, 1974; and van Elst et al., 2000).

Proposed Hypothesis

A testable hypothesis may be proposed: anger-inducing sensory stimuli cause anger via the substance P projection from the medial amygdala to medial hypothalamus. Kovach et al. (2001) describe a naturalistic way to induce aggression in dogs, which could be modified for other species to test this hypothesis.

Modulatory Structures: BNST

BNST receives inputs from the hypothalamus and amygdala and sends outputs to areas involved in emotion (Siegel and Brutus, 1990). BNST lesions increase aggression (Thomas and Van Atta, 1972; Albert and Brayley, 1979; *but see* Miczek et al., 1974). In the cat, electrical stimulation of the BNST causes a facilitation of rage elicited by medial hypothalamic stimulation (Siegel and Brutus, 1990). Davis and Whalen (2001) suggested that the BNST is involved in anxiety, while the amygdala is involved in fear. It may be speculated that the BNST also allows levels of ambient light to influence the emotional systems of the brain (Cooper et al., 1993; Grillon et al., 1997).

Modulatory Structures: Prefrontal Cortex

The prefrontal cortex is involved in behavioral inhibition, the inhibition of inappropriate responses in choice tasks, working memory, and personality. In humans and monkeys, the prefrontal cortex possesses a well-defined granular layer (layer IV), which receives extensive input from the mediodorsal nucleus of the thalamus. The mediodorsal nucleus receives projections from the amygdala and other areas. The mediodorsal nucleus projects to the prefrontal cortex in cats and rodents, but there is not a well-defined granular layer in these species (Fuster, 1997). Generally, layer IV of the cortex is a crucial synapse in the pathway by which sensory stimuli reach consciousness.

Electrical stimulation of the cat's prefrontal cortex suppresses medial–hypothalamically elicited defensive rage (reviewed in Siegel and Brutus, 1990). Additionally, in one cat, a lesion of the mediodorsal thalamic nucleus eliminated this suppressive effect (Siegel et al., 1977). Recent studies have linked reduced prefrontal functioning to human violence (Barrash et al., 2000; Davidson, 2001; Brower and Price, 2001). However, lesions of the prefrontal cortex do not lead to permanent increases in aggression in monkeys (Deets et al., 1970; Lucchetti et al., 1998).

Modulatory Structures: Nucleus Accumbens

The nucleus accumbens plays a role in motivation and the performance of rewarded and goal-directed behaviors (reviewed by Stern and Passingham, 1996; Robinson et al., 2001). It is thought that dopaminergic innervation by cells in the ventral tegmental area (VTA) plays a crucial role.

As mentioned above, most studies have found that nucleus accumbens lesions enhance aggression in rats (Paxinos, 1976; Albert and Walsh, 1984; Albert et al., 1993) and monkeys (Stern and Passingham, 1996; *but see* Miczek et al., 1974). Three studies found that electrical stimulation of the nucleus accumbens and/or ventral tegmental area (VTA) suppressed aggression in the cat (Goldstein and Siegel, 1980, 1981; Block et al., 1980). Another study suggested that endogenous opioids in the nucleus accumbens affect aggression (Brutus et al., 1989).

The papers by Goldstein and Siegel deserve particular attention because they showed that stimulation of the nucleus accumbens or VTA differentially affects the hiss component and pawstriking component of rage behavior. The hiss response, as well as other undirected somatic components, and autonomic components of the rage response, were unaffected by this stimulation. However, the goal-directed attack with claws and teeth was suppressed (Goldstein and Siegel, 1980), which is a result consistent with the hypothesis that nucleus accumbens is involved with goal-directed behavior. However, Goldstein and Siegel later concluded that this result may have been obtained because the stimulation altered the cat's sensory fields (Goldstein and Siegel, 1981). This example illustrates a difficulty in measuring aggression; namely, that it is difficult to know whether a particular experimental manipulation changes the animal's level of aggressiveness or merely affects the sensory systems in a way that makes aggression more or less likely to occur.

A new technique has been used to study the functions of the nucleus accumbens. Robinson et al. (2001) used in vivo voltammetry to measure dopamine concentrations in the nucleus accumbens in awake, freely-moving rats. By using a video camera in conjunction with high-speed computerized data collection, they were able to correlate dopamine activity with particular components of behavior during ethanol exposure and sexual behavior in rats. This technique shows great promise if it can be implemented in other paradigms, such as models of aggression.

Limbic-Motor Integration

It has recently been proposed that the pathway from nucleus accumbens to ventral pallidum to the mesencephalic locomotor region (MLR) is important for limbic control of motor activity, and, in particular, locomotor activity, such as that exhibited during flight or predation (Mogenson, 1987; Brudzynski and Wang, 1996; *see also* Bannerman et al., 2001). Although this proposed pathway seems promising, there are other pathways by which limbic activity might influence motor systems (Nauta et al., 1978; Holstege, 1991; Groenewegen et al., 1996).

Modulatory Structures: Lateral Hypothalamus

The lateral hypothalamus projects to the pedunculopontine nucleus (PPN) (the "locomotor region") (Swanson, 1987) and other areas.

Predatory Attack

Stimulation of the hypothalamus in the cat produces two different forms of attack on an anesthetized rat. Medial hypothalamic stimulation evokes defensive rage, in which the cat becomes aggressive toward any animal that happens to be in its visual field. This causes the cat to strike the rat with a paw. Lateral hypothalamic stimulation, on the other hand, elicits a completely different behavior: stealthy, quiet stalking of the rat with circular locomotion, with almost no autonomic or emotional activation, and the cat generally attacks the rat with a bite to the head or neck region (Gregg and Siegel, 2001). The response is almost identical to that seen in cats stalking prey under normal circumstances and can be elicited by electrical stimulation in individual cats that do not normally attack rats (Wasman and Flynn, 1962). Predatory aggression appears to be positively reinforcing, as opposed to defensive rage, which is not. The neural substrates for predatory aggression are less well-understood than those for defensive rage (Bandler and Halliday, 1982; Siegel and Brutus, 1990; compare with Brudnias-Graczyk and Fonberg, 1987).

Lateral Hypothalamic Modulation of Defensive Rage

Experimental manipulations which enhance rage will inhibit predation and vice versa. An explanation for this phenomenon was recently suggested, when two studies found that the lateral hypothalamus sends inhibitory GABAergic projections to the medial hypothalamus, and the medial hypothalamus reciprocates with its own GABAergic projections (Gregg and Siegel, 2001). The reciprocal inhibition makes sense, since, obviously, a cat would not succeed at catching prey if it were to alert its prey by vocalizing; similarly, predatory behavior would not be appropriate during circumstances requiring defensive rage.

Human Relevance?

The predatory response is stereotyped and nearly identical in every cat, and this behavior can be elicited in cats that have never been observed to stalk or attack. Therefore, the ontogenetic development of the neural circuit responsible for this behavior must be governed by genes. The fact that this behavior has never been elicited by electrical stimulation in any animals other than cats makes one suspect that the behavior, and the neural circuit governing it, may be unique to cats. There might be a human parallel to predatory attack (reviewed in Gregg and Siegel, 2001), but it is unclear whether the "predatory–affective" aggression described by some authors in humans is instinctual or learned (*see also* Albert et al., 1993). Still, it is possible that the ancestor common to both cats and humans possessed a neural circuit centered in the lateral hypothalamus, and this circuit divergently evolved into a circuit for predatory attack in cats and a circuit for stealthy hunting or attack behavior in humans. In this context, it may be noted that no research has focused on cooperative aggression, i.e., the "mobbing" behavior that animals (Lorenz, 1963), and perhaps humans (Ehrenreich, 1997) exhibit toward predators.

ALZHEIMER'S DISEASE AND AGGRESSION

The aggression occurring in some Alzheimer's patients may be linked to damage of the amygdala, septal nuclei, or prefrontal cortex (Chow and Cummings, 2000; Tekin et al., 2001; Callen et al., 2001).

SUMMARY AND CONCLUSIONS

1. Defensive rage behavior, which is an intense instinctive response to inescapable threat, occurs in both animals and humans. Defensive rage occurs when the dorsal, rostral PAG is activated and is still elicitable when all structures rostral to the PAG have been eliminated.
2. The motor and autonomic components of defensive rage are generated by areas that the PAG projects to.
3. Limbic and cortical influences on the PAG are sometimes direct, but are usually funneled through the hypothalamus.
4. Activation of the medial hypothalamus or medial amygdala facilitates defensive rage. Activation of the central nucleus of the amygdala, nucleus accumbens, lateral hypothalamus, or prefrontal cortex inhibits defensive rage. Activation of the cell bodies of the septal nuclei probably enhances aggression, but the evidence is not conclusive. The BNST is also involved in rage behavior.
5. Different parts of the amygdala, hypothalamus, and, perhaps, septal nuclei have different effects upon defensive rage.
6. Experiments on the VMH area have found paradoxical results.

ACKNOWLEDGMENTS

This work was supported by National Institutes of Health (NIH) grant NS 07941-31 to Allan Siegel.

REFERENCES

Adamec, R. E. (1991) Individual differences in temporal lobe sensory processing of threatening stimuli in the cat. *Physiol. Behav.* **49**, 455–464.
Adams, D. B. (1968) Cells related to fighting behavior recorded from midbrain central gray neuropil of cat. *Science* **159**, 894–896.
Aggleton, J. P. and Saunders, R. C. (2000) The amygdala—what's happened in the last decade? in *The Amygdala: A Functional Analysis.* (Aggleton, J., ed.), Wiley-Liss, New York, pp. 1–30.

Albert, D. J. and Brayley, K. N. (1979) Mouse killing and hyperreactivity following lesions of the medial hypothalamus, the lateral septum, the bed nucleus of the stria terminalis, or the region ventral to the anterior septum. *Physiol. Behav.* **23,** 439–443.

Albert, D. J. and Walsh, M. L. (1984) Neural systems and the inhibitory modulation of agonistic behavior: a comparison of mammalian species. *Neurosci. Biobehav. Rev.* **8,** 5–24.

Albert, D. J., Walsh, M. L., and Jonik, R. H. (1993) Aggression in humans: what is its biological foundation. *Neurosci. Biobehav. Rev.* **17,** 405–425.

Albert, D. J., Walsh, M. L., and Longley, W. (1985) Group rearing abolishes hyperdefensiveness induced in weanling rats by lateral septal or medial accumbens lesions but not by medial hypothalamic lesions. *Behav. Neural Biol.* **44,** 101–109.

Amorim, M. C. P. and Hawkins A. D. (2000) Growling for food: acoustic emissions during competitive feeding of the streaked gunnard. *J. Fish Biol.* **57,** 895–907.

Arnold, J. C., Topple, A. N., Mallet, P. E., Hunt, G. E., and McGregor, I. S. (2001) The distribution of cannabinoid-induced Fos expression in rat brain: differences between the Lewis and Wistar strain. *Brain Res.* **921,** 240–255.

Bailey, P. and Davis, E. W. (1942) Effects of lesions of the periaqueductal gray matter in the cat. *Proc. Soc. Exp. Biol.* **51,** 305–306.

Baklavadzhyan, O. G., Pogosyan, N. L., Arshakyan, A. V., Darbinyan, A. G., Khachatryan, A. V., and Nikogosyan, T. G. (2000) Studies of the role of the central nucleus of the amygdala in controlling cardiovascular functions. *Neurosci. Behav. Physiol.* **30,** 231–236.

Bandler, R. and Halliday, R. (1982) Lateralized loss of biting attack patterned reflexes following induction of contralateral sensory neglect in the cat: a possible role for the striatum in centrally elicited aggressive behaviour. *Brain Res.* **242,** 165–177.

Bandler, R., Keay, K. A., Floyd, N., and Price, J. (2000) Central circuits mediating patterned autonomic activity during active vs. passive emotional coping. *Brain Res. Bull.* **53,** 95–104.

Bannerman, D. M., Gilmour, G., Norman, G., Lemaire, M., Iversen, S. D., and Rawlins, J. N. (2001) The time course of the hyperactivity that follows lesions or temporary inactivation of the fimbria-fornix. *Behav. Brain Res.* **120,** 1–11.

Bard, P. and Mountcastle, V. B. (1948) Some forebrain mechanisms involved in expression of rage with special reference to suppression of angry behavior. *Res. Publ. Assoc. Res. Nerv. Ment. Dis.* **27,** 362–399.

Barrash, J., Tranel, D., and Anderson, S. W. (2000) Acquired personality disturbances associated with bilateral damage to the ventromedial prefrontal region. *Dev. Neuropsychol.* **18,** 355–381.

Bazett, H. C. and Penfield, W. G. (1922) A study of the Sherrington decerebrate animal in the chronic as well as the acute condition. *Brain* **45,** 188–265.

Behbehani, M. M. (1995) Functional characteristics of the midbrain periaqueductal gray. *Prog. Neurobiol.* **46,** 575–605.

Bernardis, L. L. and Bellinger, L. L. (1987) The dorsomedial hypothalamic nucleus revisited: 1986 update. *Brain Res. Brain Res. Rev.* **12,** 321–381.

Berntson, G. G. (1972) Blockade and release of hypothalamically and naturally elicited aggressive behaviors in cats following midbrain lesions. *J. Comp. Physiol. Psychol.* **81,** 541–554.

Blanchard, D. C., Blanchard, R. J., Lee, E. M. C., and Nakamura, S. (1979) Defensive behaviors in rats following septal and septal-amygdala lesions. *J. Comp. Physiol. Psychol.* **93,** 378–390.

Blanchard, D. C., Blanchard, R. J., Lee, E. M. C., and Williams, G. (1981) Taming of wild Rattus norvegicus by lesions of the mesencephalic central gray. *Physiol. Psychol.* **9,** 157–163.

Blanchard, D. C., Blanchard, R. J., Takahashi, L. K., and Takahashi, T. (1977) Septal lesions and aggressive behavior. *Behav. Biol.* **21,** 157–161.

Block, C. H., Siegel, A., and Edinger, H. M. (1980) Effects of stimulation of the substantia innominata upon attack behavior elicited from the hypothalamus in the cat. *Brain Res.* **197,** 57–74.

Brady, J. V. and Nauta, W. J. H. (1954) Subcortical mechanisms in emotional behavior: the duration of affective changes following septal and habenular lesions in the albino rat. *J. Comp. Physiol. Psychol.* **48**, 412–420.

Brower, M. C. and Price, B. H. (2001) Neuropsychiatry of frontal lobe dysfunction in violent and criminal behaviour: a critical review. *J. Neurol. Neurosurg. Psychiatry* **71**, 720–726.

Brudnias-Graczyk, Z. and Fonberg, E. (1987) Comparison of the effects of lateral and ventroposterior hypothalamic damage on the predatory behavior of cats. *Acta Neurobiol. Exp.* **47**, 189–198.

Brudzynski, S. M., Kadishevitz, L., and Fu, X. W. (1998) Mesolimbic component of the ascending cholinergic pathways: electrophysiological-pharmacological study. *J. Neurophysiol.* **79**, 1675–1686.

Brudzynski, S. M. and Wang, D. (1996) C-Fos immunohistochemical localization of neurons in the mesencephalic locomotor region in the rat brain. *Neuroscience* **75**, 793–803.

Brutus, M., Zuabi, S., and Siegel, A. (1989) Microinjection of D-ala2-met5-enkephalinamide placed into the nucleus accumbens suppresses feline affective defense behavior. *Exp. Neurol.* **104**, 55–61.

Budney, A. J., Hughes, J. R., Moore, B. A., and Novy, P. L. (2001) Marijuana abstinence effects in marijuana smokers maintained in their home environment. *Arch. Gen. Psychiatry* **58**, 917–924.

Callen, D. J., Black, S. E., Gao, F., Caldwell, C. B., and Szalai, J. P. (2001) Beyond the hippocampus: MRI volumetry confirms widespread limbic atrophy in AD. *Neurology* **57**, 1669–1674.

Camargo, L. A. and Saad, W. A. (2001) Role of the alpha(1)- and alpha(2)-adrenoceptors of the paraventricular nucleus on the water and salt intake, renal excretion, and arterial pressure induced by angiotensin II injection into the medial septal area. *Brain Res. Bull.* **54**, 595–602.

Cherek, D. R. and Dougherty, D. M. (1995) Provocation frequency and its role in determining the effects of smoked marijuana on human aggressive responding. *Behav. Pharmacol.* **6**, 405–412.

Chow, T. W. and Cummings, J. L. (2000) The amygdala and Alzheimer's disease, in *The Amygdala: A Functional Analysis.* (Aggleton, J. P., ed.), Wiley-Liss, New York, pp. 655–680.

Chung, S. K., Cohen, R. S., and Pfaff, D. W. (1990) Transneuronal degeneration in the midbrain central gray following chemical lesions in the ventromedial nucleus: a qualitative and quantitative analysis. *Neuroscience* **38**, 409–426.

Clemente, C. D. and Chase, M. H. (1973) Neurological substrates of aggressive behavior. *Ann. Rev. Physiol.* **35**, 329–356.

Comoli, E., Ribeiro-Barbosa, E. R., and Canteras, N. S. (2000) Afferent connections of the dorsal premammillary nucleus. *J. Comp. Neurol.* **423**, 83–98.

Cooper, H. M., Herbin, M., and Nevo, E. (1993) Visual system of a naturally microphthalmic mammal: the blind mole rat, Spalax ehrenbergi. *J. Comp. Neurol.* **328**, 313–350.

Davidson, R. J. (2001) Toward a biology of personality and emotion. *Ann. NY Acad. Sci.* **935**, 191–207.

Davis, E. W. and Bailey, P. (1944) Effects of lesions of the periaqueductal gray matter on the macaca mulatta. *J. Neuropathol. Exp. Neurol.* **3**, 69–72.

Davis, M. and Whalen, P. J. (2001) The amygdala: vigilance and emotion. *Mol. Psychiatry* **6**, 13–34.

Deets, A. C., Harlow, H. F., Singh, S. D., and Blomquist, A. J. (1970) Effects of bilateral lesions of the frontal granular cortex on the social behavior of rhesus monkeys. *J. Comp. Physiol. Psychol.* **72**, 452–461.

Dielenberg, R. A., Hunt, G. E., and McGregor, I. S. (2001) "When a rat smells a cat": the distribution of Fos immunoreactivity in rat brain following exposure to a predatory odor. *Neuroscience* **104**, 1085–1097.

Edwards, S. B. and Flynn, J. P. (1972) Corticospinal control of striking in centrally elicited attack behavior. *Brain Res.* **41**, 51–65.

Ehrenreich, B. (1997) *Blood Rites: Origins and History of the Passions of War.* Henry Holt, New York.

Ellison, G. D. (1967) Behavior after neural isolation of the hypothalamus. *Proc. Am. Psychol. Assoc.* 123–124.

Ellison, G. D. and Flynn, J. P. (1968) Organized aggressive behavior in cats after surgical isolation of the hypothalamus. *Arch. Ital. Biol.* **106,** 1–20.

Ennis, M., Xu, S. J., and Rizvi, T. A. (1997) Discrete subregions of the rat midbrain periaqueductal gray project to nucleus ambiguus and the periambigual region. *Neuroscience* **80,** 829–845.

Fanselow, M.S. (1991) The midbrain periaqueductal gray as a coordinator of action in response to fear and anxiety, in *The Midbrain Periaqueductal Gray Matter: Functional, Anatomical, and Neurochemical Organization.* (Depaulis, A. and Bandler, R. J., eds.), Plenum Press, New York, pp. 151–174.

Fernandez De Molina, A. and Hunsperger, R. W. (1962) Organization of the subcortical system governing defence and flight reactions in the cat. *J. Physiol.* **160,** 200–213.

Flannelly, K. J., Kemble, E. D., Blanchard, D. C., and Blanchard, R. J. (1986) Effects of septalforebrain lesions on maternal aggression and maternal care. *Behav. Neural Biol.* **45,** 17–30.

Flynn, J. P. (1967) The neural basis of aggression in cats, in *Neurophysiology and Emotion.* (Glass, D. C., ed.), Rockefeller University Press, Russell Sage Foundation, New York, pp. 40–60.

Flynn, W. E., Hawkins, M. F., and Tedford, W. H., Jr. (1979) The role of hyperdipsia in aggression following septal lesions in rats. *J. Comp. Physiol Psychol.* **93,** 1053–1066.

Fried, P. A. (1972) Septum and behavior: a review. *Psychol. Bull.* **78,** 292–310.

Fuchs, S. A. G., Edinger, H. M., and Siegel, A. (1985) The role of the anterior hypothalamus in affective defense behavior elicited from the ventromedial hypothalamus of the cat. *Brain Res.* **330,** 93–108.

Fuster, J. M. (1997) *The Prefrontal Cortex: Anatomy, Physiology and Neuropsychology of the Frontal Lobe, 3rd ed.*, Lippincott-Raven, Philadelphia.

Gage, F. H., Olton, D. S., and Bolanowski, D. (1978) Activity, reactivity, and dominance following septal lesions in rats. *Behav. Biol.* **22,** 203–210.

Gastaut, H., Naquet, R., Vigouroux, R., and Corriol, J. (1952) Provocation de comportements emotionnels divers par stimulation rhinencephalique chez le chat avec electrodes a demeure. *Rev. Neurol.* **86,** 319–327.

Glusman, M., Won, W., Burdock, E. I., and Ransohoff, J. (1962) Effects of midbrain lesions on "savage" behavior induced by hypothalamic lesions in the cat. *Trans. Am. Neurol. Assoc.* **87,** 216–218.

Goddard, G. V. (1964) Functions of the amygdala. *Psychol. Bull.* **62,** 89–109.

Goldstein, J. M. and Siegel, J. (1980) Suppression of attack behavior in cats by stimulation of ventral tegmental area and nucleus accumbens. *Brain Res.* **183,** 181–192.

Goldstein, J. M. and Siegel, J. (1981) Stimulation of ventral tegmental area and nucleus accumbens reduce receptive fields for hypothalamic biting reflex in cats. *Exp. Neurol.* **72,** 239–246.

Goodlett, C. R., Engellenner, W. J., Burright, R. G., and Donovick, P. J. (1982) Influence of environmental rearing history and postsurgical environmental change on the septal rage syndrome in mice. *Physiol. Behav.* **28,** 1077–1081.

Gregg, T. R. and Siegel, A. (2001) Brain structures and neurotransmitters regulating aggression in cats: implications for human aggression. *Prog. Neuropsychopharmacol. Biol. Psychiatry* **25,** 91–140.

Grillon, C., Pellowski, M., Merikangas, K. R., and Davis, M. (1997) Darkness facilitates the acoustic startle reflex in humans. *Biol. Psychiatry* **42,** 453–460.

Groenewegen, H. J., Wright, C. I., and Beijer, A. V. (1996) The nucleus accumbens: gateway for limbic structures to reach the motor system? *Prog. Brain Res.* **107,** 485–511.

Grossman, S. P. (1976) Behavioral functions of the septum: a re-analysis, in *The Septal Nuclei (Advances in Behavioral Biology, Vol. 20).* (DeFrance, J. F., ed.), Plenum Press, New York, pp. 361–422.

Grossman, S. P. (1978) An experimental dissection of the septal syndrome, in *Functions of the Septal-Hippocampal System (Ciba Found. Symp., Vol. 58).* Elsevier, Amsterdam. pp. 227–273.

Holstege, G. (1989) Anatomical study of the final common pathway for vocalization in the cat. *J. Comp. Neurol.* **284,** 242–252.

Holstege, G. (1991) Descending motor pathways and the spinal motor system: limbic and non-limbic components. *Prog. Brain Res.* **87,** 307–421.

Hunsperger, R. W. (1956) Role of substantia grisea centralis mesencephali in electrically-induced rage reactions, in *Progress in Neurobiology.* (Kappers, A. J., ed.), Elsevier, New York, pp. 289–294.

Iwata, J., Chida, K., and LeDoux, J. E. (1987) Cardiovascular responses elicited by stimulation of neurons in the central amygdaloid nucleus in awake but not anesthetized rats resemble conditioned emotional responses. *Brain Res.* **418,** 183–188.

Johansson, A. K., Bergvall, A. H., and Hansen, S. (1999) Behavioral disinhibition following basal forebrain excitotoxin lesions: alcohol consumption, defensive aggression, impulsivity and serotonin levels. *Behav. Brain Res.* **102,** 17–29.

Jolkkonen, E. and Pitkänen, A. (1998) Intrinsic connections of the rat amygdaloid complex: projections originating in the central nucleus. *J. Comp. Neurol.* **395,** 53–72.

Jurgens, U. (1994) The role of the periaqueductal grey in vocal behaviour. *Behav. Brain Res.* **62,** 107–117.

Jurgens, U. (1998) Neuronal control of mammalian vocalization, with special reference to the squirrel monkey. *Naturwissenschaften* **85,** 376–388.

Kaada, B. (1967) Brain mechanisms related to aggressive behavior, in *Aggression and Defense. Neural Mechanisms and Social Patterns (Brain Function, Vol. V).* (Clemente, C. D. and Lindsley, D. B. eds.), University of California Press, Berkeley and Los Angeles, pp. 95–133.

Keller, A. D. (1932) Autonomic discharges elicited by physiological stimuli in mid-brain preparations. *Am. J. Physiol.* **100,** 576–586.

Kling, A. and Hutt, P. J. (1958) Effect of hypothalamic lesions on the amygdala syndrome in the cat. *Arch. Neurol. Psychiatry* **79,** 511–517.

Kobayashi, M., Ikeda, K., Kinoshita, M., and Iwasaki, Y. (1999) Amyotrophic lateral sclerosis with supranuclear ophthalmoplegia and rigidity. *Neurol. Res.* **21,** 661–664.

Koprowska, M. and Romaniuk, A. (1997) Behavioral and biochemical alterations in median and dorsal raphe nuclei lesioned cats. *Pharmacol. Biochem. Behav.* **56,** 529–540.

Kovach, J. A., Nearing, B. D., and Verrier, R. L. (2001) Angerlike behavioral state potentiates myocardial ischemia-induced T-wave alternans in canines. *J. Am. Coll. Cardiol.* **37,** 1719–1725.

Lau, P. and Miczek, K. A. (1977) Differential effects of septal lesions on attack and defensive-submissive reactions during intraspecies aggression in rats. *Physiol. Behav.* **18,** 479–485.

Le Roy, I., Pothion, S., Mortaud, S., et al. (2000) Loss of aggression, after transfer onto a C57BL/6J background, in mice carrying a targeted disruption of the neuronal nitric oxide synthase gene. *Behav. Genet.* **30,** 367–373.

LeDoux, J. E. (2000) Emotion circuits in the brain. *Annu. Rev. Neurosci.* **23,** 155–184.

Lester, B. M. and Dreher, M. (1989) Effects of marijuana use during pregnancy on newborn cry. *Child Dev.* **60,** 765–771.

Li, J. H. and Mitchell, J. H. (2000) c-Fos expression in the midbrain periaqueductal gray during static muscle contraction. *Am. J. Physiol. Heart Circ. Physiol.* **279,** H2986–H2993.

Lipp, H. P. and Hunsperger, R. W. (1978) Threat, attack and flight elicited by electrical stimulation of the ventromedial hypothalamus of the marmoset monkey Callithrix jacchus. *Brain Behav. Evol.* **15,** 260–293.

Lorenz, K. (1966) *On Aggression.* Methuen, London.

Lu, C. L., Shaikh, M. B., and Siegel, A. (1992) Role of NMDA receptors in hypothalamic facilitation of feline defensive rage elicited from the midbrain periaqueductal gray. *Brain Res.* **581,** 123–132.

Lucchetti, C., Lui, F., and Bon, L. (1998) Neglect syndrome for aversive stimuli in a macaque monkey with dorsomedial frontal cortex lesion. *Neuropsychologia* **36,** 251–257.

Luthe, L., Hausler, U., and Jurgens, U. (2000) Neuronal activity in the medulla oblongata during vocalization. A single-unit recording study in the squirrel monkey. *Behav. Brain Res.* **116,** 197–210.

Maeda, H. and Maki, S. (1987) Dopamine agonists produce functional recovery from septal lesions which affect hypothalamic defensive attack in cats. *Brain Res.* **407,** 381–385.

Mark, V. H. and Sweet, W. H. (1974) The role of limbic brain dysfunction in aggression. *Res. Publ. Assoc. Nerv. Ment. Dis.* **52,** 186–200.

Masserman, J. (1941) Is the hypothalamus a center of emotion? *Psychosom. Med.* **3,** 3–25.

McBride, R. L. and Sutin, J. (1975) Amygdaloid and pontine projections to the ventromedial nucleus of the hypothalamus. *J. Comp. Neurol.* **174,** 377–396.

Melia, K. R., Sananes, C. B., and Davis, M. (1992) Lesions of the central nucleus of the amygdala block the excitatory effects of septal ablation on the acoustic startle reflex. *Physiol. Behav.* **51,** 175–180.

Miczek, K. A., Brykczynski, T., and Grossman, S. P. (1974) Differential effects of lesions in the amygdala, periamygdaloid cortex, and stria terminalis on aggressive behaviors in rats. *J. Comp. Physiol. Psychol.* **87,** 760–771.

Millhouse, O. E. (1973) Certain ventromedial hypothalamic afferents. *Brain Res.* **55,** 89–105.

Mogenson, G. L. (1987) Limbic-motor integration. *Prog. Psychobiol. Physiol. Psychol.* **12,** 117–170.

Mos, J., Olivier, B., Poth, M., Van Oorschot, R., and Van Aken, H. (1993) The effects of dorsal raphe administration of eltoprazine, TFMPP and 8-OH-DPAT on resident intruder aggression in the rat. *Eur. J. Pharmacol.* **238,** 411–415.

Mouton, L. J. and Holstege, G. (2000) Segmental and laminar organization of the spinal neurons projecting to the periaqueductal gray (PAG) in the cat suggests the existence of at least five separate clusters of spino-PAG neurons. *J. Comp. Neurol.* **428,** 389–410.

Munoz, C. and Grossman, S. P. (1980) Behavioral consequences of selective destruction of neuron perikarya in septal area of rats. *Physiol. Behav.* **24,** 779–788.

Nauta, W. J. H., Smith, G. P., Faull, R. L. M., and Domesick, V. B. (1978) Efferent connections and nigral afferents of the nucleus accumbens septi in the rat. *Neuroscience* **3,** 385–401.

Nishiyama, N., Shimazaki, T., Ikeyaga, Y., and Matsuki, N. (2001) A GABA decrease in the dorsal raphe nucleus plays a causative role in the aggressive behaviors induced by ventromedial hypothalamus lesion. *Soc. Neurosci. Abstr.* **28,** Program No. 894. CD-ROM.

Olivier, B., Olivier-Aardema, R., and Wiepkema, P. R. (1983) Effect of anterior hypothalamic and mammillary area lesions on territorial aggressive behavior in male rats. *Behav. Brain Res.* **9,** 59–81.

Omori, N., Ishimoto, T., Mutoh, F., and Chiba, S. (2001) Kindling of the midbrain periaqueductal gray in rats. *Brain Res.* **903,** 162–167.

Paxinos, G. (1975) The septum: neural systems involved in eating, drinking, irritability, nuricide, copulation, and activity in rats. *J. Comp. Physiol. Psychol.* **89,** 1154–1168.

Paxinos, G. (1976) Interruption of septal connections: effects on drinking, irritability, and copulation. *Physiol. Behav.* **17,** 81–88.

Perachio, A. A. and Alexander, M. (1975) The Neural bases of aggression and sexual behavior in the Rhesus monkey, in *The Rhesus Monkey.* (Bourne, R. C., ed.), Academic Press, New York, pp. 381–409.

Phillips, A. G. and Lieblich, I. (1972) Developmental and hormonal aspects of hyperemotionality produced by septal lesions in male rats. *Physiol. Behav.* **9,** 237–242.

Pitkanen, A. (2000) Connectivity of the rat amygdaloid complex, in *The Amygdala: A Functional Analysis.* (Aggleton, J., ed.), Wiley-Liss, New York, pp. 31–115.

Poplawsky, A. and Johnson, D. A. (1973) Open-field social behavior of rats following lateral or medial septal lesions. *Physiol. Behav.* **11,** 845–854.

Powell, E. W. and Hines, G. (1974) The limbic system: an interface. *Behav. Biol.* **12,** 149–164.

Raisman, G. (1966) Neural connexions of hypothalamus. *Br. Med. Bull.* **22,** 197–201.

Richardson, J. (1973) The Amygdala: historical and functional analysis. *Acta Neurobiol. Exp.* **33,** 623–648.

Risold, P. Y. and Swanson, L. W. (1997) Connections of the rat lateral septal complex. *Brain Res. Brain Res. Rev.* **24,** 115–195.

Robinson, D. L., Phillips, P. E., Budygin, E. A., Trafton, B. J., Garris, P. A., and Wightman, R. M. (2001) Sub-second changes in accumbal dopamine during sexual behavior in male rats. *Neuroreport* **12,** 2549–2552.

Rollins, B. L., Stines, S. G., McGuire, H. B., and King, B. M. (2001) Effects of amygdala lesions on body weight, conditioned taste aversion, and neophobia. *Physiol Behav.* **72,** 735–742.

Rothfield, L. and Harman, P. J. (1954) On the relation of the hippocampal-fornix system to the control of rage responses in cats. *J. Comp. Neurol.* **101,** 265–282.

Schubert, K., Shaikh, M. B., and Siegel, A. (1996) NMDA receptors in the midbrain periaqueductal gray mediate hypothalamically evoked hissing behavior in the cat. *Brain Res.* **726,** 80–90.

Sheehan, T., Paul, M., Amaral, E., Numan, M. J., and Numan, M. (2001) Evidence that the medial amygdala projects to the anterior/ventromedial hypothalamic nuclei to inhibit maternal behavior in rats. *Neuroscience* **106,** 341–356.

Siegel, A. and Brutus, M. (1990) Neural substrates of aggression and rage in the cat, in *Progress in Psychobiology and Physiological Psychology, Vol. 14.* (Epstein, A. N. and Morrison, A. R., eds.), Academic Press, San Diego, CA, pp. 135–233.

Siegel, A., Edinger, H., and Koo, A. (1977) Suppression of attack behavior in the cat by the prefrontal cortex: role of the mediodorsal thalamic nucleus. *Brain Res.* **127,** 185–190.

Siegel, A., Roeling, T. A. P., Gregg, T. R., and Kruk, M. R. (1999) Neuropharmacology of brain-stimulation-evoked aggression. *Neurosci. Biobehav. Rev.* **23,** 359–389.

Skultety, F. M. (1963) Stimulation of periaqueductal gray and hypothalamus. *Arch. Neurol.* **8,** 38–50.

Stern, C. E. and Passingham, R. E. (1996) The nucleus accumbens in monkeys (*Macaca fascicularis*): II. Emotion and motivation. *Behav. Brain Res.* **75,** 179–193.

Stoddard-Apter, S. L. and MacDonnell, M. F. (1980) Septal and amygdalar efferents to the hypothalamus which facilitate hypothalamically elicited intraspecific aggression and associated hissing in the cat. An autoradiographic study. *Brain Res.* **193,** 19–32.

Swanson, L. W. (1987) The hypothalamus, in *Handbook of Chemical Neuroanatomy, Vol. 5: Integrated Systems of the CNS, Part I: Hypothalamus, Hippocampus, Amygdala, Retina.* (Björklund, A., Hokfelt, T. H., and Swanson, L. W., ed.), Elsevier, New York, pp. 1–124.

Swanson, L. W. and Lind, R. W. (1986) Neural projections subserving the initiation of a specific motivated behavior in the rat: new projections from the subfornical organ. *Brain Res.* **379,** 399–403.

Sweet, W. H., Ervin, F., and Mark, V. H. (1969) The relationship of violent behaviour to focal cerebral disease, in *Aggressive Behavior.* (Garratini, S. and Sigg, E. B., eds.), Excerta Medica Foundation, Amsterdam, pp. 336–352.

Sweeten, T. L., Posey, D. J., Shekhar, A., and McDougle, C. J. (2002) The amygdala and related structures in the pathophysiology of autism. *Pharmacol. Biochem. Behav.* **71,** 449–455.

Tekin, S., Mega, M. S., Masterman, D. M., et al. (2001) Orbitofrontal and anterior cingulate cortex neurofibrillary tangle burden is associated with agitation in Alzheimer disease. *Ann. Neurol.* **49,** 355–361.

Thomas, J. B. and Van Atta, L. (1972) Hyperirritability, lever-press avoidance, and septal lesions in the albino rat. *Physiol Behav.* **8,** 225–232.

Thorne, B. M. (1993) "By the way, rats with olfactory bulb lesions are vicious." *Ann. NY Acad. Sci.* **702,** 131–147.

Treiman, D. M. (1991) Psychobiology of ictal aggression. *Adv. Neurol.* **55,** 341–356.

Tsukahara, S. and Yamanouchi, K. (2001) Neurohistological and behavioral evidence for lordosis-inhibiting tract from lateral septum to periaqueductal gray in male rats. *J. Comp. Neurol.* **431,** 293–310.

Ursin, H. (1965) The effect of amygdaloid lesions on flight and defense behavior in cats. *Exp. Neurol.* **11,** 61–79.

Ursin, H. (1981) Neuroanatomical basis of aggression, in *Multidisciplinary Approaches to Aggression Research.* (Brain, P. F. and Benton, D., eds.), Elsevier/North Holland Biomedical Press, Amsterdam, pp. 269–293.

Van Elst, L. T., Woermann, F. G., Lemieux, L., Thompson, P. J., and Trimble, M. R. (2000) Affective aggression in patients with temporal lobe epilepsy—a quantitative MRI study of the amygdala. *Brain* **123,** 234–243.

VanderHorst, V. G. J. M., Terasawa, E., Ralston, H. J., III, and Holstege, G. (2000) Monosynaptic projections from the lateral periaqueductal gray to the nucleus retroambiguus in the rhesus monkey: implications for vocalization and reproductive behavior. *J. Comp. Neurol.* **424,** 251–268.

Vochteloo, J. D. and Koolhaas, J. M. (1987) Medial amygdala lesions in male rats reduce aggressive behavior: interference with experience. *Physiol. Behav.* **41,** 99–102.

Walker, D. L., Cassella, J. V., Lee, Y., De Lima, T. C., and Davis, M. (1997) Opposing roles of the amygdala and dorsolateral periaqueductal gray in fear-potentiated startle. *Neurosci. Biobehav. Rev.* **21,** 743–753.

Wallace, T. and Thorne, B. M. (1978) The effect of lesions in the septal region on muricide, irritability, and activity in the Long-Evans rat. *Physiol. Psychol.* **6,** 36–42.

Wasman, M. and Flynn, J. P. (1962) Directed attack elicited from hypothalamus. *Arch. Neurol.* **6,** 220–227.

Woods, J. W. (1964) Behavior of chronic decerebrate rats. *J. Neurophysiol.* **27,** 635–644.

Yokosuka, M. and Hayashi, S. (1996) Colocalization of neuronal nitric oxide synthase and androgen receptor immunoreactivity in the premammillary nucleus in rats. *Neurosci. Res.* **26,** 309–314.

Zagrodzka, J., Hedberg, C. E., Mann, G. L., and Morrison, A. R. (1998) Contrasting expressions of aggressive behavior released by lesions of the central nucleus of the amygdala during wakefulness and rapid eye movement sleep without atonia in cats. *Behav. Neurosci.* **112,** 589–602.

2
Emotion Regulation
An Affective Neuroscience Approach

R. James R. Blair and Dennis S. Charney

INTRODUCTION

Emotion regulation has been defined as the "extrinsic and intrinsic processes responsible for monitoring, evaluating, and modifying emotional reactions ... to accomplish one's goals" (Thompson, 1994). Poor emotion regulation skills have been posited to be involved in most forms of childhood psychopathology (Cicchetti et al., 1995). The goal of this chapter is to take an affective cognitive neuroscience approach to emotional regulation. However, we will not discuss all neural systems involved in the regulation of all forms of emotional expression and experience. In this chapter, we will concentrate on the neural systems involved in the regulation of those basic threat systems that mediate what is termed affective (or reactive) aggression.

Affective aggression involves unplanned, "enraged" attacks on the object perceived to be the source of the threat and/or frustration. An affective aggression episode is typically preceded by a frustrating or threatening event that occurs shortly before the aggressive act. The aggression displayed is initiated without regard for any potential goal (such as accruing the victim's financial resources or increasing status within the hierarchy), but is frequently accompanied by a negative emotion, usually anger. Animals initiating affective aggression exhibit pilo-erection, autonomic arousal, hissing, and growling during their attack pattern (Flynn, 1976). Affective aggression is the eventual fight response of an animal approached by a threat. Most animals show a graded response to an approaching threat: they will freeze, then attempt to escape, and, if this is impossible, attack the threat (Blanchard et al., 1977). Thus, affective aggression can be considered to be an appropriate part of an individual's emotional behavioral repertoire. There can be problems, however, if there are difficulties in emotion regulation. These can occur if, due to psychiatric disturbance or highly aversive life events, the individual's threat system is elevated and regulatory systems are insufficient to suppress this elevated activity. Alternatively, they can occur if the regulatory systems themselves are damaged, resulting in threatening stimuli overactivating the basic stress response, such that a stimulus that might have resulted in avoidance initiates affective aggression.

THE NEUROBIOLOGY OF AFFECTIVE AGGRESSION

Considerable work has been conducted characterizing the neural circuitry involved in affective aggression in nonhuman mammals (Gregg and Siegel, 2001; Panksepp, 1998). It is this circuitry that is responsible for the animal's response to threat. Low-level stimulation of the circuitry initiates freezing, higher-level stimulation elicits escape behaviors, and still higher-level stimulation, elicits affective aggression. The circuit has been identified that runs from the medial amygdaloidal areas downward, largely via the stria terminalis to the medial hypothalamus, and from there to the dorsal half of the periaqueductal gray (PAG). The system is organized in a hierarchical manner, such that aggression evoked from the amygdala is dependent on functional integrity of the medial hypothalamus and PAG, but that aggression evoked from the PAG is not dependent on the functional integrity of the amygdala (Bandler, 1988; Gregg and Siegel, 2001; Panksepp, 1998). Moreover, this circuit has been shown to be modulated by neurons in the frontal cortex, in particular, medial frontal and orbital frontal cortex (Gregg and Siegel, 2001; Panksepp, 1998).

There are several reasons for believing that human affective aggression is mediated by this circuitry also. Thus, hamartomas of the hypothalamus can lead to attacks of rage and affective aggression (Reeves and Plum, 1969; Weissenberger et al., 2001). Moreover, there are suggestions that in patients with post-traumatic stress disorder (PTSD), there is elevated responsiveness of the subcortical threat response neural architecture. In line with this suggestion, patients with PTSD show an elevated startle to aversive stimuli in comparison to healthy individuals (Morgan et al., 1996, 1997). Unsurprisingly, given this elevated responsiveness, the display of affective aggression has been linked to PTSD (Silva et al., 2001). Other psychiatric disorders, such as depression and generalized anxiety disorder, have also been linked to overactivity in some of the elements of this basic threat circuitry (Drevets, 2001; Thomas et al., 2001). In these disorders also, there is a heightened risk of affective aggression (Galovski et al., 2002; Kendell, 1970; Knox et al., 2000; Pine et al., 2000).

NEURAL SYSTEMS THAT REGULATE EMOTION

Three main neural systems have been implicated in the regulation of emotion. These are the amygdala and two regions of prefrontal cortex, the medial frontal cortex (including anterior cingulate) and the orbital frontal cortex (Davidson et al., 2000). We will concentrate here on their role in the regulation of affective aggression. Each will be considered in turn.

AMYGDALA AND THE REGULATION OF AFFECTIVE AGGRESSION

The amygdala modulates the probability that the organism will initiate affective aggression. In rodents, stimulation of the medial amygdala elicits affective aggression (Gregg and Siegel, 2001; Panksepp, 1998). This suggests that an amygdala lesion, preventing the impact of potential stimulation on the subcortical system responding to threat, should reduce the probability that an individual will engage in affective aggression. There are data consistent with this position (Lilly et al., 1983). More specifically, in an open retrospective study of 481 cases of bilateral amygdalectomies performed for the control of conservatively untreatable aggressiveness, moderate to excellent improvement of aggressive behavior was reported in 70–76% of cases (Ramamurthi, 1988). However, unilateral damage to the central nucleus of the amygdala in cats can actually increase the expression of affective aggression (Zagrodzka et al., 1998). In addition, the impact of amygdalectomy on aggression in monkeys is a function of the animal's position in the dominance hierarchy; submissive monkeys show increased aggression following amygdalectomies (Rosvold et al., 1954). Moreover, a significant subgroup (20%) of aggressive patients with temporal lobe epilepsy, present with very severe amygdalar atrophy (van Elst et al., 2000). Therefore, it would appear that while stimulation of neurons in the central nucleus of the amygdala can elicit affective aggression, lesions of the amygdala can either increase or decrease the probability of the emergence of affective aggression.

The amygdala is known to react to reinforcing as well as aversive stimuli (Everitt et al., 2000). This suggests that the amygdala would be in a position to both upgrade (as a response to an aversive stimulus) or downgrade (as a response to reinforcement) the responsiveness of the subcortical systems that respond to threat. Indeed, this suggestion is supported by findings that, while aversive visual threat primes augment the magnitude of the startle reflex relative to neutral primes, appetitive visual primes reduce the magnitude of the startle reflex (Lang et al., 1990). The amygdala is known to be involved in the modulation of the startle reflex (Davis, 2000). This suggests that amygdala lesions might, therefore, reduce the probability of affective aggression in threatening circumstances by reducing the patient's sensitivity to a learned threat. Learned threats would not activate the amygdala, and through the amygdala, the subcortical system. The subcortical threat system would therefore not respond to learnt threats. However, amygdala lesions might also increase the probability of affective aggression in nonthreatening circumstances. The amygdala lesion would prevent the suppression of affective aggression as a function of amygdala activation by appetitive stimuli in the environment.

FRONTAL CORTEX AND THE REGULATION OF AFFECTIVE AGGRESSION

The animal literature suggests that frontal neurons are involved in the modulation of the subcortical circuit mediating affective aggression (Gregg and Siegel, 2001; Panksepp, 1998). In one particularly elegant study, the latency of aggression initiated by electrical stimulation of the hypothalamus in the cat was increased by the stimulation of anterior cingulate (Brutus et al., 1984). In humans, there are clear data that damage to the anterior cingulate and orbital frontal cortex is associated with an increased risk for the display of affective aggression, whether the lesion occurs in childhood (Anderson et al., 1999; Pennington and Bennetto, 1993) or adulthood (Grafman et al., 1996).

In addition to the data from patients with neurological lesions, a series of neuroimaging studies have provided evidence of frontal dysfunction in aggressive individuals (Critchley et al., 2000; Goyer et al., 1994; Raine et al., 1997; Raine et al., 1994; Raine et al., 1998; Soderstrom et al., 2000; Volkow and Tancredi, 1987; Volkow et al., 1995; Wong et al., 1997). Unfortunately, only a subset of these studies have characterized the predominant form of the aggression shown by their patients.

Volkow and colleagues in two major studies examined cerebral blood flow (CBF) under rest conditions using positron emission tomography (PET) in affectively violent psychiatric patients and comparison individuals. In both studies, the affectively aggressive individuals were found to show significantly less CBF in medial temporal and frontal cortex than the comparison individuals (Volkow and Tancredi, 1987; Volkow et al., 1995). Soderstrom and colleagues reported very similar findings, again implicating hypoperfusion in the temporal and/or frontal lobes in 16 out of 21 affectively aggressive individuals using hexamethyl propylene amine oxime single photon emission computed tomography (HMPAO-SPECT)-CBF (Soderstrom et al., 2000).

Raine and colleagues examined CBF using PET during performance of a continuous performance task in murderers pleading not guilty by reason of insanity and matched comparison individuals ($n = 22$ and $n = 41$ in both groups in the 1994 and 1997 studies, respectively). Both studies reported reduced CBF in the prefrontal cortex (Raine et al., 1997; Raine et al., 1994). While the predominant nature of the aggression presented by the participants was not described in these studies, in later work, Raine demonstrated that the reduced prefrontal cortex functioning was only shown by affectively aggressive individuals and not those who present with instrumental aggression (Raine et al., 1998).

Thus, the neuroimaging studies on these individuals presenting with predominantly affective aggression have provided evidence of frontal and temporal lobe dysfunction. It is necessary to be cautious, however, concerning much of the above neuroimaging data. First, it should be noted that all of the above studies included individuals with known organic brain damage and/or patients with schizophrenia. Thus, the atypical blood flow reported may reflect organic damage or schizophrenia that may, or may not, be related to their aggressive behavior. Indeed, Wong and colleagues investigated CBF under rest conditions, using PET, and structural abnormalities, using magnetic resonance imaging (MRI), in 20 repetitive violent schizophrenic offenders and a matched group of 19 nonrepetitive violent schizophrenics. This study reported no group differences in functioning or structure (Wong et al., 1997).

Secondly, the existing neuroimaging literature has placed little emphasis on considering the separable regions of frontal cortex. However, neuropsychological, neuroimaging, and animal lesion data all suggest that different aspects of executive functions are dissociable and mediated by distinct neural systems subserved by different regions of the prefrontal cortex (Baddeley and Della Sala, 1998; Fuster, 1980; Luria, 1966). Indeed, neuropsychological data have shown that only lesions to orbital frontal and medial frontal cortex, rather than dorsolateral prefrontal cortex, increase risk for aggression (Grafman et al., 1996). One of the few studies to dissociate regions of frontal cortex was conducted by Goyer and colleagues who examined the CBF under rest conditions using PET of 17 patients with personality disorder (antisocial, borderline, dependent, and narcissistic) and 43 comparison individuals (Goyer et al., 1994). The patients' aggression was predominantly affective. Interestingly, they found, in line with the neuropsychological data, that it was lower normalized CBF in lateral orbitofrontal cortex (Brodmann's Area [BA 47]) that correlated with a history of aggression.

Thus, and in agreement with the animal literature (Gregg and Siegel, 2001; Panksepp, 1998), it is likely that it is dysfunction in medial frontal and orbital frontal cortex, rather than frontal cortex more generally, that can lead to disinhibition of the subcortical circuitry mediating affective aggression.

MEDIAL FRONTAL ORBITAL FRONTAL CORTEX: DIFFERENTIAL CONTRIBUTIONS TO EMOTION REGULATION?

Medial frontal cortex and orbital frontal cortex are massively interconnected and frequently function in tandem. However, there are reasons to believe that medial and orbital frontal cortex have differential roles and that they achieve the regulation of emotion through differential computational processes. Thus, recent animal work has identified medial frontal cortex involvement in the representation of the animal's goals during instrumental learning paradigms (Balleine and Dickinson, 1998). The maintenance of the animal's goals must necessitate the suppression of information, including emotional representations, that are likely to interfere with the achievement of the goal. Alternatively, of course, it would be advantageous if the sudden presence of an unexpected aversive stimulus, which is associated with threat, interferes with current goal-directed behavior; the individual should not maintain food-gathering behavior if a predator is observed. Interestingly, recent data have demonstrated that conditioned aversive tones depress medial frontal cortex neurons as a function of the degree to which the conditioned aversive tone activates the amygdala (Garcia et al., 1999).

Orbitofrontal cortex is involved in at least two processes that are likely to result in the modulation of the subcortical systems allowing affective aggression. The first is to allow the disinhibition of the subcortical systems under conditions of frustration (Panksepp, 1998). Frustration has long been linked to the display of affective aggression (Berkowitz, 1993). It occurs following the initiation of a behavior in expectation of, and to achieve, a specific reward when the reward is not consequent on the action. Orbitofrontal cortex is crucially involved in the computation of expectations of reward and identifying if these expectations have been violated (Rolls, 2000). Thus, orbitofrontal cortex may increase neuronal activity in the circuit under conditions when an expected reward has not been achieved and suppress neuronal activity when the expected reward is achieved.

The second process is in social response reversal (SRR) (Blair and Cipolotti, 2000). It has been suggested that neurons in orbitofrontal cortex are recruited by a system that is crucial for social cognition and the modulation of affective aggression, but which is separable from the system computing violations of reward expectancies (Blair and Cipolotti, 2000). The position stresses the role of social cues in modulating social behavior (Blair, 2001; Blair and Cipolotti, 2000). Thus, angry expressions are known to curtail the behavior of others in situations where social rules or expectations have been violated (Averill, 1982). The SRR system is thought to be activated by several classes of stimuli: (*i*) another individual's angry expressions; (*ii*) other negative valence expressions (e.g., staring that can precede a sense of embarrassment and perhaps others' disgusted expressions); and (*iii*) situations associated with social disapproval. The suggestion is that this system modulates current behavioral responding, in particular the modulation of affective aggression, but that this modulation is a function of the position in the dominance hierarchy of the other individual. Thus, for example, the angry expression of an individual higher in the dominance hierarchy will suppress affective aggression and lead to alterations in current instrumental behavior. In contrast, the angry expression of an individual lower in the dominance hierarchy will lead to activation of the subcortical circuitry for affective aggression. In line with this, there are data from work with primates demonstrating that affective aggression is modulated by the individual's position in the dominance hierarchy. Thus, neural stimulated animals will vent their rage on more submissive animals and avoid confrontations with more dominant ones (Alexander and Perachio, 1973).

This hypothesis has drawn support from findings that orbitofrontal cortex (BA 47) is activated by negative emotional expressions; in particular, anger but also fear and disgust (Blair et al., 1999; Kesler-West et al., 2001; Sprengelmeyer et al., 1998). In addition, this region is also activated if an individual is induced to feel angry (Dougherty et al., 1999). Moreover, if this region is lesioned, patients are impaired in the ability to recognize facial expressions, particularly anger (Blair and Cipolotti, 2000; Hornak et al., 1996). Such patients have also been found to show impairment in both appropriately attributing anger and embarrassment to story protagonists (Blair and Cipolotti, 2000). In addition, such patients have a deficit in identifying the sorts of violations of social norms that induce anger in others (Blair and Cipolotti, 2000; Stone et al., 1998). In line with the suggestion that this region of lateral orbitofrontal cortex (BA 47) is crucially involved in responding to expectancies of social anger, prompting avoidance of violations of social norms, a recent functional MRI (fMRI) study has revealed its activation when individuals read descriptions of social norm violations (Berthoz et al., 2002). Finally, both alcohol and diazepam, two pharmacological agents associated with increased risk for affective aggression (Bond et al., 1995), selectively impair the ability of healthy individuals to process angry expressions (Blair and Curran, 1999; Borrill et al., 1987).

Thus, in conclusion, affective aggression is mediated by the circuit depicted in Fig. 1, which includes medial amygdaloid areas, the medial hypothalamus, and the dorsal half of the PAG. The amygdala and frontal cortex, in particular medial frontal cortex and lateral orbital frontal cortex (BA 47), modulate this circuit. The amygdala's modulation occurs as a function of the presence of threat or appetitive cues in the environment. The modulation by frontal cortex occurs as a function of social emotional cues, representations of social norms and knowledge of the other individuals' position in the dominance hierarchy.

NEUROTRANSMITTER INVOLVEMENT IN EMOTION REGULATION

In addition to the neural systems described above, there are neurotransmitters that have been implicated in the regulation of emotion and, particularly, the regulation of affective aggression (Gregg and Siegel, 2001). In this chapter, we will concentrate on two: serotonin (5-HT) and γ-amino butyric acid (GABA). The involvement of both of these neurotransmitters in emotion regulation, with particular reference to the regulation of affective aggression, will be discussed in turn.

SEROTONIN (5-HT)

5-HT has long been implicated in the regulation of emotion and the modulation of affective aggression (Davidson et al., 2000; Lesch and Merschdorf, 2000; Nelson and Chiavegatto, 2001). Generally, experimental manipulations, which increase 5-HT receptor activation, decrease affective aggression, and those that decrease receptor activation increase affective aggression (Bell et al., 2001; Shaikh et al., 1997). Thus, the selective destruction of 5-HT neurons in the raphe complex in cats and rats leads to increases in affective aggression (File and Deakin, 1980). In humans, there have been consistent reports

Emotion Regulation 27

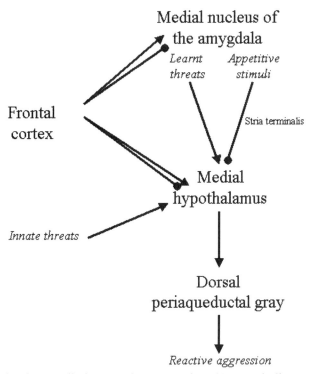

Fig. 1. Neural circuitry mediating reactive aggression. (Arrows indicate excitatory connections. Circled connections indicate inhibitory connections.)

linking affective aggression with low cerebrospinal fluid (CSF) concentrations of 5-hydroxyindoleacetic acid (5-HIAA) (Lesch and Merschdorf, 2000).

Pharmacological challenge studies also suggest that 5-HT plays a role in the modulation of affective aggression. Thus, the prolactin elevation in response to a single dose of a 5-HT agonist can be used to index central 5-HT activity. Peak prolactin responses to the 5-HT-releasing agent, fenfluramine, significantly correlate inversely with an interview-assessed life history of aggression in males (although not females) (Manuck et al., 1998). Interestingly, regarding the frontal cortical structures implicated in the regulation of emotion and affective aggression above, the administration of fenfluramine and meta-chlorophenylpiperazine (m-CPP), a nonspecific 5-HT agonist (Fiorella et al., 1995), both increase CBF in orbital frontal cortex and medial frontal cortex in healthy individuals (Hommer et al., 1997; Mann et al., 1996). However, increases in these regions following the administration of either fenfluramine or m-CPP are not found in patients presenting with affective aggression (New et al., 2002; Siever et al., 1999). This suggests that one potential role in which 5-HT might modulate affective aggression, by innervating orbital frontal cortex and medial frontal cortex, does not occur in patients presenting with affective aggression.

Tryptophan depletion increases laboratory aggression in both men and women (Bjork et al., 2000; Bond et al., 2001). Moreover, there is now data suggesting that the aggressive effect of tryptophan depletion is mediated via the $5-HT_{1A}$ receptor (Cleare and Bond, 2000). Thus, participants in whom aggression can be provoked or inhibited by tryptophan depletion or enhancement, respectively, show a blunted hypothermic response to ipsapirone (Cleare and Bond, 2000). Since ipsapirone acts specifically to stimulate 5-HT_{1A} receptors, the blunted hypothermic response is likely to represent impaired 5-HT_{1A} receptor function in the aggressive individuals. Moreover, animal work has demonstrated a selective suppressive action of 5-HT_{1A} receptors on the PAG neurons mediating affective aggression (Gregg and Siegel, 2001).

Thus, disturbed serotonergic functioning is likely to lead to reduced emotional and affective aggression regulation through two main routes. First, by reducing the functional efficacy of those frontal regions, orbital and medial frontal cortex, implicated in the regulation of affective aggression. Second, by preventing the selective suppressive action of 5-HT_{1A} receptors on the PAG neurons mediating affective aggression (Gregg and Siegel, 2001).

GABA

Alcohol and benzodiazepines have been consistently shown to increase aggression (Fish et al., 2001). In vitro, alcohol- and receptor-specific positive modulators of the $GABA_A$ receptor complex (i.e., $GABA_A$ receptor agonists, benzodiazepines, barbiturates, and certain neurosteroids) share the action of increasing Cl flux through the ion channel in the presence of GABA. Behaviorally, alcohol and the $GABA_A$ positive modu-lators exert similar muscle relaxant and aggression heightening effects that can be potentiated by the co-administration of these agents (Fish et al., 2001).

In humans, various clinical reports have detailed incidences of aggression following treatment with benzodiazepines varying from 10%, in a mixed diagnostic group (Rosenbaum et al., 1984), to 58%, in patients with borderline personality disorder (Gardner and Cowdry, 1985). Experimental studies have shown that administration of benzodiazepines is associated with increased behavioral aggression following provocation, not only in anxious patients undergoing benzodiazepine treatment (Bond et al., 1995), but even in healthy volunteers administered only one dose of a benzodiazepine (Bond and Silveira, 1993).

Interestingly, both alcohol and diazepam appear to affect the functioning of the lateral region of the orbitofrontal cortex, which has been linked to mediating anger and angry behavior, and is presumably involved in the control of affective aggression, i.e., BA 47 (Blair, 2001). Thus, both alcohol and diazepam impair recognition of the angry expressions of others (Blair and Curran, 1999; Borrill et al., 1987). Moreover, in recent neuroimaging work, we have shown that this impairment is associated with a reduction, following the ingestion of 15 mg of diazepam, in the response in BA 47 to angry expressions. Strikingly, the amygdala's response to fearful expressions is not affected by the ingestion of this dose of diazepam.

In conclusion, both 5-HT and GABA appear to modulate aggression. Increases in 5-HT suppress affective aggression, while increases in GABA increase the probability of affective aggression.

IN SUMMARY

Considerable progress has been made in recent years in understanding the neural systems involved in the regulation of emotion and, in particular, human affective aggression. The subcortical structures involved in mediating affective aggression have been well specified. Moreover, there is now far more understanding of the role of the amygdala and medial frontal and orbital frontal cortex in modulating these subcortical structures and their role in emotion and affective aggression. There is now also data on the role of neurotransmitters, such as 5-HT and GABA, in modulating this subcortical circuitry also. Such neurochemical understanding holds, of course, the promise of successful interventions in the future. If this promise is realized, there is a strong possibility that the impact of affective aggression on victims and society in the future will be considerably ameliorated.

REFERENCES

Alexander, M. and Perachio, A. A. (1973) The influence of target sex and dominance on evoked attack in Rhesus monkeys. *Am. J. Phys. Anthropol.* **38,** 543–547.

Anderson, S. W., Bechara, A., Damasio, H., Tranel, D., and Damasio, A. R. (1999) Impairment of social and moral behaviour related to early damage in human prefrontal cortex. *Nat. Neurosci.* **2,** 1032–1037.

Averill, J. R. (1982) *Anger and Aggression: An Essay on Emotion.* Springer-Verlag, New York.

Baddeley, A. and Della Sala, S. (1998) Working memory and executive control, in *The Prefrontal Cortex.* (Roberts, A. C., Robbins, T. W., Weiskrantz, L., eds.), Oxford University Press, New York, pp. 9–21.

Balleine, B. W. and Dickinson, A. (1998) Goal-directed instrumental action: contingency and incentive learning and their cortical substrates. *Neuropharmacology* **37,** 407–419.

Bandler, R. (1988) Brain mechanisms of aggression as revealed by electrical and chemical stimulation: suggestion of a central role for the midbrain periaqueductal gray region, in *Progress in Psychobiology and Physiological Psychology.* (Epsein, A. N. and Morrison, A. R., eds.), Academic Press, San Diego, pp. 135–233.

Bell, C., Abrams, J., and Nutt, D. (2001) Tryptophan depletion and its implications for psychiatry. *Br. J. Psychiatry* **178,** 399–405.

Berkowitz, L. (1993) *Aggression: Its Causes, Consequences, and Control.* Temple University Press, Philadelphia.

Berthoz, S., Armony, J., Blair, R. J. R., and Dolan, R. (2002) Neural correlates of violation of social norms and embarrassment. *Brain* **125,** 1696–1708.

Bjork, J. M., Dougherty, D. M., Moeller, F. G., and Swann, A. C. (2000) Differential behavioral effects of plasma tryptophan depletion and loading in aggressive and nonaggressive men. *Neuropsychopharmacology* **22,** 357–369.

Blair, R. J. R. (2001) Neuro-cognitive models of aggression, the antisocial personality disorders and psychopathy. *J. Neurol. Neurosurg. Psychiatry* **71,** 727–731.

Blair, R. J. R. and Cipolotti, L. (2000) Impaired social response reversal: a case of "acquired sociopathy." *Brain* **123,** 1122–1141.

Blair, R. J. R. and Curran, H. V. (1999) Selective impairment in the recognition of anger induced by diazepam. *Psychopharmacology* **147,** 335–338.

Blair, R. J. R., Morris, J. S., Frith, C. D., Perrett, D. I., and Dolan, R. (1999) Dissociable neural responses to facial expressions of sadness and anger. *Brain* **122,** 883–893.

Blanchard, R. J., Blanchard, D. C., and Takahashi, L. K. (1977) Attack and defensive behaviour in the albino rat. *Animal Beh.* **25,** 197–224.

Bond, A. J., Curran, H. V., Bruce, M., O'Sullivan, G., and Shine, P. (1995) Behavioural aggression in panic disorder after 8 weeks of treatment with alprazolam. *J. Affect. Disord.* **35,** 117–123.

Bond, A. J. and Silveira, J. C. (1993) The combination of alprazolam and alcohol on behavioral aggression. *J. Stud. Alcohol Suppl.* **11,** 30–39.

Bond, A. J., Wingrove, J., and Critchlow, D. G. (2001) Tryptophan depletion increases aggression in women during the premenstrual phase. *Psychopharmacology (Berl)* **156,** 477–480.

Borrill, J. A., Rosen, B. K., and Summerfield, A. B. (1987) The influence of alcohol on judgment of facial expressions of emotion. *Br. J. Med. Psychol.* **60**, 71–77.

Brutus, M., Shaikh, M. B., Siegel, H. E., and Siegel, A. (1984) An analysis of the mechanisms underlying septal area control of hypothalamically elicited aggression in the cat. *Brain Res.* **310**, 235–248.

Cicchetti, D., Ackerman, B. P., and Izard, C. E. (1995) Emotions and emotion regulation in developmental psychopathology. *Dev. Psychopathol.* **7**, 1–10.

Cleare, A. J. and Bond, A. J. (2000) Experimental evidence that the aggressive effect of tryptophan depletion is mediated via the 5-HT1A receptor. *Psychopharmacology (Berl)* **147**, 439–441.

Critchley, H. D., Simmons, A., Daly, E. M., et al. (2000) Prefrontal and medial temporal correlates of repetitive violence to self and others. *Biol. Psychiatry* **47**, 928–934.

Davidson, R. J., Putnam, K. M., and Larson, C. L. (2000) Dysfunction in the neural circuitry of emotion regulation—a possible prelude to violence. *Science* **289**, 591–594.

Davis, M. (2000) The role of the amygdala in conditioned and unconditioned fear and anxiety, in *The Amygdala: A Functional Analysis.* (Aggleton, J. P., eds.), Oxford University Press, Oxford, pp. 289–310.

Dougherty, D. D., Shin, L. M., Alpert, N. M., et al. (1999) Anger in healthy men: a PET study using script-driven imagery. *Biol. Psychiatry* **46**, 466–472.

Drevets, W. C. (2001) Neuroimaging and neuropathological studies of depression: implications for the cognitive-emotional features of mood disorders. *Curr. Opin. Neurobiol.* **11**, 240–249.

Everitt, B. J., Cardinal, R. N., Hall, J., Parkinson, J. A., and Robbins, T. W. (2000) Differential involvement of amygdala subsystems in appetitive conditioning and drug addiction, in *The Amygdala: A Functional Analysis.* (Aggleton, J. P., eds.), Oxford University Press, Oxford, pp. 289–310.

File, S. E. and Deakin, J. F. (1980) Chemical lesions of both dorsal and median raphe nuclei and changes in social and aggressive behaviour in rats. *Pharmacol. Biochem. Behav.* **12**, 855–859.

Fiorella, D., Helsley, S., Rabin, R. A., and Winter, J. C. (1995) 5-HT2C receptor-mediated phosphoinositide turnover and the stimulus effects of m-chlorophenylpiperazine. *Psychopharmacology (Berl)* **122**, 237–243.

Fish, E. W., Faccidomo, S., DeBold, J. F., and Miczek, K. A. (2001) Alcohol, allopregnanolone and aggression in mice. *Psychopharmacology (Berl)* **153**, 473–483.

Flynn, J. P. (1976) Neural basis of threat and attack, in *Biological Foundations of Psychiatry.* (Grenell, R. G. and Abau, S. G., eds.), Raven Press, New York, pp. 275–295.

Fuster, J. M. (1980) *The Prefrontal Cortex.* Raven Press, New York.

Galovski, T., Blanchard, E. B., and Veazey, C. (2002) Intermittent explosive disorder and other psychiatric comorbidity among court-referred and self-referred aggressive drivers. *Behav. Res. Ther.* **40**, 641–651.

Garcia, R., Vouimba, R. M., Baudry, M., and Thompson, R. F. (1999) The amygdala modulates prefrontal cortex activity relative to conditioned fear. *Nature* **402**, 294–296.

Gardner, D. L. and Cowdry, R. W. (1985) Alprazolam-induced dyscontrol in borderline personality disorder. *Am. J. Psychiatry* **146**, 98–100.

Goyer, P. F., Andreason, P. J., Semple, W. E., et al. (1994) Positron-emission tomography and personality disorders. *Neuropsychopharmacology* **10**, 21–28.

Grafman, J., Schwab, K., Warden, D., Pridgen, B. S., and Brown, H. R. (1996) Frontal lobe injuries, violence, and aggression: a report of the Vietnam head injury study. *Neurology* **46**, 1231–1238.

Gregg, T. R. and Siegel, A. (2001) Brain structures and neurotransmitters regulating aggression in cats: implications for human aggression. *Prog. Neuropsychopharmacol. Biol. Psychiatry* **25**, 91–140.

Hommer, D., Andreasen, P., Rio, D., et al. (1997) Effects of m-chlorophenylpiperazine on regional brain glucose utilization: a positron emission tomographic comparison of alcoholic and control subjects. *J. Neurosci.* **17**, 2796–2806.

Hornak, J., Rolls, E. T., and Wade, D. (1996) Face and voice expression identification in patients with emotional and behavioural changes following ventral frontal damage. *Neuropsychologia* **34**, 247–261.

Kendell, R. E. (1970) Relationship between aggression and depression. Epidemiological implications of a hypothesis. *Arch. Gen. Psychiatry* **22**, 308–318.

Kesler-West, M. L., Andersen, A. H., Smith, C. D., et al. (2001) Neural substrates of facial emotion processing using fMRI. *Cognit. Brain Res.* **11**, 213–226.

Knox, M., King, C., Hanna, G. L., Logan, D., and Ghaziuddin, N. (2000) Aggressive behavior in clinically depressed adolescents. *J. Am. Acad. Child Adolesc. Psychiatry* **39**, 611–618.

Lang, P. J., Bradley, M. M., and Cuthbert, B. N. (1990) Emotion, attention, and the startle reflex. *Psychol. Rev.* **97**, 377–398.

Lesch, K. P. and Merschdorf, U. (2000) Impulsivity, aggression, and serotonin: a molecular psychobiological perspective. *Behav. Sci. Law* **18**, 581–604.

Lilly, R., Cummings, J., Benson, F., and Frankel, M. (1983) The human Klüver-Bucy syndrome. *Neurology* **33**, 1141–1145.

Luria, A. (1966) *Higher Cortical Functions in Man.* Basic Books, New York.

Mann, J. J., Malone, K. M., Diehl, D. J., Perel, J., Nichols, T. E., and Mintun, M. A. (1996) Positron emission tomographic imaging of serotonin activation effects on prefrontal cortex in healthy volunteers. *J. Cereb. Blood Flow Metab.* **16**, 418–426.

Manuck, S. B., Flory, J. D., McCaffery, J. M., Matthews, K. A., Mann, J. J., and Muldoon, M. F. (1998) Aggression, impulsivity, and central nervous system serotonergic responsivity in a nonpatient sample. *Neuropsychopharmacology* **19**, 287–299.

Morgan, C. A., III, Grillon, C., Lubin, H., and Southwick, S. M. (1997) Startle reflex abnormalities in women with sexual assault-related posttraumatic stress disorder. *Am. J. Psychiatry* **154**, 1076–1080.

Morgan, C. A., III, Grillon, C., Southwick, S. M., Davis, M., and Charney, D. S. (1996) Exaggerated acoustic startle reflex in Gulf War veterans with posttraumatic stress disorder. *Am. J. Psychiatry* **153**, 64–68.

Nelson, R. J. and Chiavegatto, S. (2001) Molecular basis of aggression. *Trends Neurosci.* **24**, 713–719.

New, A. S., Hazlett, E. A., Buchsbaum, M. S., et al. (2002) Blunted prefrontal cortical 18fluorodeoxyglucose positron emission tomography response to meta-chlorophenylpiperazine in impulsive aggression. *Arch. Gen. Psychiatry* **59**, 621–629.

Panksepp, J. (1998) *Affective Neuroscience: The Foundations of Human and Animal Emotions.* Oxford University Press, New York.

Pennington, B. F. and Bennetto, L. (1993) Main effects or transaction in the neuropsychology of conduct disorder? Commentary on "The neuropsychology of conduct disorder." *Dev. Psychopathol.* **5**, 153–164.

Pine, D. S., Cohen, E., Cohen, P., and Brook, J. S. (2000) Social phobia and the persistence of conduct problems. *J. Child Psychol. Psychiatry* **41**, 657–665.

Raine, A., Buchsbaum, M. S., and LaCasse, L. (1997) Brain abnormalities in murderers indicated by positron emission tomography. *Biol. Psychiatry* **42**, 495–508.

Raine, A., Buchsbaum, M. S., Stanley, J., Lottenberg, S., Abel, L., and Stoddard, J. (1994) Selective reductions in prefrontal glucose metabolism in murderers. *Biol. Psychiatry* **15**, 365–373.

Raine, A., Meloy, J. R., Birhle, S., Stoddard, J., LaCasse, L., and Buchsbaum, M. S. (1998) Reduced prefrontal and increased subcortical brain functioning assessed using positron emission tomography in predatory and affective murderers. *Behav. Sci. Law* **16**, 319–332.

Raine, A., Phil, D., Stoddard, J., Bihrle, S., and Buchsbaum, M. (1998) Prefrontal glucose deficits in murderers lacking psychosocial deprivation. *Neuropsychiatry Neuropsychol. Behav. Neurol.* **11**, 1–7.

Ramamurthi, B. (1988) Stereotactic operation in behaviour disorders. Amygdalotomy and hypothalamotomy. *Acta Neurochir. Suppl. (Wien)* **44**, 152–157.

Reeves, A. and Plum, F. (1969) Hyperphagia, rage, and dementia accompanying a ventromedial hypothalamic neoplasm. *Arch. Neurol.* **20,** 616–624.

Rolls, E. T. (2000) The orbitofrontal cortex and reward. *Cereb. Cortex* **10,** 284–294.

Rosenbaum, J. F., Woods, S. W., Groves, J. E., and Klerman, G. L. (1984) Emergence of hostility during alprazolam treatment. *Am. J. Psychiatry* **141,** 793.

Rosvold, H., Mirsky, A., and Pribram, K. (1954) Influence of amygdalectomy on social behavior in monkeys. *J. Comp. Physiol. A* **47,** 173–178.

Shaikh, M. B., De Lanerolle, N. C., and Siegel, A. (1997) Serotonin 5-HT1A and 5-HT2/1C receptors in the midbrain periaqueductal gray differentially modulate defensive rage behavior elicited from the medial hypothalamus of the cat. *Brain Res.* **765,** 198–207.

Siever, L. J., Buchsbaum, M. S., New, A. S., et al. (1999) d,l-Fenfluramine response in impulsive personality disorder assessed with [18F]fluorodeoxyglucose positron emission tomography. *Neuropsychopharmacology* **20,** 413–423.

Silva, J. A., Derecho, D. V., Leong, G. B., Weinstock, R., and Ferrari, M. M. (2001) A classification of psychological factors leading to violent behavior in posttraumatic stress disorder. *J. Forensic. Sci.* **46,** 309–316.

Soderstrom, H., Tullberg, M., Wikkelso, C., Ekholm, S., and Forsman, A. (2000) Reduced regional cerebral blood flow in non-psychotic violent offenders. *Psychiatry Res.* **98,** 29–41.

Sprengelmeyer, R., Rausch, M., Eysel, U. T., and Przuntek, H. (1998) Neural structures associated with the recognition of facial basic emotions. *Proc. R. Soc. Lond. B* **265,** 1927–1931.

Stone, V. E., Baron-Cohen, S., and Knight, R. T. (1998) Frontal lobe contributions to theory of mind. *J. Cogn. Neurosci.* **10,** 640–656.

Thomas, K. M., Drevets, W. C., Dahl, R. E., et al. (2001) Amygdala response to fearful faces in anxious and depressed children. *Arch. Gen. Psychiatry* **58,** 1057–1063.

Thompson, R. A. (1994) Emotion regulation: a theme in search of definition, in *The Development of Emotion Regulation.* (Fox, N. A., eds.), Monographs of the Society for Research in Child Development, Ann Arbor, MI, pp. 25–52.

van Elst, L. T., Woermann, F. G., Lemieux, L., Thompson, P. J., and Trimble, M. R. (2000) Affective aggression in patients with temporal lobe epilepsy: a quantitative MRI study of the amygdala. *Brain* **123,** 234–243.

Volkow, N. D. and Tancredi, L. (1987) Neural substrates of violent behaviour. A preliminary study with positron emission tomography. *Br. J. Psychiatry* **151,** 668–673.

Volkow, N. D., Tancredi, L. R., Grant, C., et al. (1995) Brain glucose metabolism in violent psychiatric patients: a preliminary study. *Psychiatry Res.* **61,** 243–253.

Weissenberger, A. A., Dell, M. L., Liow, K., et al. (2001) Aggression and psychiatric comorbidity in children with hypothalamic hamartomas and their unaffected siblings. *J. Am. Acad. Child Adolesc. Psychiatry* **40,** 696–703.

Wong, M., Fenwick, P., Fenton, G., Lumsden, J., Maisey, M., and Stevens, J. (1997) Repetitive and non-repetitive violent offending behaviour in male patients in a maximum security mental hospital—clinical and neuroimaging findings. *Med. Sci. Law* **37,** 150–160.

Zagrodzka, J., Hedberg, C. E., Mann, G. L., and Morrison, A. R. (1998) Contrasting expressions of aggressive behavior released by lesions of the central nucleus of the amygdala during wakefulness and rapid eye mocvement sleep without atonia in cats. *Behav. Neurosci.* **112,** 589–602.

3
The Serotonergic Dimension of Aggression and Violence

Klaus Peter Lesch

INTRODUCTION

Investigations of the mechanisms underlying impulsivity (individual tendency toward impulsive behavior) and aggressiveness (individual tendency toward aggressive behavior) seek to determine the neurobehavioral factors that underlie violence or other destructive behaviors in humans. Behavioral, imaging, pharmacologic, and genetic studies indicate that anatomically and functionally distinct neural circuits, as well as numerous neurotransmitters, growth factors, hormones, and their intracellular signaling effectors, influence impulsivity and aggressiveness in animal models and humans (Lesch and Merschdorf, 2000). Aggression is an umbrella term for various facets of aggressive behavior, which ranges from the establishment of hierarchies and dominance to antisocial behavior and delinquency. In both humans and animals, the term aggression comprises a variety of behaviors that are heterogeneous for neurobiological features and clinical phenomenology. Although most neurobiological studies of aggression and violence typically do not differentiate between premeditated and impulsive aggression, which is relatively unplanned and spontaneous but often culminates in physical violence, this distinction is likely relevant in understanding their genetic, functional neuroanatomical, and neurochemical bases.

It is believed that the propensity for impulsive aggression is associated with a low threshold for activating negative affect (a combination of emotion and mood including anger, distress, and agitation) and with a failure to respond appropriately to the anticipated negative consequences of behaving aggressively (Davidson et al., 2000). Social psychological research underscores the relation between aggression and emotion: negative affect frequently precipitates and accentuates aggressive behavior. While the impact of complex cultural variables on behavior impedes simple extrapolation of animal subtypes to humans, clinical observation, experimental paradigms in the laboratory, and cluster-factor analytic statistics have been used in attempts to subdivide aggression.

Based on a spectrum of different approaches, human aggression may be differentiated into several subtypes depending on the presence or absence of causes or motivation (spontaneous–impulsive or reactive–hostile, offensive or defensive, proactive–instrumental), nature of trigger (e.g., conditioned, response to narcissistic insult), characteristics of mediators (physiologic, biochemical, gender-specific, arousal–anger–affect-related, injurious),

From: *Neurobiology of Aggression: Understanding and Preventing Violence*
Edited by: M. Mattson © Humana Press Inc., Totowa, NJ

form of manifestation (cognitive, symbolic, verbal, physical, direct vs indirect, open vs concealed), direction (outward vs inward), and function (e.g., intentional harm, injury, or damage to subjects or objects, expression of an emotional–affective reaction, compensation of hypoarousal) (Archer and Browne, 1989; Buss, 1961; Crick and Dodge, 1996; Loeber and Stouthamer-Loeber, 1998; Scarpa and Raine, 1997). In a review on qualitatively distinct subtypes of human aggression, the dichotomy between an impulsive–reactive–hostile–affective subtype and a controlled–proactive–instrumental–predatory subtype has emerged as the most promising construct (Vitiello and Stoff, 1997).

The prevalence of aggression and violence in our society has stimulated both social and biological scientists to search for the predictors and causes. While serotonin (5-HT) stands out as the primary modulator of aggression, the other factors appear to indirectly involve 5-HT neurotransmission (for review, *see* Coccaro, 1992; Lesch et al., 2002; Nelson and Chiavegatto, 2001). Changes in 5-HT concentrations and metabolism, or in receptor subtype activation, density, and binding affinity, modify aggression-related behaviors. Here, I focus on the nature of deficits in central 5-HT function that might predispose an individual to these aberrant, often destructive forms of human behavior. Studies on the role of 5-HT in impulsivity and aggression, focuses on gene effects in inbred and knock-out strains of mice with increased aggression-related behavior are reviewed, and the relevance of several serotonergic gene variations in humans, which include high aggressiveness as part of the phenotype, are assessed. Although emphasis is put on the molecular psychobiology of 5-HT in aggression-related behavior, pertinent problems in the search for candidate genes for impulsivity and aggressiveness in humans and conceptual paradigms that have been applied to the interpretation of aggression-related phenotypes in knock-out mice are also discussed.

SEROTONERGIC SYSTEM IN EMOTION REGULATION

A neural circuit composed of several regions of the prefrontal cortex, amygdala, hippocampus, medial preoptic area, hypothalamus, anterior cingulate cortex, insular cortex, ventral striatum, and other interconnected structures and involving multiple neurotransmitter systems, such as 5-HT, has been implicated in emotion regulation including the associated affective phenomena of anger, negative affect, and impulsive aggression (Fig. 1) (for review, *see* Davidson et al., 2000). Although the brain systems mediating aggression appear to be fairly constant among mammals, several details of the regulatory pathways are species-specific. While genetic and environmental factors contribute to the structure and function of this circuitry, the amygdala is central to processes of learning to associate stimuli with events that are either punishing or rewarding.

The function of the amygdala in emotion regulation is highly complex. Perceptions of danger or threat are transmitted to the lateral nucleus of the amygdala, which projects to the basal nuclei where information regarding the social context derived from orbitofrontal projections is integrated with the perceptual information. Behavioral responses can then be initiated via activation of projections from the basal nuclei to various association cortices, while physiological responses can be produced via projections from the basal nuclei to the central nucleus and then to the hypothalamus and brainstem. Excessive or insufficient activation of the amygdaloid complex leads to either disproportionate negative affect or impaired sensitivity to social signals. The orbitofrontal cortex,

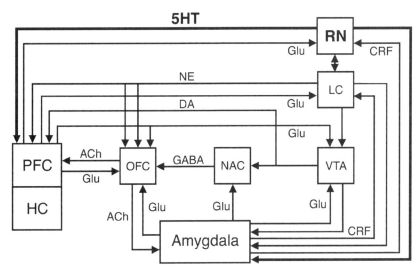

Fig. 1. Serotonergic modulation of the cortico-cingulo-amygdaloid circuit and emotion regulation (modified from Coupland, 2000). 5-HT and acetylcholine (ACh) induce direct prefrontal cortex activation, whereas norepinephrine (NE) operates indirectly via the forebrain cholinergic neurons. Dopamine (DA) acts via inhibition of γ-aminobutyric acid (GABA) input into the orbitofrontal cortex. Emotional responses activate the ACh, NE, and DA projections to the forebrain, thus facilitating attention, learning, and memory processes in conjunction with the septohippocampal system, whereas 5-HT suppresses them. DA influences response selection and increases motivation in relation to emotional stimuli, while 5-HT mediates response inhibition. Stress-induced sensitization of corticotropin-releasing hormone (CRH) activity influences monoamine function. Glu, glutamatergic projections; PFC, prefrontal cortex; OFC, orbitofrontal cortex; HC, hippocampus; NAC, nucleus accumbens; VTA, ventral tegmental area; LC, locus coeruleus; RN, raphe complex.

through its connections with other domains of the prefrontal cortex and with the amygdala, plays a critical role in limiting impulsive outbursts, and the anterior cingulate cortex recruits other neural systems during anger arousal and other negative emotions. Functional or structural abnormalities in these regions or in their interconnections can modify negative affect and aggressiveness, and it has therefore been suggested that impulsive aggression and violence arise as a consequence of a deficiency in emotion regulation (Davidson et al., 2000).

In humans, nonhuman primates, and other mammals, preclinical and clinical studies have accumulated substantial evidence that serotonergic signaling is a major modulator of emotional behavior, including anxiety, impulsivity, and aggression, and integrates complex brain functions such as cognition, sensory processing, and motor activity. This diversity of these functions is because 5-HT orchestrates the activity and interaction of several other neurotransmitter systems. The central serotonergic system, which originates in the midbrain and brainstem raphe complex, is widely distributed throughout the brain, and its chemical messenger is viewed as a master control neurotransmitter within this highly elaborate system of neural communication mediated by 14+ pre- and postsynaptic receptor subtypes with a multitude of isoforms (e.g., functionally relevant splice variants) and subunits. The prefrontal cortex receives major serotonergic input, which appears dysfunctional in individuals who are impulsively violent. Individuals vulnerable to faulty regulation of negative emotion are therefore at risk for aggression and violence.

The action of 5-HT as a messenger is regulated by 5-HT synthesizing and metabolizing enzymes and the 5-HT transporter. Serotonergic raphe neurons diffusely project to all brain regions implicated in aggressive behavior, while neurons in aggression-mediating areas are rich in both 5-HT1 and 5-HT2 receptor subtypes. In addition to its role as a neurotransmitter, 5-HT is an important regulator of morphogenetic activities during early brain development, as well as during adult neurogenesis and plasticity, including cell proliferation, migration, differentiation, and synaptogenesis (Azmitia and Whitaker-Azmitia, 1997; Lauder, 1993).

INDICES OF CENTRAL 5-HT FUNCTION

5-Hydroxyindoleacetic Acid in Cerebrospinal Fluid

5-HT has been hypothesized to exert inhibitory control over aggressiveness. Dysregulation of central serotonergic neurotransmission, whether physiologically occurring or pharmacologically induced, leads to deficits in waiting for reinforcement and withholding behavior to avoid punishment and, thus, has been linked to impulsive aggression and violence. Cerebrospinal fluid (CSF) concentrations of the major 5-HT metabolite 5-hydroxyindoleacetic acid (5-HIAA) is believed to reflect presynaptic serotonergic activity in the brain. Reduced CSF 5-HIAA has been found in aggressive patients with behavioral and neuropsychiatric disorders, impulsive violent men, and victims of violent suicide (Asberg, 1997; Linnoila and Virkkunen, 1992; Mann et al., 2001; Virkkunen and Linnoila, 1993). Furthermore, low 5-HIAA concentrations have been found to predict explosive aggression and impulsive violence in conduct-disordered boys and have been implicated in social incompetence (Siever and Trestman, 1993). In comparison with non-impulsive violent offenders, lower CSF 5-HIAA levels have been reported in impulsive violent offenders and impulsive fire setters. Included among these psychopathological patterns are high rates of criminality, alcohol dependence, and premature mortality from murder (Berman and Coccaro, 1998; Lewis, 1991).

Given their evolutionary similarity, the relationship between CSF 5-HIAA concentrations and behavior has been studied extensively in nonhuman primates. Male rhesus macaques with low concentrations of CSF 5-HIAA develop patterns of antisocial behavior that parallel deleterious impulse control deficits observed in their human male counterparts. Among these patterns are severe unrestrained aggression, impulsive risk-taking, excessive alcohol intake, and most severe forms of violence leading to trauma and early mortality (for details see below).

The synthesis of 5-HT depends on the availability of the amino acid tryptophan. By limiting dietary tryptophan, brain concentrations of 5-HT are reduced. Tryptophan depletion increased aggression in normal men in a laboratory setting (Cleare and Bond, 2000a; Moeller et al., 1996), induces behavioral disinhibition in aggressive adolescent males and in nonalcoholic young men with multigenerational family histories of paternal alcoholism (LeMarquand et al., 1998, 1999).

Imaging

In an investigation designed to evaluate the functional impact of a serotonergic challenge on regional cerebral glucose metabolism assessed with positron emission tomography (PET), the increase in glucose metabolic rate in prefrontal cortex and anterior cingulate cortex of normal subjects in response to the indirect 5-HT agonist fenfluramine was attenuated or absent in patients with aggressive impulsive personality disorder (Siever et al., 1999). This finding suggests that abnormalities in 5-HT function in regions of the prefrontal and anterior cingulate cortical regions may be especially important, however, other neurotransmitters, neuromodulators, and hormones including norepinephrine, dopamine, as well as corticotropin-releasing factor and testosterone are also likely involved. At the morphological level, patients with antisocial personality disorder who exhibit a propensity for impulsive aggression exhibit a reduction of prefrontal gray matter volume by magnetic resonance imaging (MRI) (Raine et al., 2000).

Studies of regional glucose metabolism assessed by PET also revealed prefrontal abnormalities in individuals liable to impulsive aggression (Raine et al., 1997, 1998). In individuals who committed murder, hypoactivation in prefrontal areas, including lateral and medial zones of the prefrontal cortex, as well as hyperactivation in the right but not the left amygdala, were reported. When these murderers were classified as those who committed affective and impulsive murder or those who committed planned predatory kill, only the affective impulsive murderers showed reductions in lateral prefrontal cortex metabolism.

Interestingly, the amygdala is activated in response to cues that connote threat (e.g., such as facial signs of fear), as well as during fear-conditioning and generalized negative affect, whereas increasing intensity of angry facial expressions or specifically induced anger is associated with increased activation of the orbitofrontal and anterior cingulate cortex (reviewed in Davidson et al., 2000. These activations are likely part of an automatic regulatory response that controls the intensity of expressed anger, and in individuals prone to aggression and violence, the increase in orbitofrontal and anterior cingulate cortex activation usually observed in such conditions would be attenuated. While normal individuals are able to voluntarily regulate their negative affect and also profit from restraint-producing cues in their environment, impulsive aggression may be the product of a failure of emotion regulation.

Pharmacological Studies

Pharmacological strategies of increasing 5-HT levels, such as the use of 5-HT precursors, 5-HT reuptake inhibitors, in addition to several 5-HT receptor subtype agonists, are able to reduce aggressive behavior in rodents. Pharmacological classification based on ligand binding experiments and on the study of functional responses to agonists–antagonists were initially utilized to define four 5-HT receptor subfamilies, 5-HT1–4. Molecular biology has subsequently both confirmed this classification and also revealed the existence of novel 5-HT receptor subtypes for which little pharmacological or functional data exist (5-HT1E, 5-HT1F, 5-HT5A, 5-HT5B, 5-HT6, and 5-HT7) (Hoyer and Martin, 1997). The recent discovery of an additional subunit B of the 5-HT3 receptor further adds to the complexity of serotonergic signaling. Current research is also focusing on participating constituents and regulatory mechanisms of gene transcription, messenger

Table 1
Serotonergic Genes Implicated
in Impulsivity and Aggression-Related Behaviors

	Rodents		Humans
	Pharmacologic studies (rats/mice)	Knock-out mice	Gene variation
5-HT receptors			
1A	↑?[a]	↓	nd
1B	↓[a], –[b]	↑	+
2A	↓[b]	nd	nd
3	↓[b]	nd	nd
7	nd		+
TPH	nd		+
MAOA	–	↑	+
5-HT transporter	–	↓	+

[a]Agonist-induced increase.
[b]Antagonist-induced decrease.
↑/↓, Increase/decrease in aggression and/or related behavior.
+, Association or linkage with impulsivity, aggression, and/or related behavior.
–, No effect.
nd, Not determined.

RNA (mRNA) processing and translation, as well as intracellular trafficking and posttranslational modification of proteins relevant to synaptic and postreceptor signaling.

In particular, the 5-HT1A and 5-HT1B receptor have been implicated in impulsivity and aggression-related behavior, and various models of rodent agonistic behavior, which differentiate between offensive aggression and defensive–flight models, have been described (Olivier and Mos, 1992; Olivier et al., 1995) (Table 1). Pharmacological challenge studies provide an in vivo method for studying central 5-HT function. The prolactin release in response to a single dose of the indirect 5-HT agonist fenfluramine and of the nonselective 5-HT agonist m-chlorophenylpiperazine (m-CPP) has been used to index central 5-HT activity. Lower prolactin responses to these enhancers of 5-HT transmission have been associated with aggressiveness (Coccaro et al., 1997), antisocial personality disorder (Moss et al., 1990), alcohol dependence (Soloff et al., 2000), and suicidal behavior (Coccaro et al., 1997). Traits of aggression in a large normative community-derived sample were associated with lower prolactin responses to fenfluramine in men (Manuck et al., 1998). Recently, pharmacological challenge of neuroendocrine and hypothermic responses to the 5-HT1A receptor partial agonist ipsapirone in aggressive men showed an inverse correlation between 5-HT1A receptor function and aggression (Cleare and Bond, 2000b; Moeller et al., 1998).

Nevertheless, documentation of the contribution of other 5-HT receptor subtypes in aggression is still constrained by the lack of selective ligands, but drugs that target the 5-HT2A, 5-HT2C, or 5-HT3 sites have generally not influenced aggressiveness. Thus, it may currently be more instructive to evaluate aggressive behavior systematically in constitutive and conditional knock-out mice of various 5-HT receptor subtypes. With the identification of 14+ 5-HT receptor subtypes, much work remains to be done to clarify the specific role of the various 5-HT receptors to discriminate between pre- and postsynaptic effects on aggression and the interactions among 5-HT receptor subtypes that underlie aggression.

GENETIC EVIDENCE

Quantitative Trait Locus and Gene-Targeting Strategies

Behaviors related to impulsivity and aggressiveness seem to delineate a biologically based model of dispositions to both normal and pathological functioning, with a continuum of genetic risk underlying personality and behavioral dimensions that extend from normal to abnormal. The analysis of genetic contributions to aggressive behavior is therefore conceptually and methodologically difficult, and consistent findings remain sparse. The documented heterogeneity of both genetic and environmental determinants suggests the futility of searching for unitary causes. This vista has therefore increasingly encouraged the pursuit of dimensional approaches to behavioral genetics (Plomin et al., 1994), and gene variants with a significant impact on the functionality of components of the 5-HT system are a rational strategy.

While systematic studies of the patterns of inheritance of impulsivity and aggressiveness indicate that these traits are likely to be influenced by many genes, making them polygenic or "quantitative" traits, behavioral genetics convincingly documents the significance of environmental factors. Individual differences in impulsivity and aggressiveness and the ultimate behavioral consequences, such as distinct types of aggression, violence, self-injurious behavior, including suicidality, and addiction, are relatively enduring, continuously distributed, as well as substantially heritable, and, therefore, are likely to result from additive or nonadditive interaction of genetic variations with environmental influences (McGue and Bouchard, 1998). Joint genetic and environmental influences on impulsivity and aggression were recently confirmed in mono- and dizygotic male twin pairs (Seroczynski et al., 1999). While the results indicate significant heritability of a nonadditive nature, additive genetic variance accounted for 47% of the individual differences.

Mice strains that have been selectively bred to display a phenotype of interest are currently being used to identify genetic loci that contribute to behavioral traits. This quantitative trait loci (QTL) approach has been applied to a trait in mice called "emotionality" (Flint et al., 1995). Such linkage analyses, however, provide only a rough chromosomal localization, whereas the next step, identifying the relevant genes by positional cloning, remains a challenging task. Since mice and humans share many orthologous genes mapped to syntenic chromosomal regions, it is conceivable that individual genes, successfully identified for one or more types of murine aggressive behavior, may be developed as animal models for human aggression. Although mice and humans share >90% of their genes in common and the underlying molecular and neurobiologic mechanisms found in aggressive mice are also reported in humans displaying aggression, aggression observed in mice is rarely directly comparable. Following chromosomal mapping of polymorphic genes and evaluation of gene function using knock-out mutants, behavioral parameters, including the type of aggression, measure of aggression, test situation, and opponent type, are investigated (Maxson, 1996). Thus, the combination of elaborate genetic and behavioral analyses provides an attractive tool to discover new candidate genes with effects on variation and development of one or more forms of murine aggressive behavior.

Notwithstanding the confounding issues, the "classic" knock-out mouse remains a powerful tool for modeling the genetic basis of behavior (Gingrich and Hen, 2000). Constitutively created mutations mimic genetic variability, in the sense that they are present during the entire developmental process and that the spectrum seen in human behavior is the result of developmental adaptation. As demonstrated in knock-out mice, the developmental impact of a mutation might be more prominent than the actual loss of function that occurs with its absence in adulthood. A major challenge is therefore the identification of neural mechanisms that underlie aggressiveness, and attempts to unravel the genetic basis of impulsivity and aggressiveness should also reflect on the complex nature of these traits, which is expressed in many different facets. Such facets can be distinguished by specific testing procedures that identify particular categories of aggression, such as isolation-induced offensive aggression, defensive aggression, predatory aggression, shock-induced or irritability-associated aggression, and infanticide (Tecott and Barondes, 1996). Despite the usefulness of the gene knock-out strategy in identifying specific gene products that may be involved in aggression, this approach is limited to known candidate genes. Owing to the complexity in the expression of aggressive behavior, it is impossible to predict which genes contribute to the variability of this trait in different populations. Thus, QTL analysis, although technically demanding, should ultimately prove to be an important complementary approach, because it is likely that identification of the particular alleles of the various genes that influence aggressiveness in inbred strains will facilitate elucidation of epistatic (gene–gene) interactions, as well as the phenomenon of pleiotropy, which is the multiple and apparently independent effects of a genes on phenotypical expression.

Serotonergic-Related Genes

5-HT Receptor Subtypes

The 5-HT1B receptor was the first subtype to have its gene inactivated by classical homologous recombination (Saudou et al., 1994). 5-HT1B receptors are expressed in the basal ganglia, central gray, lateral septum, hippocampus, amygdala, and raphe nuclei, either at presynaptic terminals inhibiting 5-HT release or as a heteroreceptor modulating the release of other neurotransmitters. Selective agonists and antagonists for 5-HT1B receptors are largely lacking, but indirect pharmacological evidence suggests that 5-HT1B activation influences food intake, sexual activity, locomotion, and aggression. Mice with a targeted disruption of the 5-HT1B gene therefore facilitated investigation of the concept of 5-HT-related impulsivity in the context of aggressive behavior. Two of the behaviors, locomotion and aggression, postulated to be modulated by 5-HT1B receptors were analyzed (Ramboz et al., 1996). Wild-type and homozygous null mutant (5-HT1B–/–) mice were found to display similar levels of locomotor activity in an open field. Impulsivity and aggression-related behavior of 5-HT1B–/– male mice was assessed by isolation and subsequent exposure to a nonisolated male wild-type intruder mouse. The latency and number of attacks displayed by the knock-out mice were used as indices of aggression. The 5-HT1B–/– mice, when compared with wild-type mice, showed more rapid, more intense, and more frequent attacks. Lactating female 5-HT1B−/− mice also attack unfamiliar male mice more rapidly and violently. In addition to increased aggression, knock-out mice acquire cocaine self-administration faster and ingest more ethanol than controls (Brunner and Hen, 1997). Thus, the 5-HT1B receptor modulates not only motor impulsivity and aggression but also addictive behavior.

Nevertheless, these results further support the notion that the 5-HT1B subtype is not the only 5-HT receptor modulating this behavior. In particular, 5-HT1A receptor activation may influence aggressive behavior. In contrast to 5-HT1B knock-out mice, 5-HT1A knock-outs are less reactive and possibly less aggressive, but show more anxiety-related behavior than control mice (for review, see Lesch and Mössner, 1999; Zhuang et al., 1999), although both 5-HT1A and 5-HT1B receptors control the tone of the serotonergic system and mediate some of the postsynaptic 5-HT effects. The regional variation of 5-HT receptor expression and the complex autoregulatory processes of 5-HT function, which are operational in different brain areas, may lead to a plausible hypothesis to explain this apparent contradiction. 5-HT1A receptor expression was assessed in male mice selected for high and low offensive aggression showed that high-aggressive mice, characterized by a short attack latency, decreased plasma corticosterone concentration, and increased levels of 5-HT1A mRNA in the dorsal hippocampus (dentate gyrus and CA1) compared to low-aggressive mice that had long attack latency and high plasma corticosterone levels (Korte et al., 1996). Increased postsynaptic 5-HT1A receptor radioligand binding was found in dentate gyrus, CA1, lateral septum, and frontal cortex, whereas no difference in ligand binding was found for the 5-HT1A autoreceptor on cell bodies in the dorsal raphe nucleus. These results suggest that high offensive aggression is associated with reduced (circadian peak) plasma corticosterone and increased postsynaptic 5-HT1A receptor availability in limbic and cortical regions. Although pharmacologic evidence has implicated other 5-HT receptor subtypes in the modulation of aggression (Rudissaar et al., 1999), few studies have employed genetic approaches, such as gene targeting.

Based on the relationship of aggression, suicide, and drug abuse in clinical samples, the association of psychopathology with 5-HT1B receptor gene and postmortem human brain 5-HT1B receptor binding was recently studied (Huang et al., 1999). Two common polymorphisms were identified in the 5-HT1B receptor gene, a silent C to T substitution at nucleotide 129 and a silent G to C substitution at nucleotide 861 of the amino acid coding region. While the C129 or G861 allele had 20% fewer 5-HT1B receptors compared to the 129T or 861C allele, no association between suicide, major depression, alcoholism, or pathological aggression with 5-HT1B receptor binding indices or genotype was identified. Lappalainen et al. (1998) have investigated whether the 5-HT1B gene (the HTR1B G861C polymorphism and the short-tandem-repeat locus D6S284) is linked to impulsive and aggressive behavior in two patient populations, Finnish sibling pairs and a large multigenerational family derived from a Southwestern American Indian tribe, with antisocial personality disorder and intermittent explosive disorder co-morbid with alcoholism. While Finnish antisocial alcohol-dependent patients had a significantly higher HTR1B-861C allele frequency than the other Finns, significant sib-pair linkage of antisocial alcoholism to HTR1B-G861C to D6S284 was observed, thus indicating that a locus predisposing to antisocial behavior and aggressiveness associated with alcoholism may be linked to HTR1B at 6q13-15.

Pesonen and coworkers (1998) have reported that a Pro279Leu amino acid substitution in the 5-HT7 receptor gene may be a predisposing allele in a subgroup of Finnish alcoholic offenders with multiple behavioral problems. The present challenge, however, is to further characterize the physiological relevance of the large variety of 5-HT receptor gene products, establish their function as endogenous receptors, find selective ligands, and determine potential therapeutic application of these compounds. However, many of these receptors are remarkably similar in their ligand-binding domains, and it has been difficult to design pharmacological compounds that will specifically interact with a single subtype. The new insights into neural plasticity and complexity of gene regulation in 5-HT subsystems will eventually provide the means for novel approaches of studying 5-HT receptor subtype-related behaviors at the molecular level.

At the next dimension of complexity, signaling through 5-HT receptors involves different transduction pathways, and each receptor subtype modulates distinct, though frequently interacting, second messenger systems and multiple effectors. The gene of the effector enzyme calcium-calmodulin kinase II (CamKII), which participates in some intracellular responses to 5-HT receptor activation, has also been implicated in aggressive behavior by a knock-out experiment (Chen et al., 1994). While CamKII−/− mutants showed global behavioral impairment, male mice heterozygous for the inactivated CamKII gene had a greater tendency to fight with each other when housed together. In detail, they showed enhanced offensive aggression, normal defensive aggression, and decreased fear-related responses.

The discovery of a considerable number of hyperaggressive mutant strains in the course of gene knock-out experiments highlights the extraordinary diversity of genes involved in the genetic influence on impulsivity and aggression. Interestingly, genetic support for a role of 5-HT in aggression also derives from mice lacking specific genes, including neuronal nitric oxide (NO) synthase (nNOS) and neural cell adhesion molecule (NCAM), that either directly or indirectly affect 5-HT turnover or 5-HT receptor sensitivity. Male nNOS−/− mice and wild-type mice, in which nNOS is pharmacologically suppressed, are highly aggressive (Chiavegatto et al., 2001). Excessive aggressiveness and impulsiveness of nNOS knock-out mice are caused by a selective decrease in 5-HT turnover and deficient 5-HT1A and 5-HT1B receptor function in brain regions regulating emotion. The NCAM plays a critical role during brain development and in adult plasticity, and NCAM−/− mice display elevated anxiety and aggression (Stork et al., 1997, 2000). Lower doses of 5-HT1A agonists are necessary to reduce anxiety and aggressiveness in the NCAM-deficient mice, suggesting a functional change in the 5-HT1A receptor, although 5-HT1A binding as well as brain 5-HT and 5-HIAA tissue concentrations were unaltered (Stork et al., 1999). Taken together, these findings indicate an involvement of nNOS and NCAM and the 5-HT system through 5-HT1A and 5-HT1B receptors, but the specific molecular mechanisms in aggression remain to be elucidated.

Tryptophan Hydroxylase

Brain serotonergic activity correlates inversely with human aggressive behavior and individual differences in aggressive disposition are influenced by genetic factors (Table 1). The first step of 5-HT biosynthesis in 5-HT neurons is catalyzed by the rate-limiting enzyme tryptophan hydroxylase (TPH). A role of L-tryptophan availability and of TPH activity in impulsivity, aggressiveness, and associated suicidality has been reported by several studies of psychiatric patients or offender populations (Dougherty et al., 1999). For example, an increase in aggressive responses on a free-operant laboratory measure of aggression following experimental tryptophan depletion in healthy males was recently shown, supporting the hypothesis that low plasma tryptophan concentration and associated decrease in brain 5-HT facilitates aggression-related behavior.

The human TPH gene located on chromosome 11p is a member of the aromatic amino acid hydroxylase family, spans a region of 29 kb, and contains at least 11 exons (transcribed DNA sequence after splicing) (Boularand et al., 1995). An unusual splicing complexity in the 5'-untranslated region (5'-UTR) results in at least four TPH mRNA species transcribed from a single transcriptional start site. Although a detailed analysis of the gene's transcriptional control region is still lacking, DNA elements important for serotonergic neuron-specific expression of TPH appear to be contained in 6.1 kb of 5'-flanking transcriptional control region of the mouse TPH gene (Huh et al., 1994; Son et al., 1996). A mouse model with a targeted disruption of the TPH gene will hopefully soon be available for an assessment of the effect of 5-HT deficiency on aggressive behavior (J. Mallet, personal communication).

Several common gene variations have been described in the 5'-flanking regulatory region (T-7180G, C-7065T, A-6526G, and G-5806T, designated as nucleotides upstream of the translation start codon) and in intron 7 (A218C and C779A) (an intron is the transcribed DNA sequence that is removed from a transcript by splicing) (Rotondo et al., 1999), while functional variants have not been reported in the coding sequence of this gene (Han et al., 1999). In a landmark study, Nielsen et al. (1994) reported that the TPH A779C polymorphism influences 5-HIAA concentrations in CSF and may predispose to suicidality, a pathophysiological mechanism that may involve impaired impulse control. This finding was subsequently replicated by the same group using a family-based design in an extended sample of Finnish alcoholic offenders (Nielsen et al., 1998).

Additional investigations indicate that the intronic polymorphism may be associated with interview and self-report measures of aggression and anger-related traits of personality, as well as central nervous system (CNS) 5-HT activity assessed by pharmacologic challenge (prolactin response to fenfluramine) in healthy volunteers (Manuck et al., 1999). Similarly, in a population of male personality disorder patients, individual differences in aggressive disposition, but not prolactin response to fenfluramine, was associated with the intronic TPH genotypes (New et al., 1998). In Finnish offenders, previously studied for the TPH intron7 C779A polymorphism, a significant association was observed between the TPH promoter polymorphism A-6526G and suicidality (Rotondo et al., 1999). Although not consistently replicated in other populations (Abbar et al., 1995; Furlong et al., 1998), these findings in conjunction with results from association studies in various psychiatric disorders, including bipolar disorder and alcoholism (Bellivier et al., 1998; Manuck et al., 1999), further support the notion that functional variant(s) in or close to the TPH gene may predispose individuals to suicidality or externally directed aggressiveness and underscore the relevance of synthesis-dependent 5-HT homeostasis in the expression of other behaviors thought to be influenced by 5-HT.

Monoamine Oxidase A

Alterations in monoamine oxidase A (MAOA) activity have been implicated in a wide range of behavioral traits and disorders (Table 1). MAOA is a mitochondrial enzyme that oxidizes 5-HT, norepinephrine, as well as dopamine, and is expressed in a cell-type selective manner. MAOA-deficient mice were generated accidentally by the replacement of exons 2 and 3 of the MAOA gene with an interferon transgene (Cases et al., 1995). Mice with a targeted inactivation of the MAOA gene display elevated brain levels of 5-HT, norepinephrine, and dopamine, increased reactivity to stress, hyperactive startle responses, violent motions during sleep, and abnormal posture and aggressive behavior. Enhanced male aggressiveness was demonstrated by resident–intruder tests and by increased injury between male cagemates. The increased aggressiveness of the MAOA mutant mice was indicated by the large percentage of mutant males that become wounded under standard group housing conditions and confirmed by enhanced offensive aggression by mutants in the resident–intruder assay (Seif and De Maeyer, 1999). The MAOA mutants also displayed increased copulatory behavior of males with nonreceptive female mice.

Since these phenotypical alterations are restrained by 5-HT synthesis inhibition but not by catecholamine synthesis suppression, the observed behavioral abnormalities are likely to be specifically the result of attenuated 5-HT degradation. MAOA-deficient mice also show disrupted formation of sensory maps in the visual and somatosensory systems (cortical barrelfields) (Cases et al., 1996; Salichon et al., 2001), underscoring the role of 5-HT as a morphogenetic factor in brain development. It remains to be elucidated, however, whether some of the behavioral abnormalities are influenced by structural abnormalities.

The aggressive phenotype of MAOA-deficient mice appears to complement the behavioral consequences of a mutation in the coding region of the human MAOA gene. This X-linked hemizygous chain termination mutation has been linked to mild mental retardation and occasional episodes of impulsive aggression, arson, and hypersexual behavior, such as attempted rape and exhibitionism, in affected males from a single large family (Brunner et al., 1993). Affected males exhibit markedly disturbed monoamine metabolism and an absence of MAOA enzymatic activity in cultured fibroblasts. A nonconservative point mutation was found in all affected males and all carrier females; the mutation introduces a stop codon (base triplet, e.g., TAA, that serves as a signal for termination of transcription) at position 296. Although inhibition of MAOA in adults leads to antidepressant effects but not aggression-related behavior, the deviate behavior in MAOA-deficient men may be due to structural or compensatory changes resulting from altered monoamine metabolism during neurodevelopment. However, screening of volunteers in the general population and from putative high-risk groups for possible MAO deficiency states suggests that marked MAO deficiency states are very rare (Schuback et al., 1999). The fact that humans with an inactive MAOA gene also show increased impulsive aggression and sexual aggressiveness demonstrates the potential relevance of mutant mouse models to human behavior, although the rarity of the human mutation indicates that other genetic and/or nongenetic influences contribute to these forms of misconduct.

The human MAOA gene is localized on chromosome Xp11.23, extends over 70 kb, and is composed of 15 exons (Shih et al., 1993; Zhu et al., 1992). Two species of MAOA mRNA, 2.1 kb and 5.0 kb, are generated by the use of two alternative polyadenylation sites. While there is considerable controversy regarding the site where mRNA synthesis is initiated, tissue-selective length variability of the 5'-UTR with multiple transcription start sites clustered primarily around a initiator element, which may also act as a negative regulatory element have been reported (Denney et al., 1994; Zhu et al., 1994). The core promoter region contains two 90-bp repeat sequences, which are further divided into four imperfect tandem repeats, each containing an Sp1 binding site in reversed orientation (Zhu and Shih, 1997).

Although the MAOA gene is a potential candidate for affective illness, none of several previously described gene variants are consistently associated with disease. A functional 30-bp repeat polymorphism was identified in the promoter region of the human MAOA gene that differentially modulates gene transcription (Deckert et al., 1999; Sabol et al., 1998) as well as enzyme activity in fibroblasts. A corresponding functional repeat polymorphism was recently found in rhesus monkeys (Y. Syagailo and K.P. Lesch, manuscript submitted). Variation in the number of repeats (three to five) of this MAOA gene-linked polymorphic region (MAOA-LPR) had different transcriptional efficiency when fused to a luciferase reporter gene and transfected into cell lines. The transcriptional efficiency of the three-repeat allele was two-fold lower than those with longer repeats, and enzyme activity is correlated with repeat length (Denney et al., 1999). Preliminary evidence indicates that length variation of the MAOA-LPR confers vulnerability to antisocial behavior in alcohol-dependent males (Samochowiec et al., 1999), is linked to impulsivity, hostility, and lifetime aggression history, as well as brain serotonergic function in a community sample of men (Manuck et al., 2000), and appears to be a risk factor for panic disorder and unipolar depression in female patients (Deckert et al., 1999; Schulze et al., 2000), but not for other psychiatric disorders (Furlong et al., 1999; Syagailo et al., 2001).

5-HT Transporter

While 5-HT controls a highly complex system of neural communication mediated by multiple pre- and postsynaptic 5-HT receptor subtypes, high-affinity 5-HT transport into the presynaptic neuron is mediated by a single protein. The 5-HT transporter (5-HTT) removes 5-HT from the synaptic cleft and determines the magnitude and duration of postsynaptic receptor-mediated signaling, thus playing a pivotal role in the fine-tuning of 5-HT neurotransmission (for review, see Lesch, 1997). The 5-HTT is also the initial target for several antidepressant drugs (e.g., clomipramine, fluoxetine), which also display anti-aggressive properties. A polymorphism in the 5'-flanking transcriptional control region of the 5-HTT gene that results in allelic variation in functional 5-HTT expression is associated with anxiety, depression, and aggression-related personality traits (Lesch et al., 1996). Advances in 5-HTT gene knock-out studies are also changing views of the relevance of adaptive 5-HT uptake function and 5-HT homeostasis in brain development and plasticity, as well as processes underlying aggressiveness and substance abuse (reviewed in Lesch, 2001).

The human 5-HTT is encoded by a single gene (*SLC6A4*) on chromosome 17q11.2 (Shen et al., 2000). It is composed of 14 to 15 exons spanning approx 35 kb (Lesch et al., 1994). Transcriptional activity of the human 5-HTT gene is modulated by a 44-bp length variation of a repetitive element, the 5-HTT gene-LPR (5-HTT-LPR) located upstream of the transcription start site. Additional variations have been described in the 5'-UTR due to alternative splicing of exon 1B, in intron 2 (variable number of a 16/17-bp tandem repeat, VNTR-17), and in the 3'-UTR (for review, *see* Lesch and Mössner, 1998). Comparison of different mammalian species confirmed the presence of the 5-HTT-LPR in platyrrhini and catarrhini (hominoids, cercopithecoids), but not in prosimian primates and other mammals (Lesch et al., 1997). Thus, the 5-HTT-LPR is unique to humans and simian primates. In humans, the majority of alleles are composed of either 14- or 16-repeat elements (short and long allele, respectively), although alleles with up to 22-repeat elements sometimes occur, as do variants with single-base insertions and/or deletions or substitutions within individual repeat elements are rare. A predominantly Caucasian population displayed allele frequencies of 57% for the long (l) allele and 43% for the short (s) allele with a 5-HTT-LPR genotype distribution of 32% l/l, 49% l/s, and 19% s/s (Lesch et al., 1996); different allele and genotype distributions are found in other populations (Gelernter et al., 1997; Kunugi et al., 1997).

The GC-rich sequence of the 5-HTT-LPR is likely to give rise to the formation of DNA secondary structure that has the potential to regulate 5-HTT gene transcription. The short (s) and long (l) 5-HTT-LPR variants differentially modulate transcriptional activity of the 5-HTT gene promoter, 5-HTT protein concentration, and 5-HT uptake activity in human lymphoblastoid cell lines (Lesch et al., 1996). Membrane preparations from l/l lymphoblasts showed higher inhibitor binding than did s/s cells. Furthermore, the rate of specific 5-HT uptake was more than two-fold higher in cells homozygous for the l form of the 5-HTT-LPR than in cells carrying one or two copies of the s variant of the promoter. Further evidence from studies of 5-HTT gene promoter activity in other cell lines (Mortensen et al., 1999), mRNA concentrations in the raphe complex of human postmortem brain (Little et al., 1998), platelet 5-HT uptake and content (Greenberg et al., 2000; Hanna et al., 1998; Nobile et al., 1999), 5-HT system responsivity elicited by pharmacologic challenge tests with clomipramine and fenfluramine (Reist et al., 2001; Whale et al., 2000), mood changes following tryptophan depletion (Neumeister et al., 2002), and in vivo single photon emission computed tomography (SPECT) imaging of human brain 5-HTT (Heinz et al., 1999) confirmed that the s variant is associated with lower 5-HTT expression and function.

Despite growing evidence for a potential role of the 5-HTT in the integration of synaptic connections in the mammalian brain during development, adult life, and old age, detailed knowledge of the molecular mechanisms involved in this fine-tuning process is just beginning to emerge. For this reason, we will draw upon examples and issues that have emerged from our and collaborating groups' work on the 5-HTT gene to illustrate several critical points.

5-HTT AND BEHAVIORAL TRAITS RELATED TO IMPULSIVITY AND AGGRESSIVENESS

Following systematic attempts to characterize genetically driven variation in 5-HT uptake function, the 5-HTT has assumed importance as a piece in the mosaic-like texture of personality traits, such as anxiety, negative emotionality, impulsivity, and aggressiveness (Table 1). The contribution of 5-HTT-LPR variability to individual phenotypic differences in temperament, personality, and behavior was explored in two independent population–family genetic studies. In our initial study, we found population and within-family associations between the low-expressing s allele and neuroticism, a trait related to anxiety, hostility, and depression, on the NEO personality inventory; revised (NEO-PI-R), a self-report inventory based on the five-factor model of personality ("Big

Five") (Costa and McCrae, 1992), in a primarily male population ($n = 505$), and that the s allele was dominant (Lesch et al., 1996). Individuals with either one or two copies of the short 5-HTT-LPR variant (group S) had significantly greater levels of neuroticism, defined as proneness to negative emotionality, including anxiety, hostility, and depression, than those homozygous for the long genotype (group L) in the sample as a whole and also within sibships. Individuals with 5-HTT-LPR S genotypes also had significantly decreased agreeableness, a dimension reflecting expression of a spectrum of traits ranging from cooperativeness to aggressiveness. Recently, this association was reassessed in a new sample ($n = 397$, 84% female, primarily sib-pairs). The findings robustly replicated the 5-HTT-LPR-neuroticism association, and the dominance of the s allele (Greenberg et al., 2000). Combined data from the two studies ($n = 902$) gave a highly significant association between the s allele and higher NEO neuroticism, both across individuals and within families, reflecting a genuine genetic influence rather than an artifact of ethnic admixture.

Another association encountered in the original study between the s allele and lower scores of NEO agreeableness, including the subscales straightforwardness, compliance, and trust, was also replicated and was stronger in the primarily female replication sample. Gender-related differences in 5-HTT-LPR-personality trait associations are possible, since several lines of evidence demonstrate gender-related differences in the 5-HT system functioning in humans and in animals (Fink et al., 1999; McQueen et al., 1997). These findings include effects of gonadal steroids on 5-HTT expression in rodent brain and differences in anxiety-related behaviors in male and female 5-HTT knockout mice. While such evidence provides a theoretical basis for possible gender-related differences in the 5-HTT-LPR-personality association, we found that the 5-HTT-LPR has a qualitatively similar influence in women and men. However, the results of the replication sample suggest a possibly stronger association between 5-HTT-LPR-S genotypes and a predisposition to lower agreeableness and related traits in women.

These findings show that the 5-HTT-LPR influences a constellation of neuroticism and agreeableness, traits of negative emotionality related to interpersonal hostility and aggression. Other efforts to detect associations between the 5-HTT-LPR and personality traits, which are discussed elsewhere (Greenberg et al., 2000; Lesch et al., 2000), have been complicated by the use of small sample sizes, heterogeneous subject populations, ethnic and sociocultural characteristics, and differing methods of personality assessment. The relationship between these two aspects of negative emotionality is not unexpected in view of the previously observed negative correlation between angry hostility, a subscale of neuroticism, and agreeableness, indicating that both dimensions assess a behavioral predisposition toward uncooperative and aggressive interpersonal behavior. The effect sizes for the 5-HTT-LPR-personality associations, which were comparable in the two samples, indicate that this polymorphism has a moderate influence on these behavioral predispositions of approx 0.30 standard deviation unit. This corresponds to 3 to 4% of the total variance and 7–9% of the genetic variance, based on estimates from twin studies using these and related measures that have consistently demonstrated that genetic factors contribute 40–60% of the variance in personality traits. Thus, the results are consistent with the view that the influence of a single common polymorphism on continuously distributed traits is likely to be small in humans, as well as different quantitative characteristics in other species (Plomin et al., 1994).

Recently, Hariri and associates (2002) reported that individuals with one or two copies of the short 5-HTT-LPR variant exhibit greater amygdala neuronal activity, as assessed by functional MRI (fMRI), in response to fearful stimuli compared with individuals homozygous for the long allele. These findings confirm that genetically driven variation of serotonergic function contributes to the response of brain regions underlying human emotional behavior and indicate that differential excitability of the amygdala to emotional stimuli may contribute to increased fear and anxiety-related responses including defensive aggression.

At first sight, association between the high-activity 5-HTT-LPR l allele with lower neuroticism and related traits seemed inconsistent with the known antidepressant, antianxiety, and anti-aggressive effects of 5-HTT inhibitors (serotonin reuptake inhibitors [SRIs]). Likewise, Knutson et al. (1998) reported that long-term inhibition of the 5-HTT by the SRI paroxetine reduced indices of hostility through a more general decrease in negative affect, which is a personality dimension related to neuroticism. The same individuals also demonstrated an increase in directly measured social cooperation after paroxetine treatment, an interesting finding in view of the replicated reciprocal association between 5-HTT-LPR genotype and agreeableness. That a drug which inhibits the 5-HTT lessened negative emotionality and increased social cooperation appears to conflict with findings that the 5-HTT-LPR long allele, which confers greater 5-HTT expression, is associated with lower NEO neuroticism and higher NEO agreeableness.

The conclusion that the 5-HTT may affect personality traits via an influence on brain development is strongly supported by recent findings in rodents. Studies in rats confirmed that the 5-HTT gene is expressed in brain regions central to emotional behavior during fetal development, but not later in life (Hansson et al., 1998, 1999), hence enduring individual differences in personality could result from 5-HTT-LPR-driven differential 5-HTT expression during pre- and perinatal life. In support of this notion, it has recently been reported that prenatal exposure of mice to a clinically relevant dose of paroxetine produced no major behavioral alterations, but increases in some anxiety-related measures in infant offspring and on aggressive behavior in adult males (Coleman et al., 1999). Altered serotonergic function associated with desensitization of 5-HT1A and 1B receptors have also been implicated in increased anxiety and attenuated aggression-related behaviors recently found in 5-HTT-deficient mice (Bengel et al., 1998; Fabre et al., 2000; la Cour et al., 2001; Li et al., 2000; Murphy et al., 2001). Male 5-HTT–/– mice are slower to attack the intruder and attacked with less frequency than control littermates, but showed no alterations of social interaction (Holmes et al., 2002). Heterozygous 5-HTT knock-out mice were as quick to attack, but made fewer overall attacks, as compared to controls. Aggression increased with repeated exposure to an intruder in 5-HTT+/– and control mice, but not in 5-HTT–/– mice.

CORTICAL DEVELOPMENT IN 5-HTT KNOCK-OUT MICE

Morphological analyses of brain structures where 5-HT has been suggested to act as a differentiation signal in development revealed a detrimental effect of 5-HTT inactivation on the formation and plasticity of cortical and subcortical structures. The timing of serotonergic innervation coincides with pronounced growth and synaptogenesis in the cortex, and perinatal manipulations of 5-HT affects cortical 5-HT receptors. The period for 5-HT action corresponds to the period when incoming axons begin to establish synaptic interactions with target neurons and to elaborate a profuse branching pattern. Investigations of 5-HT participation in neocortical development and plasticity have been concentrated on the rodent somatosensory cortex (SSC), due to its one-to-one correspondence between each whisker and its cortical barrel-like projection area. The processes underlying patterning of projections in the SSC have been intensively studied with a widely held view that the formation of somatotopic maps does not depend on neural activity. While pharmacologically induced 5-HT depletion at birth yields smaller barrels, but does not prevent the formation of the barrel pattern itself (Bennett-Clarke et al., 1994; Osterheld-Haas et al., 1994), excess of extracellular 5-HT, as demonstrated in MAOA knock-out, results in a complete absence of cortical barrel patterns (Cases et al., 1996). Additional evidence for a role of 5-HT in the development of neonatal rodent SSC derives from the transient barrel-like distribution of 5-HT, 5-HT1B, and 5-HT2A receptors, and of the 5-HTT (Lebrand et al., 1996; Mansour-Robaey et al., 1998). The transient barrel-like 5-HT pattern visualized in layer IV of the SSC of neonatal rodents apparently stems from 5-HT uptake and vesicular storage in thalamocortical neurons, which express both the 5-HTT and the vesicular monoamine transporter (VMAT2) at this developmental stage.

Inactivation of the 5-HTT gene profoundly disturbs formation of the SSC with altered cytoarchitecture of cortical layer IV, which is the layer that contains synapses between thalamocortical terminals and their postsynaptic target neurons (Persico et al., 2001) Brains of 5-HTT knock-out mice display no or only very few barrels. Cell bodies as well as terminals, typically more dense in barrel septa, appear homogeneously distributed in layer IV of adult 5-HTT knock-out brains. Injections of a 5-HT synthesis inhibitor within a narrow time window of 2 d postnatally completely rescued formation of SSC barrel fields. Of note, heterozygous knock-out mice develop all SSC barrel fields, but frequently present irregularly shaped barrels and less defined cell gradients between septa and barrel hollows. These findings demonstrate that excessive concentrations of extracellular 5-HT are deleterious to SSC development and suggest that transient 5-HTT expression in thalamocortical neurons is responsible for barrel patterns in neonatal rodents, and its permissive action is required for normal barrel pattern formation, presumably by maintaining extracellular 5-HT concentrations below a critical threshold. Because normal synaptic density in SSC layer IV of 5-HTT knock-out mice was shown, it is more likely that 5-HT affects SSC cytoarchitecture by promoting dendritic growth toward the barrel hollows, as well as by modulating cytokinetic movements of cortical granule cells, similar to concentration-dependent 5-HT modulation of cell migration described in other tissues. Because the gene–dose-dependent reduction in 5-HTT availability in heterozygous knock-out mice, which leads to a modest delay in 5-HT uptake but distinctive irregularities in barrel and septum shape, is similar to those reported in humans carrying low activity allele of the 5-HTT-LPR, it may be speculated that allelic variation in 5-HTT function also affects the human brain during development with due consequences for disease liability and therapeutic response.

These findings demonstrate that excessive amounts of extracellular 5-HT are detrimental to SSC development and suggest that transient 5-HTT expression and its permissive action is required for barrel pattern formation, presumably by maintaining extracellular

5-HT concentrations below a critical threshold. Two key players of serotonergic neurotransmission appear to mediate the deleterious effects of excess 5-HT: the 5-HTT and the 5-HT1B receptor. Both molecules are expressed in primary sensory thalamic nuclei during the period when the segregation of thalamocortical projections occurs (Bennett-Clarke et al., 1996; Hansson et al., 1998; Lebrand et al., 1996). 5-HT is internalized via 5-HTT in thalamic neurons and is detectable in axon terminals (Cases et al., 1998; Lebrand et al., 1996). The presence of the VMAT2 within the same neurons allows internalized 5-HT to be stored in vesicles and used as a co-transmitter of glutamate. Lack of 5-HT degradation in MAOA knock-out mice, as well as severe impairment of 5-HT clearance in mice with an inactivation of the 5-HTT, results in an accumulation of 5-HT and overstimulation of 5-HT receptors all along thalamic neurons (Cases et al., 1998). Since 5-HT1B receptors are known to inhibit the release of glutamate in the thalamocortical somatosensory pathway, excessive activation of 5-HT1B receptors could prevent activity-dependent processes involved in the patterning of afferents and barrel structures. This hypothesis is supported by a recent study using a strategy of combined knock-out of MAOA, 5-HTT, and 5-HT1B receptor genes (Salichon et al., 2001). While only partial disruption of the patterning of somatosensory thalamocortical projections was observed in 5-HTT knock-out, MAOA/5-HTT double-knock-out mice showed that 5-HT accumulation in the extracellular space causes total disruption of the patterning of these projections. Moreover, the removal of 5-HTB receptors in MAOA and 5-HTT knock-out, as well as in MAOA/5-HTT double-knock-out mice, allows a normal segregation of the somatosensory projections. These findings point to an essential role of the 5-HT1B receptor in mediating the deleterious effects of excess 5-HT in the somatosensory system.

The evidence that changes of 5-HT system homeostasis exerts long-term effects on cortical development and adult brain plasticity may be an important step forward in establishing the psychobiological groundwork for a neurodevelopmental hypothesis of negative emotionality, aggressiveness, and violence. Although there is converging evidence that serotonergic dysfunction contributes to anxiety-related behavior, the precise mechanism that renders 5-HTT-deficient mice less aggressive remains to be elucidated.

GENE–ENVIRONMENT INTERACTION AND BEHAVIOR IN NONHUMAN PRIMATES

Personality defines the framework to adapt to other people as a crucial task in long-term reproductive success. Extraversion and agreeableness are important to the formation of social structures ranging from pair-bonds to coalitions of group. Emotional stability and conscientiousness are critical to the endurance of these structures, while openness may reflect the capacity for innovation. Although studies in nonhuman primates have yielded models of aggressive behaviors, insights into the biological mechanisms that underlie these behaviors are beginning to emerge, and the concept of aggression as an antisocial instinct is being replaced by a framework that considers it a tool of competition and negotiation (de Waal, 2000). Since the genetic basis of present-day temperamental, personality, and behavioral traits may reflect selective forces among our remote ancestors (Loehlin, 1992), research efforts have recently been focused on rhesus macaques. In this nonhuman primate model, environmental influences are probably less complex, can be more easily controlled for, and thus, less likely to confound associations between temperament and genes. All forms of aggression in rhesus monkeys—major categories are defensive and offensive aggression—appear to be modulated by environmental factors, and marked disruptions to the mother–infant relationship likely confer increased risk (Kalin, 1999).

One of the most replicated findings in psychobiology is the observation of lower 5-HIAA in the brain and CSF of subjects with impulsive aggression and suicidal behavior (for review, see Asberg, 1997). Human and nonhuman primate behavior is similarly modified by deficits in 5-HT function. In rhesus monkeys, 5-HT turnover, as measured by cisternal CSF 5-HIAA concentrations, shows a strong heritable component and is traitlike, with demonstrated stability over an individual's lifespan (Higley et al., 1991, 1993; Kraemer et al., 1989). Low or lower than average CSF 5-HIAA concentrations have been reported in individuals who display inappropriate aggression as children, engage in frequent impulsive and violent criminal behavior, exhibit excessive alcohol abuse and dependence, and attempt suicide.

Recently, a study has assessed its generalizability across primates by making simultaneous comparisons between and within closely related species. Between-species analyses indicated higher CSF 5-HIAA concentrations in pigtailed macaques (*Macaca nemestrina*) and higher rates of high-intensity aggression, escalated aggression, and wounds requiring medical treatment in rhesus macaques (*Macaca mulatta*) (Westergaard et al., 1999). Within-species analyses indicated that inter-individual differences in CSF 5-HIAA concentrations were inversely correlated with escalated aggression and positively correlated with social dominance rank, further supporting the notion that 5-HT functioning plays an important role in controlling impulsivity that regulate severe impulsive aggression and social dominance relationships in nonhuman primates, and that between-species differences in agonistic temperament can be predicted by species-typical brain 5-HT function.

Not unexpectedly, CSF 5-HIAA concentrations are also subject to the long-lasting influence of deleterious events early in life as well as by situational stressors. Monkeys separated from their mother and reared in absence of conspecific adults (peer-reared) have altered serotonergic function and exhibit behavioral deficits throughout their lifetimes when compared to their mother-reared counterparts. Comparison of different mammalian species indicates that the 5-HTT-LPR is unique to humans and simian primates, although the composition and number of repeats display intraspecies and interspecies variability. In hominoids, all alleles originate from variation at a single locus (polymorphic locus 1, PL1), whereas a corresponding locus for a 21-bp length variation (PL2) was found in the 5-HTT-LPR of rhesus monkeys (rh5-HTT-LPR) (Lesch et al., 1997). The 5-HTT-LPR sequence may be informative in the comparison of closely related species and reflects the phylogeny of the old world monkeys, great apes, and humans. The presence of an analogous rh5-HTT-LPR and resulting allelic variation of 5-HTT activity in rhesus monkeys provides a unique model to dissect the relative contribution of genes and environmental sources to central serotonergic function and related behavioral outcomes.

Genotype–environment interaction was recently studied by testing associations between central 5-HT turnover and rh5-HTT-LPR genotype in rhesus monkeys with well-characterized environmental histories. The monkeys' rearing fell into one of the following categories: (*i*) mother-reared, either reared with the biological mother or cross-fostered; or (*ii*) peer-reared, either with a peer group of three to four monkeys or with an inanimate surrogate and daily contact with a playgroup of peers. Peer-reared monkeys were separated from their mothers, placed in the nursery at birth, and given access to peers at 30 d of age either continuously or during daily play sessions. Mother-reared and cross-fostered monkeys remained with the mother, typically within a social group. At roughly 7 mo of age, mother-reared monkeys were weaned and placed together with their peer-reared cohort in large mixed-gender social groups. Because the monkey population encompassed two groups that received dramatically different social and rearing experience early in life, the interactive effects of environmental experience and the rh5-HTT-LPR on cisternal CSF 5-HIAA levels and 5-HT-related behavior was assessed (Bennett et al., 2002). CSF 5-HIAA concentrations were significantly influenced by genotype for peer-reared, but not for mother-reared, subjects. Peer-reared rhesus monkeys with the low-activity rh5-HTT-LPR s allele had significantly lower concentrations of CSF 5-HIAA than their homozygous l/l counterparts. Low 5-HT turnover in monkeys with the s allele is congruent with in vitro studies that show reduced binding and transcriptional efficiency of the 5-HTT gene associated with the 5-HTT-LPR s allele (Lesch et al., 1996). This suggests that the rh5-HTT-LPR genotype is predictive of CSF 5-HIAA concentrations, but that early experiences make unique contributions to variation in later 5-HT functioning. This finding is the first to provide evidence of an environment-dependent association between a polymorphism in the 5'-regulatory region of the 5-HTT gene and a direct measure of 5-HT functioning, cisternal CSF 5-HIAA concentration, thus revealing an interaction between rearing environment and rh5-HTT-LPR genotype. Similar to the 5-HTT-LPR's influence on NEO neuroticism in humans, however, the effect size is small, with 4.7% of variance in CSF 5-HIAA accounted for by the rh5-HTT-LPR-rearing environment interaction.

Previous work has shown that monkeys' early experiences have long-term consequences for the functioning of the central 5-HT system, as indicated by robustly altered CSF 5-HIAA levels, as well as anxiety, depression, aggression-related behavior in monkeys deprived of their parents at birth and raised only with peers (Higley et al., 1991, 1993; Kraemer et al., 1989). Intriguingly, the biobehavioral results of deleterious early experiences of social separation are consistent with the notion that the 5-HTT-LPR may influence the risk for affective spectrum disorders. Evolutionary preservation of two prevalent 5-HTT-LPR variants and the resulting allelic variation in 5-HTT expression may be part of the genetic mechanism resulting in the emergence of temperamental traits that facilitate adaptive functioning in the complex social worlds most primates inhabit. The uniqueness of the 5-HTT-LPR among humans and simian nonhuman primates, but not among prosimians or other mammals, along with the role 5-HT plays in complex primate sociality, form the basis for the hypothesized relationship between the 5-HTT function and personality traits that mediate individual differences in social behavior. This conclusion concurs with an increasing body of evidence for a complex interaction between individual differences in the central 5-HT system and social success. In monkeys, lowered 5-HT functioning, as indicated by decreased CSF 5-HIAA levels, is associated with lower rank within a social group, less competent social behavior, and greater impulsive aggression (Higley et al., 1992, 1996; Mehlman et al., 1994,

1995). It is well established that, while subjects with low CSF 5-HIAA concentrations are no more likely to engage in competitive aggression than other monkeys, when they do engage in aggression, it frequently escalates to violent and hazardous levels.

Association between the rh5-HTT-LPR genotype and behavior was studied by analyzing the joint effects of genotype and early rearing environment on social play and aggression (Barr et al. 2002, manuscript submitted). Infant rhesus monkeys homozygous for the l variant were more likely to engage in rough play than were l/s individuals with a significant interaction between 5-HTT genotype and rearing condition. Peer-reared infants carrying the s variant were less likely to play with peers than those homozygous for the l allele, whereas the rh5-HTT-LPR genotype had no effect on the incidence of social play among mother-reared monkeys. Socially dominant mother-reared monkeys were more likely than their peer-reared counterparts to engage in aggression. In contrast, peer-reared, but not mother-reared, monkeys with the low-activity s allele exhibited more aggressive behaviors than their l/l counterparts. This genotype by rearing interaction for aggressive behavior indicates that peer-reared subjects with the s allele, while unlikely to win in a competitive encounter, are more inclined to persist in aggression once it begins. Moreover, high composite scores for alcohol intake and alcohol-elicited aggression are associated with the rh5-HTT-LPR s variant in male rhesus monkeys, a potential model for type II alcoholism (S. Suomi, personal communication). A role of s allele-dependent low 5-HTT function in nonhuman primate anxious and aggressive behavior is in remarkable agreement with the association of NEO subscales neuroticism (increased angry hostility) and agreeableness (decreased compliance equal to increased aggressiveness and hostility) and genetically influenced low 5-HTT function.

As the focus of studies is increasingly extended to the neonatal period, a time in early development when environmental influences are modest and least likely to confound gene–behavior associations, complementary approaches have recently been applied to nonhuman primates. Rhesus macaque infants heterozygous for the s and l variant of the rh5-HTT-LPR (l/s) displayed higher behavioral stress reactivity compared to infants homozygous for the long variant of the allele (l/l) (Champoux et al., in press). Mother-reared and peer-reared monkeys were assessed on a standardized primate neurobehavioral test designed to measure orienting, motor maturity, reflex functioning, and temperament. Main effects of genotype and, in some cases, interactions between rearing condition and genotype were demonstrated for items indicative of orienting, attention, and temperament. In general, heterozygote animals demonstrated diminished orientation, lower attentional capabilities, and increased affective responding relative to l/l homozygotes. However, the genotype effects were more pronounced for animals raised in the neonatal nursery than for animals reared by their mothers. These findings support the notion that the comparatively recent appearance of the 5-HTT-LPR-associated genetic variation may have helped permit more sophisticated modulation of social behaviors during the evolution of higher-order primates (Buss, 1991, 1995; Lesch et al., 1997).

Another well-defined behavioral pattern among nonhuman primates that seems to be directly related to CSF 5-HIAA concentrations is dispersal from the natal group (Mehlman et al., 1995). Most male rhesus monkeys leave their natal group and either visit or join other social groups or form small transient all-male groups before returning to their birth group. While natal dispersal is occurring at a highly variable age and is almost always associated with loss of social status and an increase in stress, injury, and mortality, the cause and intention remains controversial. Interestingly, a recent study by Trefilov and coworkers (2000) showed a gene–dose effect of the rh5-HTT-LPR s variant on the age of dispersal, with s/s homozygotes leaving earlier than carriers of the l allele. This finding further supports the notion that impaired 5-HTT function resulting in low 5-HT turnover is associated with impulsive behavior together with a high tendency toward risk-taking activity that leads to early dispersal.

Taken together, these findings provide evidence of an environment-dependent association between allelic variation of 5-HTT expression and central 5-HT function and illustrate the possibility that specific genetic factors play a role in 5-HT-mediated social competence in primates. Because rhesus monkeys exhibit temperamental and behavioral traits that parallel anxiety, depression, and aggression-related personality dimensions associated in humans with the low-activity 5-HTT-LPR variant, it may be possible to search for evolutionary continuity in this genetic mechanism for individual differences. Nonhuman primate studies may also be useful to help identify environmental factors that either compound the vulnerability conferred by a particular genetic makeup or, conversely, act to improve the behavioral outcome associated with a distinct genetic makeup.

SUMMARY AND FUTURE DIRECTIONS

Aggression and violence are complex social behaviors that arise out of multiple causes involving biologic, psychological, and social forces, and different forms of violent antisocial behavior may each result from different biopsychosocial pathways. The expression of impulsivity and associated aggressive behavior must be carefully modulated to ensure the success of individuals, small groups, and large societies, especially within the current framework of rapid globalization. Even though individual differences in impulsivity and the behavioral consequences, such as aggression, addiction, and suicidality, are substantially heritable, they ultimately result from an interplay between genetic variations and environmental factors. Multiple neural networks, including the orbital frontal cortex, amygdala, anterior cingulate cortex, and several other interconnected regions, whose formation, function, and integration depend on the actions of several classical neurotransmitters, such as 5-HT, have been linked to aggression. Several lines of evidence suggest that deficits in serotonergic modulation of this cortico-cingulo-amygdaloid circuit modify an individual's liability to impulsive aggression. The 5-HT system generally attenuates aggression in animal models and violent behavior in humans. Impulsivity and inappropriate aggressiveness are correlated with low CSF concentrations of the 5-HT metabolite, 5-HIAA, in humans and nonhuman primates and are modified by drugs that influence 5-HT system function and their associated behaviors throughout development and adult life. Moreover, genetically influenced variation of 5-HT system function, in conjunction with other predisposing genetic factors and with inadequate adaptive responses to environmental stressors, is also likely to contribute to impulsivity and aggression-related behavior emerging from compromised brain development and from neuroadaptive processes.

More functionally relevant polymorphisms in genes within a single neurotransmitter system, or in genes which comprise a functional unit in their concerted actions, need to be identified and assessed in both large-population and family-based association studies to avoid stratification artifacts and to elucidate complex interactions of multiple loci. Even pivotal regulatory proteins of neurotransmission, such as receptors, transporters, and modifying enzymes, will have only a modest impact, while noise from nongenetic mechanisms may seriously obstruct identification of relevant genes. Although current methods for the detection of gene–environment interaction in behavioral genetics are largely indirect, the most relevant consequence of gene identification for personality and behavioral traits related to aggression may be that it will provide the tools required to systematically clarify the effects of gene–environment interaction.

Based on the remarkable progress in technologies that allow the alteration or elimination of individual genes to create transgenic animal models, gene knock-out strategies are likely to increase our knowledge about which gene products are involved in aggression-associated traits. However, because a missing gene might affect many developmental processes throughout ontogeny and compensatory mechanisms may be activated in knock-outs, behavioral data from mice with targeted gene deletions should be interpreted with caution. It is becoming increasingly evident that many neurotransmitters and their receptors are expressed at early periods of neural development, and it is increasingly appreciated that they participate in the structural organization of the brain. An additional shortcoming of current knock-out experiments is the inability to provide region-specific control of the disruption. The ability to use native and exogenous promoters to control the expression of specifically targeted genes may allow region-specific and temporal control of protein expression. Systems that may prove useful include tetracycline or tamoxifen-responsive inducible promoters and the loxP-cre approach of inducibly deleting sections of DNA. The development of conditional knock-outs, in which a specific gene can be inactivated tissue-specifically any time during ontogeny, are, therefore, likely to avoid these imperfections associated with behavioral data from constitutive knock-outs.

REFERENCES

Abbar, M., Courtet, P., Amadeo, S., et al. (1995) Suicidal behaviors and the tryptophan hydroxylase gene. *Arch. Gen. Psychiatry* **52**, 846–849.

Archer, J. and Browne, K. (1989) Concepts and approaches to the study of aggression, in *Human Aggression: Naturalistic Approaches.* (Archer, J. and Browne, K., eds.), Routledge, London, pp. 3–24.

Asberg, M. (1997) Neurotransmitters and suicidal behavior. The evidence from cerebrospinal fluid studies. *Ann. NY Acad. Sci.* **836**, 158–181.

Azmitia, E. C. and Whitaker-Azmitia, P. M. (1997) Development and adult plasticity of serotonergic neurons and their target cells, in *Serotonergic Neurons and 5-HT Receptors in the CNS, Vol. 129.* (Baumgarten, H. G. and Göthert, M., eds.), Springer, Berlin, Heidelberg, New York, pp. 1–39.

Bellivier, F., Leboyer, M., Courtet, P., et al. (1998) Association between the tryptophan hydroxylase gene and manic-depressive illness [see comments]. *Arch. Gen. Psychiatry* **55**, 33–37.

Bengel, D., Murphy, D. L., Andrews, A. M., et al. (1998) Altered brain serotonin homeostasis and locomotor insensitivity to 3, 4-methylenedioxymethamphetamine ("Ecstasy") in serotonin transporter-deficient mice. *Mol. Pharmacol.* **53**, 649–655.

Bennett, A. J., Lesch, K. P., Heils, A., et al. (2002) Early experience and serotonin transporter gene variation interact to influence primate CNS function. *Mol. Psychiatry* **7**, 118–122.

Bennett-Clarke, C. A., Chiaia, N. L., and Rhoades, R. W. (1996) Thalamocortical afferents in rat transiently express high-affinity serotonin uptake sites. *Brain Res.* **733,** 301-306.

Bennett-Clarke, C. A., Leslie, M. J., Lane, R. D., and Rhoades, R. W. (1994) Effect of serotonin depletion on vibrissa-related patterns of thalamic afferents in the rat's somatosensory cortex. *J. Neurosci.* **14,** 7594–7607.

Berman, M. E. and Coccaro, E. F. (1998) Neurobiologic correlates of violence: relevance to criminal responsibility. *Behav. Sci. Law* **16,** 303–318.

Boularand, S., Darmon, M. C., and Mallet, J. (1995) The human tryptophan hydroxylase gene. An unusual splicing complexity in the 5'-untranslated region. *J. Biol. Chem.* **270,** 3748–3756.

Brunner, D. and Hen, R. (1997) Insights into the neurobiology of impulsive behavior from serotonin receptor knockout mice. *Ann. NY Acad. Sci.* **836,** 81–105.

Brunner, H. G., Nelen, M., Breakefield, X. O., Ropers, H. H., and van Oost, B. A. (1993) Abnormal behavior associated with a point mutation in the structural gene for monoamine oxidase A. *Science* **262,** 578–580.

Buss, A. H. (1961) *The Psychology of Aggression.* Wiley & Sons, New York.

Buss, D. M. (1991) Evolutionary personality psychology. *Annu. Rev. Psychol.* **42,** 459–491.

Buss, D. M. (1995) Evolutionary psychology: a new paradigm for psychological science. *Psychol. Inquiry* **6,** 1–30.

Cases, O., Lebrand, C., Giros, B., et al. (1998) Plasma membrane transporters of serotonin, dopamine, and norepinephrine mediate serotonin accumulation in atypical locations in the developing brain of monoamine oxidase A knock-outs. *J. Neurosci.* **18,** 6914–6927.

Cases, O., Seif, I., Grimsby, J., et al. (1995) Aggressive behavior and altered amounts of brain serotonin and norepinephrine in mice lacking MAOA [see comments]. *Science* **268,** 1763–1766.

Cases, O., Vitalis, T., Seif, I., De Maeyer, E., Sotelo, C., and Gaspar, P. (1996) Lack of barrels in the somatosensory cortex of monoamine oxidase A-deficient mice: role of a serotonin excess during the critical period. *Neuron* **16,** 297–307.

Champoux, M., Bennett, A. J., Shannon, C., Higley, J. D., Lesch, K. P., and Suomi, S. J. (2002) Serotonin transporter gene polymorphism, differential early rearing, and behavior in rhesus monkey neonates. *Mol. Psychiatry* **7,** 1058–1063.

Chen, C., Rainnie, D. G., Greene, R. W., and Tonegawa, S. (1994) Abnormal fear response and aggressive behavior in mutant mice deficient for calcium–calmodulin kinase II. *Science* **288,** 291–294.

Chiavegatto, S., Dawson, V. L., Mamounas, L. A., Koliatsos, V. E., Dawson, T. M., and Nelson, R. J. (2001) Brain serotonin dysfunction accounts for aggression in male mice lacking neuronal nitric oxide synthase. *Proc. Natl. Acad. Sci. USA* **98,** 1277–1281.

Cleare, A. J. and Bond, A. J. (2000a) Experimental evidence that the aggressive effect of tryptophan depletion is mediated via the 5-HT1A receptor. *Psychopharmacology (Berl)* **147,** 439–441.

Cleare, A. J. and Bond, A. J. (2000b) Ipsapirone challenge in aggressive men shows an inverse correlation between 5-HT1A receptor function and aggression. *Psychopharmacology (Berl)* **148,** 344–349.

Coccaro, E. F. (1992) Impulsive aggression and central serotonergic system function in humans: an example of a dimensional brain-behavior relationship. *Int. Clin. Psychopharmacol.* **7,** 3–12.

Coccaro, E. F., Kavoussi, R. J., Trestman, R. L., Gabriel, S. M., Cooper, T. B., and Siever, L. J. (1997) Serotonin function in human subjects: intercorrelations among central 5-HT indices and aggressiveness. *Psychiatry Res.* **73,** 1–14.

Coleman, F. H., Christensen, H. D., Gonzalez, C. L., and Rayburn, W. F. (1999) Behavioral changes in developing mice after prenatal exposure to paroxetine (Paxil). *Am. J. Obstet. Gynecol.* **181,** 1166–1171.

Costa, P. and McCrae, R. (1992) *Revised NEO Personality Inventory (NEO PI-R) and NEO Five Inventory (NEO-FFI) professional manual.* Psychological Assessment Resources.

Coupland, N. J. (2000) Brain mechanisms and neurotransmitters, in *Post-Traumatic Stress Disorder: Diagnosis, Management, and Treatment*. (Nutt, D., Davidson, J. R. T., and Zohar, J., eds.), Martin Dunitz Publishers, London, pp. 69–99.

Crick, N. R. and Dodge, K. A. (1996) Social information-processing mechnisms in reactive and proactive aggression. *Child Dev.* **67,** 993–1002.

Davidson, R. J., Putnam, K. M., and Larson, C. L. (2000) Dysfunction in the neural circuitry of emotion regulation—a possible prelude to violence. *Science* **289,** 591–594.

de Waal, F. B. (2000) Primates-a natural heritage of conflict resolution. *Science* **289,** 586–590.

Deckert, J., Catalano, M., Syagailo, Y.V., et al. (1999) Excess of high activity monoamine oxidase A gene promoter alleles in female patients with panic disorder. *Hum. Mol. Genet.* **8,** 621–624.

Denney, R. M., Koch, H., and Craig, I. W. (1999) Association between monoamine oxidase A activity in human male skin fibroblasts and genotype of the MAOA promoter-associated variable number tandem repeat. *Hum. Genet.* **105,** 542–551.

Denney, R. M., Sharma, A., Dave, S. K., and Waguespack, A. (1994) A new look at the promoter of the human monoamine oxidase A gene: mapping transcription initiation sites and capacity to drive luciferase expression. *J. Neurochem.* **63,** 843–856.

Dougherty, D. M., Moeller, F. G., Bjork, J. M., and Marsh, D. M. (1999) Plasma L-tryptophan depletion and aggression. *Adv. Exp. Med. Biol.* **467,** 57–65.

Fabre, V., Beaufour, C., Evrard, A., et al. (2000) Altered expression and functions of serotonin 5-HT1A and 5-HT1B receptors in knock-out mice lacking the 5-HT transporter. *Eur. J. Neurosci.* **12,** 2299–2310.

Fink, G., Sumner, B., Rosie, R., Wilson, H., and McQueen, J. (1999) Androgen actions on central serotonin neurotransmission: relevance for mood, mental state and memory [in process citation]. *Behav. Brain Res.* **105,** 53–68.

Flint, J., Corley, R., DeFries, J. C., et al. (1995) A simple genetic basis for a complex psychological trait in laboratory mice. *Science* **269,** 1432–1435.

Furlong, R. A., Ho, L., Rubinsztein, J. S., Walsh, C., Paykel, E. S., and Rubinsztein, D. C. (1998) No association of the tryptophan hydroxylase gene with bipolar affective disorder, unipolar affective disorder, or suicidal behaviour in major affective disorder. *Am. J. Med. Genet.* **81,** 245–247.

Furlong, R. A., Ho, L., Rubinsztein, J. S., Walsh, C., Paykel, E. S., and Rubinsztein, D. C. (1999) Analysis of the monoamine oxidase A (MAOA) gene in bipolar affective disorder by association studies, meta-analyses, and sequencing of the promoter. *Am. J. Med. Genet.* **88,** 398–406.

Gelernter, J., Kranzler, H., and Cubells, J. F. (1997) Serotonin transporter protein (SLC 6A4) allele and haplotype frequencies and linkage disequilibria in African- and European-American and Japanese populations and in alcohol-dependent subjects. *Hum. Genet.* **101,** 243–246.

Gingrich, J. A. and Hen, R. (2000) The broken mouse: the role of development, plasticity and environment in the interpretation of phenotypic changes in knockout mice. *Curr. Opin. Neurobiol.* **10,** 146–152.

Greenberg, B. D., Li, Q., Lucas, F. R., et al. (2000) Association between the serotonin transporter promoter polymorphism and personality traits in a primarily female population sample. *Am. J. Med. Genet.* **96,** 202–216.

Han, L., Nielsen, D.A., Rosenthal, N.E., et al. (1999) No coding variant of the tryptophan hydroxylase gene detected in seasonal affective disorder, obsessive-compulsive disorder, anorexia nervosa, and alcoholism. *Biol. Psychiatry* **45,** 615–619.

Hanna, G. L., Himle, J. A., Curtis, G. C., et al. (1998) Serotonin transporter and seasonal variation in blood serotonin in families with obsessive-compulsive disorder. *Neuropsychopharmacology* **18,** 102–111.

Hansson, S. R., Mezey, E., and Hoffman, B. J. (1998) Serotonin transporter messenger RNA in the developing rat brain: early expression in serotonergic neurons and transient expression in non-serotonergic neurons. *Neuroscience* **83**, 1185–1201.

Hansson, S. R., Mezey, E., and Hoffman, B. J. (1999) Serotonin transporter messenger RNA expression in neural crest-derived structures and sensory pathways of the developing rat embryo. *Neuroscience* **89**, 243–265.

Hariri, A. R., Mattay, V. S., Tessitore, A., et al. (2002) Serotonin transporter genetic variation and the response of the human amygdala. *Science* **297**, 400–403.

Heinz, A., Jones, D. W., Mazzanti, C., et al. (1999) A relationship between serotonin transporter genotype and in vivo protein expression and alcohol neurotoxicity. *Biol. Psychiatry* **47**, 643–649.

Higley, J. D., King, S. T., Jr., Hasert, M. F., Champoux, M., Suomi, S. J., and Linnoila, M. (1996) Stability of interindividual differences in serotonin function and its relationship to severe aggression and competent social behavior in rhesus macaque females. *Neuropsychopharmacology* **14**, 67–76.

Higley, J. D., Mehlman, P. T., Taub, D. M., et al. (1992) Cerebrospinal fluid monoamine and adrenal correlates of aggression in free-ranging rhesus monkeys. *Arch. Gen. Psychiatry* **49**, 436–441.

Higley, J. D., Suomi, S. J., and Linnoila, M. (1991) CSF monoamine metabolite concentrations vary according to age, rearing, and sex, and are influenced by the stressor of social separation in rhesus monkeys. *Psychopharmacology* **103**, 551–556.

Higley, J. D., Thompson, W. W., Champoux, M., et al. (1993) Paternal and maternal genetic and environmental contributions to cerebrospinal fluid monoamine metabolites in rhesus monkeys (Macaca mulatta). *Arch. Gen. Psychiatry* **50**, 615–623.

Holmes, A., Murphy, D. L., and Crawley, J. N. (2002) Reduced aggression in mice lacking the serotonin transporter. *Psychopharmacology (Berl)* **161**, 160–167.

Hoyer, D. and Martin, G. (1997) 5-HT receptor classification and nomenclature: towards a harmonization with the human genome. *Neuropharmacology* **36**, 419–428.

Huang, Y. Y., Grailhe, R., Arango, V., Hen, R., and Mann, J. J. (1999) Relationship of psychopathology to the human serotonin1B genotype and receptor binding kinetics in postmortem brain tissue. *Neuropsychopharmacology* **21**, 238–246.

Huh, S. O., Park, D. H., Cho, J. Y., Joh, T. H., and Son, J. H. (1994) A 6.1 kb 5' upstream region of the mouse tryptophan hydroxylase gene directs expression of E. coli lacZ to major serotonergic brain regions and pineal gland in transgenic mice. *Brain Res. Mol. Brain Res.* **24**, 145–152.

Kalin, N. H. (1999) Primate models to understand human aggression. *J. Clin. Psychiatry* **60**, 29–32.

Knutson, B., Wolkowitz, O. M., Cole, S. W., et al. (1998) Selective alteration of personality and social behavior by serotonergic intervention. *Am. J. Psychiatry* **155**, 373–379.

Korte, S. M., Meijer, O. C., de Kloet, E. R., et al. (1996) Enhanced 5-HT1A receptor expression in forebrain regions of aggressive house mice. *Brain Res.* **736**, 338–343.

Kraemer, G. W., Ebert, M. H., Schmidt, D. E., and McKinney, W. T. (1989) A longitudinal study of the effect of different social rearing conditions on cerebrospinal fluid norepinephrine and biogenic amine metabolites in rhesus monkeys. *Neuropsychopharmacology* **2**, 175–189.

Kunugi, H., Hattori, M., Kato, T., et al. (1997) Serotonin transporter gene polymorphisms: ethnic difference and possible association with bipolar affective disorder. *Mol. Psychiatry* **2**, 457–462.

la Cour, C. M., Boni, C., Hanoun, N., Lesch, K. P., Hamon, M., and Lanfumey, L. (2001) Functional consequences of 5-HT transporter gene disruption on 5-HT(1a) receptor-mediated regulation of dorsal raphe and hippocampal cell activity. *J. Neurosci.* **21**, 2178–2185.

Lappalainen, J., Long, J. C., Eggert, M., et al. (1998) Linkage of antisocial alcoholism to the serotonin 5-HT1B receptor gene in 2 populations. *Arch. Gen. Psychiatry* **55**, 989–994.

Lauder, J. M. (1993) Neurotransmitters as growth regulatory signals: role of receptors and second messengers. *Trends Neurosci.* **16,** 233–240.

Lebrand, C., Cases, O., Adelbrecht, C., et al. (1996) Transient uptake and storage of serotonin in developing thalamic neurons. *Neuron* **17,** 823–835.

LeMarquand, D. G., Benkelfat, C., Pihl, R. O., Palmour, R. M., and Young, S. N. (1999) Behavioral disinhibition induced by tryptophan depletion in nonalcoholic young men with multigenerational family histories of paternal alcoholism. *Am. J. Psychiatry* **156,** 1771–1779.

LeMarquand, D. G., Pihl, R. O., Young, S. N., et al. (1998) Tryptophan depletion, executive functions, and disinhibition in aggressive, adolescent males. *Neuropsychopharmacology* **19,** 333–341.

Lesch, K. P. (1997) Molecular biology, pharmacology, and genetics of the serotonin transporter: psychobiological and clinical implications, in *Serotonergic Neurons and 5-HT Receptors in the CNS, Vol. 129.* (Baumgarten, H. G. and Göthert, M., eds.), Springer, Berlin, Heidelberg, New York, pp. 671–705.

Lesch, K. P. (2001) Serotonin transporter: from genomics and knockouts to behavioral traits and psychiatric disorders, in *Molecular Genetics of Mental Disorders.* (Briley, M. and Sulser, F., eds.), Martin Dunitz Publishers, London, pp. 221–267.

Lesch, K. P., Balling, U., Gross, J., et al. (1994) Organization of the human serotonin transporter gene. *J. Neural Transm. Gen. Sect.* **95,** 157–162.

Lesch, K. P., Bengel, D., Heils, A., et al. (1996) Association of anxiety-related traits with a polymorphism in the serotonin transporter gene regulatory region. *Science* **274,** 1527–1531.

Lesch, K. P., Greenberg, B. D., Higley, J. D., and Murphy, D. L. (2002) Serotonin transporter, personality, and behavior: toward dissection of gene-gene and gene-environment interaction, in *Molecular Genetics and the Human Personality.* (Benjamin, J., Ebstein, R., and Belmaker, R. H., eds.), American Psychiatric Press, Washington, DC, pp. 109–135.

Lesch, K. P. and Merschdorf, U. (2000) Impulsivity, aggression, and serotonin: a molecular psychobiological perspective. *Behav. Sci. Law* **18,** 581–604.

Lesch, K. P., Meyer, J., Glatz, K., et al. (1997) The 5-HT transporter gene-linked polymorphic region (5-HTTLPR) in evolutionary perspective: alternative biallelic variation in rhesus monkeys. Rapid communication. *J. Neural Transm.* **104,** 1259–1266.

Lesch, K. P. and Mössner, R. (1998) Genetically driven variation in serotonin uptake: is there a link to affective spectrum, neurodevelopmental, and neurodegenerative disorders. *Biol. Psychiatry* **44,** 179–192.

Lesch, K. P. and Mössner, R. (1999) 5-HT1A receptor inactivation: anxiety or depression as a murine experience. *Int. J. Neuropsychopharmacol.* **2,** 327–331.

Lewis, C. E. (1991) Neurochemical mechanisms of chronic antisocial behavior (psychopathy). A literature review. *J. Nerv. Ment. Dis.* **179,** 720–727.

Li, Q., Wichems, C., Heils, A., Lesch, K. P., and Murphy, D. L. (2000) Reduction in the density and expression, but not G-protein coupling, of serotonin receptors (5-HT1A) in 5-HT transporter knock-out mice: gender and brain region differences. *J. Neurosci.* **20,** 7888–7895.

Linnoila, V. M. and Virkkunen, M. (1992) Aggression, suicidality, and serotonin. *J. Clin. Psychiatry* **53,** 46–51.

Little, K. Y., McLaughlin, D. P., Zhang, L., et al. (1998) Cocaine, ethanol, and genotype effects on human midbrain serotonin transporter binding sites and mRNA levels. *Am. J. Psychiatry* **155,** 207–213.

Loeber, R. and Stouthamer-Loeber, M. (1998) Development of juvenile aggression and violence. Some common misconceptions and controversies. *Am. Psychol.* **53,** 242–259.

Loehlin, J. C. (1992) *Genes and Environment in Personality Development.* Sage Publications, Newburg Park, CA.

Mann, J. J., Brent, D. A., and Arango, V. (2001) The neurobiology and genetics of suicide and attempted suicide: a focus on the serotonergic system. *Neuropsychopharmacology* **24,** 467–477.

Mansour-Robaey, S., Mechawar, N., Radja, F., Beaulieu, C., and Descarries, L. (1998) Quantified distribution of serotonin transporter and receptors during the postnatal development of the rat barrel field cortex. *Dev. Brain Res.* **107,** 159–163.

Manuck, S. B., Flory, J. D., Ferrell, R. E., Dent, K. M., Mann, J. J., and Muldoon, M. F. (1999) Aggression and anger-related traits associated with a polymorphism of the tryptophan hydroxylase gene. *Biol. Psychiatry* **45,** 603–614.

Manuck, S. B., Flory, J. D., Ferrell, R. E., Mann, J. J., and Muldoon, M. F. (2000) A regulatory polymorphism of the monoamine oxidase-A gene may be associated with variability in aggression, impulsivity, and central nervous system serotonergic responsivity. *Psychiatry Res.* **95,** 9–23.

Manuck, S. B., Flory, J. D., McCaffery, J. M., Matthews, K. A., Mann, J. J., and Muldoon, M. F. (1998) Aggression, impulsivity, and central nervous system serotonergic responsivity in a nonpatient sample. *Neuropsychopharmacology* **19,** 287–299.

Maxson, S. C. (1996) Issues in the search for candidate genes in mice as potential animal models of human aggression. *Ciba Found. Symp.* **194,** 21–30.

McGue, M. and Bouchard, T. J. (1998) Genetic and environmental influences on human behavioral differences. *Annu. Rev. Neurosci.* **21,** 1–24.

McQueen, J. K., Wilson, H., and Fink, G. (1997) Estradiol-17 beta increases serotonin transporter (SERT) mRNA levels and the density of SERT-binding sites in female rat brain. *Brain Res. Mol. Brain Res.* **45,** 13–23.

Mehlman, P. T., Higley, J. D., Faucher, I., et al. (1994) Low CSF 5-HIAA concentrations and severe aggression and impaired impulse control in nonhuman primates [see comments]. *Am. J. Psychiatry* **151,** 1485–1491.

Mehlman, P. T., Higley, J. D., Faucher, I., et al. (1995) Correlation of CSF 5-HIAA concentration with sociality and the timing of emigration in free-ranging primates. *Am. J. Psychiatry* **152,** 907–913.

Moeller, F. G., Allen, T., Cherek, D. R., Dougherty, D. M., Lane, S., and Swann, A. C. (1998) Ipsapirone neuroendocrine challenge: relationship to aggression as measured in the human laboratory. *Psychiatry Res.* **81,** 31–38.

Moeller, F. G., Dougherty, D. M., Swann, A. C., Collins, D., Davis, C. M., and Cherek, D. R. (1996) Tryptophan depletion and aggressive responding in healthy males. *Psychopharmacology (Berl)* **126,** 97–103.

Mortensen, O. V., Thomassen, M., Larsen, M. B., Whittemore, S. R., and Wiborg, O. (1999) Functional analysis of a novel human serotonin transporter gene promoter in immortalized raphe cells. *Brain Res. Mol. Brain Res.* **68,** 141–148.

Moss, H. B., Yao, J. K., and Panzak, G. L. (1990) Serotonergic responsivity and behavioral dimensions in antisocial personality disorder with substance abuse. *Biol. Psychiatry* **28,** 325–338.

Murphy, D. L., Li, Q., Engel, S., et al. (2001) Genetic perspectives on the serotonin transporter. *Brain Res. Bull.* **56,** 487–494.

Nelson, R. J. and Chiavegatto, S. (2001) Molecular basis of aggression. *Trends Neurosci.* **24,** 713–719.

Neumeister, A., Konstantinidis, A., Stastny, J., et al. (2002) Association between serotonin transporter gene promoter polymorphism (5HTTLPR) and behavioral responses to tryptophan depletion in healthy women with and without family history of depression. *Arch. Gen. Psychiatry* **59,** 613–620.

New, A. S., Gelernter, J., Yovell, Y., et al. (1998) Tryptophan hydroxylase genotype is associated with impulsive-aggression measures: a preliminary study. *Am. J. Med. Genet.* **81,** 13–17.

Nielsen, D. A., Goldman, D., Virkkunen, M., Tokola, R., Rawlings, R., and Linnoila, M. (1994) Suicidality and 5-hydroxyindoleacetic acid concentration associated with a tryptophan hydroxylase polymorphism. *Arch. Gen. Psychiatry* **51,** 34–38.

Nielsen, D. A., Virkkunen, M., Lappalainen, J., et al. (1998) A tryptophan hydroxylase gene marker for suicidality and alcoholism. *Arch. Gen. Psychiatry* **55**, 593–602.

Nobile, M., Begni, B., Giorda, R., et al. (1999) Effects of serotonin transporter promoter genotype on platelet serotonin transporter functionality in depressed children and adolescents [in process citation]. *J. Am. Acad. Child Adolesc. Psychiatry* **38**, 1396–1402.

Olivier, B. and Mos, J. (1992) Rodent models of aggressive behavior and serotonergic drugs. *Prog. Neuropsychopharmacol. Biol. Psychiatry* **16**, 847–870.

Olivier, B., Mos, J., van Oorschot, R., and Hen, R. (1995) Serotonin receptors and animal models of aggressive behavior. *Pharmacopsychiatry* **28(Suppl. 2)**, 80–90.

Osterheld-Haas, M. C., Van der Loos, H., and Hornung, J. P. (1994) Monoaminergic afferents to cortex modulate structural plasticity in the barrelfield of the mouse. *Dev. Brain Res.* **77**, 189–202.

Persico, A. M., Revay, R. S., Mössner, R., et al. (2001) Barrel pattern formation in somatosensory cortical layer IV requires serotonin uptake by thalamocortical endings, while vesicular monoamine release is necessary for development of supragranular layers. *J. Neurosci.* **21**, 6862–6873.

Pesonen, U., Koulu, M., Bergen, A., et al. (1998) Mutation screening of the 5-hydroxytryptamine7 receptor gene among Finnish alcoholics and controls. *Psychiatry Res.* **77**, 139–145.

Plomin, R., Owen, M. J., and McGuffin, P. (1994) The genetic basis of complex human behaviors. *Science* **264**, 1733–1739.

Raine, A., Buchsbaum, M., and LaCasse, L. (1997) Brain abnormalities in murderers indicated by positron emission tomography. *Biol. Psychiatry* **42**, 495–508.

Raine, A., Lencz, T., Bihrle, S., LaCasse, L., and Colletti, P. (2000) Reduced prefrontal gray matter volume and reduced autonomic activity in antisocial personality disorder. *Arch. Gen. Psychiatry* **57**, 119–129.

Raine, A., Meloy, J. R., Bihrle, S., Stoddard, J., LaCasse, L., and Buchsbaum, M. S. (1998) Reduced prefrontal and increased subcortical brain functioning assessed using positron emission tomography in predatory and affective murderers. *Behav. Sci. Law* **16**, 319–332.

Ramboz, S., Saudou, F., Amara, D. A., et al. (1996) 5-HT1B receptor knock out–behavioral consequences. *Behav. Brain Res.* **73**, 305–312.

Reist, C., Mazzanti, C., Vu, R., Tran, D., and Goldman, D. (2001) Serotonin transporter promoter polymorphism is associated with attenuated prolactin response to fenfluramine. *Am. J. Med. Genet.* **105**, 363–368.

Rotondo, A., Schuebel, K., Bergen, A., et al. (1999) Identification of four variants in the tryptophan hydroxylase promoter and association to behavior. *Mol. Psychiatry* **4**, 360–368.

Rudissaar, R., Pruus, K., Skrebuhhova, T., Allikmets, L., and Matto, V. (1999) Modulatory role of 5-HT3 receptors in mediation of apomorphine-induced aggressive behaviour in male rats. *Behav. Brain Res.* **106**, 91–96.

Sabol, S. Z., Hu, S., and Hamer, D. (1998) A functional polymorphism in the monoamine oxidase A gene promoter. *Hum. Genet.* **103**, 273–279.

Salichon, N., Gaspar, P., Upton, A. L., et al. (2001) Excessive activation of serotonin (5-HT) 1B receptors disrupts the formation of sensory maps in monoamine oxidase a and 5-ht transporter knock-out mice. *J. Neurosci.* **21**, 884–896.

Samochowiec, J., Lesch, K. P., Rottmann, M., et al. (1999) Association of a regulatory polymorphism in the promoter region of the monoamine oxidase A gene with antisocial alcoholism [in process citation]. *Psychiatry Res.* **86**, 67–72.

Saudou, F., Amara, D. A., Dierich, A., et al. (1994) Enhanced aggressive behavior in mice lacking 5-HT1B receptor. *Science* **265**, 1875–1878.

Scarpa, A. and Raine, A. (1997) Psychophysiology of anger and violent behavior. *Psychiatr. Clin. N. Am.* **20**, 375–394.

Schuback, D. E., Mulligan, E. L., Sims, K. B., et al. (1999) Screen for MAOA mutations in target human groups. *Am. J. Med. Genet.* **88,** 25–28.

Schulze, T. G., Müller, D. J., Krauss, H., et al. (2000) Association between a functional polymorphism in the monoamine oxidase A gene promoter and major depressive disorder. *Am. J. Med. Genet.* **96,** 801–803.

Seif, I. and De Maeyer, E. (1999) Knockout corner: knockout mice for monoamine oxidase A. *Int. J. Neuropsychopharmacol.* **12,** 241–243.

Seroczynski, A. D., Bergeman, C. S., and Coccaro, E. F. (1999) Etiology of the impulsivity/aggression relationship: genes or environment? *Psychiatry Res.* **86,** 41–57.

Shen, S., Battersby, S., Weaver, M., Clark, E., Stephens, K., and Harmar, A. J. (2000) Refined mapping of the human serotonin transporter (SLC6A4) gene within 17q11 adjacent to the CPD and NF1 genes. *Eur. J. Hum. Genet.* **8,** 75–78.

Shih, J. C., Grimsby, J., Chen, K., and Zhu, Q. S. (1993) Structure and promoter organization of the human monoamine oxidase A and B genes. *J. Psychiatry Neurosci.* **18,** 25–32.

Siever, L. and Trestman, R. L. (1993) The serotonin system and aggressive personality disorder. *Int. Clin. Psychopharmacol.* **8(Suppl. 2),** 33–39.

Siever, L. J., Buchsbaum, M. S., New, A. S., et al. (1999) d,l-fenfluramine response in impulsive personality disorder assessed with [18F]fluorodeoxyglucose positron emission tomography. *Neuropsychopharmacology* **20,** 413–423.

Soloff, P. H., Lynch, K. G., and Moss, H. B. (2000) Serotonin, impulsivity, and alcohol use disorders in the older adolescent: a psychobiological study. *Alcohol Clin. Exp. Res.* **24,** 1609–1619.

Son, J. H., Chung, J. H., Huh, S. O., et al. (1996) Immortalization of neuroendocrine pinealocytes from transgenic mice by targeted tumorigenesis using the tryptophan hydroxylase promoter. *Brain Res. Mol. Brain Res.* **37,** 32–40.

Stork, O., Welzl, H., Cremer, H., and Schachner, M. (1997) Increased intermale aggression and neuroendocrine response in mice deficient for the neural cell adhesion molecule (NCAM). *Eur. J. Neurosci.* **9,** 1117–1125.

Stork, O., Welzl, H., Wolfer, D., et al. (2000) Recovery of emotional behaviour in neural cell adhesion molecule (NCAM) null mutant mice through transgenic expression of NCAM180. *Eur. J. Neurosci.* **12,** 3291–3306.

Stork, O., Welzl, H., Wotjak, C. T., et al. (1999) Anxiety and increased 5-HT1A receptor response in NCAM null mutant mice. *J. Neurobiol.* **40,** 343–355.

Syagailo, Y. V., Stober, G., Grassle, M., et al. (2001) Association analysis of the functional monoamine oxidase A gene promoter polymorphism in psychiatric disorders. *Am. J. Med. Genet.* **105,** 168–171.

Tecott, L. H. and Barondes, S. H. (1996) Genes and aggressiveness. Behavioral genetics. *Curr. Biol.* **6,** 238–240.

Trefilov, A., Berard, J., Krawczak, M., and Schmidtke, J. (2000) Natal dispersal in rhesus macaques is related to serotonin transporter gene promoter variation. *Behav. Genet.* **30,** 295–301.

Virkkunen, M. and Linnoila, M. (1993) Brain serotonin, type II alcoholism and impulsive violence. *J. Stud. Alcohol Suppl.* **11,** 163–169.

Vitiello, B. and Stoff, D. M. (1997) Subtypes of aggression and their relevance to child psychiatry. *J. Am. Acad. Child Adolesc. Psychiatry* **36,** 307–315.

Westergaard, G. C., Suomi, S. J., Higley, J. D., and Mehlman, P. T. (1999) CSF 5-HIAA and aggression in female macaque monkeys: species and interindividual differences. *Psychopharmacology (Berl)* **146,** 440–446.

Whale, R., Quested, D. J., Laver, D., Harrison, P. J., and Cowen, P. J. (2000) Serotonin transporter (5-HTT) promoter genotype may influence the prolactin response to clomipramine. *Psychopharmacology (Berl)* **150,** 120–122.

Zhu, Q. and Shih, J. C. (1997) An extensive repeat structure down-regulates human monoamine oxidase A promoter activity independent of an initiator-like sequence. *J. Neurochem.* **69,** 1368–1373.

Zhu, Q. S., Chen, K., and Shih, J. C. (1994) Bidirectional promoter of human monoamine oxidase A (MAO A) controlled by transcription factor Sp1. *J. Neurosci.* **14**, 7393–7403.

Zhu, Q. S., Grimsby, J., Chen, K., and Shih, J. C. (1992) Promoter organization and activity of human monoamine oxidase (MAO) A and B genes. *J. Neurosci.* **12**, 4437–4446.

Zhuang, X., Gross, C., Santarelli, L., Compan, V., Trillat, A. C., and Hen, R. (1999) Altered emotional states in knockout mice lacking 5-HT1A or 5-HT1B receptors. *Neuropsychopharmacology* **21**, 52S–60S.

4
The Neurochemical Genetics of Serotonin in Aggression, Impulsivity, and Suicide

Mark D. Underwood and J. John Mann

INTRODUCTION

The serotonin (5-HT) neurotransmitter system is implicated in the etiology of aggression, impulsivity, and suicide (for a recent review, see Oquendo and Mann, 2000). Data increasingly indicate that the level of serotonergic transmission dictates behavior and not *vice versa*. A serotonergic abnormality may underlie the aggression–impulsivity and may be the diathesis for suicidal behavior. The serotonergic abnormality can be anatomically restricted to a discrete brain region or can affect brain "circuits," including the ventral prefrontal cortex.

CEREBROSPINAL FLUID

Despite technical limitations and inherent flaws of studies of the cerebrospinal fluid (CSF), insight into relationships between brain chemistry and behavior have been gleaned. In the mid to late 1970s Åsberg and Träskman-Benz (1976) reported that CSF 5-hydroxyindoleacetic acid (5-HIAA) was low in depressed subjects who had a history of a serious suicide attempt. This was the first published evidence suggesting a 5-HT abnormality involving suicidal behavior independent of the 5-HT abnormality associated with depression. The finding of reduced 5-HIAA in the CSF of suicide attempters has been replicated by many researchers (*see* Mann et al., 1996 for a complete review).

Linnoila and colleagues (1983) brought attention to a possible connection between 5-HT and impulsivity. CSF was measured in 36 impulsive and nonimpulsive violent offenders. Low CSF 5-HIAA was found in impulsive violent offenders compared to those who premeditated their acts. Importantly, other monoamines or metabolites were not different, providing an early indication that 5-HT was perhaps the transmitter most associated with the impulsivity. The lowest 5-HIAA values were found in violent offenders who attempted suicide.

Goodwin and Post (1983) wrote a review article assessing the literature to date regarding the relationship among 5-HT, aggression, suicide, and impulsivity. Autopsy data, CSF, 5-HT reuptake in platelets, and imipramine binding in platelets were reviewed. Fewer imipramine sites were generally reported in brains of suicide victims. The data further indicated a relationship between low 5-HIAA and suicide, aggression, and impulsivity.

A 5-HT deficit(s) appears separately associated with both aggressive behavior and suicide behavior. A study by Stanley and colleagues (2000) examined the role of 5-HT in aggression, but not suicide, by specifically excluding subjects with a history of suicide attempt. The aggressive group had a lower concentration of CSF 5-HIAA than the nonaggressive group. Also, the aggressive group had higher scores on impulsiveness subscores. These findings suggest reduced serotonergic function in aggression and/or impulsivity, independent of the serotonergic alterations associated with suicide behavior.

Stein and colleagues (1993) have advanced the notion that impulsivity is a failure of serotonergically mediated behavioral restraint. The authors point out that impulsivity has been characterized in both dimensional and categorical terms. The Diagnostic and Statistical Manual of Mental Disorders, 3rd edition, revised (DSMIII-R) classifies impulsivity and impulse control disorders as discrete entities. The argument is made that impulsivity and impulse control disorders have an underlying behavioral dimension, and the evidence regarding 5-HT is most consistent with the hypothesis of reduced serotonergic function being associated with a breakdown of normal behavioral restraint mechanisms.

More recently, efforts have been made attempting to sort out the biological effects of impulsivity from those that may be associated with violence (Cremniter et al., 1999). While CSF 5-HIAA has repeatedly been found to be lower in suicides than controls, Cremniter and colleagues found that the reduced CSF 5-HIAA in suicides was entirely due to the impulsive attempters. An inverse correlation between 5-HIAA and the Impulsivity Rating Scale was observed in the suicide group, with only a trend for the dopamine metabolite homovanillic acid (HVA). CSF 5-HIAA was lower in impulsive violent attempters than nonimpulsive violent attempters. Similarly, Mann and colleagues (1996) did not find any difference in CSF 5-HIAA between violent and nonviolent suicide attempters. Low CSF 5-HIAA, but not other monoamines, was found to be associated with a history of planned suicide attempt. Low 5-HIAA was also associated with suicide attempts with more medical damage (Mann et al., 1996). Elsewhere, Mann et al. (1989) have argued that the suicide method is related more to availability than to biology. However, Mann and Malone (1997) did not observe CSF 5-HIAA differences associated with impulsivity. Taken together, the findings suggest that low serotonergic function is related more to impulsivity than to violence; but also that impulsivity is associated more with the serotonergic system than the dopaminergic or noradrenergic system.

In summary, although measurement of transmitters in the CSF is an indirect and limited method for determining the level of neurotransmission and the relationship with behavior, over the years and decades, CSF studies implicate reduced serotonergic neurotransmission as a significant contributor to impulsivity.

The indirect indication of reduced 5-HT in aggression and suicide from studies of CSF need not be considered alone. For example, prolactin release in response to administration of the 5-HT releasing agent fenfluramine has been used as an index of the integrity of the serotonergic system (*see* Mann et al., 1995, for an extensive review of the application of the fenfluramine challenge test in studies of major depression). Coccaro et al. (1989) found an inverse correlation between the peak prolactin response to fenfluramine and measures of aggression (Brown-Goodwin Assessment for Life History of Aggression) and impulsivity (Baratt Impulsiveness Scale). In the same study, patients with a history of suicide attempt had lower peak change in prolactin values than either patients without a suicide history or normal controls (Coccaro et al., 1989).

Neither CSF nor neuroendocrine studies provided anatomical information about which brain regions are affected. Oquendo and colleagues (in press) recently examined the brain regions affected in the fenfluramine challenge in depressed suicide attempters using brain imaging of cerebral metabolism by positron emission tomography (PET). Regional cerebral metabolism was lower in ventral, medial, and lateral prefrontal cerebral cortex in high lethality suicide attempters than low lethality attempters, particularly after fenfluramine challenge. Low glucose utilization in the ventromedial prefrontal cortex was associated with low impulsivity, high suicide planning, and high attempt lethality, indicating a relative importance of this brain region. Additional evidence for a role of the ventral prefrontal cortex in suicide can be found in postmortem studies. Arango and colleagues (1995) found localized alterations in 5-HT reuptake sites and postsynaptic $5-HT_{1A}$ receptors in the brain of suicide victims. Mann et al. (2000) confirmed the finding of localized ventral prefrontal cortex reduction in the concentration of 5-HT transporter sites in suicide and, furthermore, found that in major depression, the reduction in 5-HT reuptake sites is widespread throughout the prefrontal cortex. The data from these and numerous other studies not discussed in this review has lead to the hypothesis that a serotonergic deficiency in the ventral prefrontal cortex has a permissive effect of allowing suicidal behavior (e.g., Arango and Underwood, 1997).

GENETIC STUDIES

Studies of the CSF, platelets, and brain tissue have contributed much in establishing 5-HT as the mediator of behavioral restraint and low 5-HT as permissive to impulsivity. These studies, however, do little to identify the underlying cause or causes of the reduced serotonergic neurotransmission and whether reduced serotonergic function is causative to behavior. The decade of the brain has attracted molecular biologists and psychiatrists alike to seek explanations of brain function, behavior, mental illness in general, and, specifically, for the purposes of this chapter, impulsivity and aggression.

Tryptophan Hydroxylase

Tryptophan hydroxylase (TPH) is the rate-limiting enzyme in the synthesis of 5-HT. Two polymorphisms have been identified in intron 7 of the TPH gene. There are seven studies showing an association between a polymorphism in the TPH gene and suicide (*see* Mann et al., 2001, for a recent review of the genetics of suicide). The relationship between TPH gene polymorphisms and aggression and impulsivity is less clear.

Neilsen and colleagues examined the relationship between TPH genotype and CSF 5-HIAA in 56 impulsive and 14 nonimpulsive alcoholic violent offenders and 20 healthy controls (Nielsen et al., 1994). An association between TPH genotype and CSF 5-HIAA was observed in the extreme impulsive group. There was no association between TPH genotype and impulsive behavior. A polymorphism associated with history of suicide attempts in violent offenders was also seen. Others have failed to find such associations. Manuck et al. (1999) found an association between the U allele of the A218C TPH polymorphism and aggressive behavior and a blunted prolactin response.

The combination of genetic variants may be critical in the determination of intermediate phenotypes. Turecki et al. (2001) examined three TPH gene polymorphisms (two promotor, one in intron 7) in 101 suicide completers vs 129 living controls. While no difference was observed at a single loci, haplotype associations were observed in suicides and more so in suicides with a violent method. These findings in suicide completers replicate a finding in suicide attempters (Rotondo et al., 1999). Abbar et al. (2001) examined seven polymorphisms in 231 suicide attempters and 281 controls. A genetic variant of the 3' part of the TPH gene was found and hypothesized to be a susceptibility factor for a phenotype combining suicide, mood disorder, and impulsive aggression.

5-HT Transporter

The 5-HT transporter (SERT) functions to take 5-HT released from the neuron back up into the cell for reuse. The SERT is the principal means by which the action of 5-HT on receptors is terminated. Major depression and suicide are independently associated with reduced SERT in the prefrontal cortex (Mann et al., 2000) and perhaps elsewhere in the brain (Parsey et al., 2001). One possible explanation for reduction in the number or concentration of SERT sites in major depression is that genetic variants in the gene reduce transcription or produce post-translational modifications resulting in decreased or less efficient reuptake of 5-HT. As in several and perhaps most genetic association studies, the results are discrepant and not conclusive. A recurring theme of this chapter is that genetic variants may contribute more to intermediate phenotypes, such as impulsivity, than to explaining major classes of mental disorders.

Zalsman et al. (2001) performed an association study of the SERT promoter in 48 suicidal adolescents. No association between the polymorphism and suicide was found. However, an association between violence and the LL LS genotype was found. Consistent with these findings, Courtet et al. (2001) found the SERT S allele and the SS genotype were more prevalent in violent suicide attempters than controls. An alternate explanation for genotype–phenotype associations is that behavior phenotypes somehow interact with genetics to increase risk of other behaviors. For example, mood disorders and alcohol abuse may interact with genetics to reduce 5-HT reuptake and increase risk for aggressive–impulsive behaviors like suicide. Gorwood et al. (2000) examined the SERT gene transporter polymorphism (5-HTT-LPR) in 110 male alcohol-dependent patients and 61 unaffected blood donors. The S allele was unrelated to alcoholism, but was associated with increased risk for suicide attempts.

A critical feature of any polymorphism is whether the variant(s) have functional significance for the amount or structure of the resulting protein. Regarding the SERT, for example, we find reduced amounts of SERTs in the ventral prefrontal cortex in suicide victims, but no relationship between SERT polymorphisms and the amount of SERT sites or association with suicide (Mann et al., 2000).

Taking the evidence together, the possibility is raised that variations in the SERT gene may be more associated with a "violence" phenotype.

$5-HT_{1B}$ Receptor

Transgenic mice are proving of heuristic value in dissecting the contributions of specific genes and gene products or proteins to complex behaviors. Of particular importance to aggressive–impulsive behaviors is the $5-HT_{1B}$ knock-out ($5-HT_{1B}$ KO) mouse. Brunner and Hen (1997) have effectively demonstrated the utility of the $5-HT_{1B}$ KO mouse as a model of impulsive behavior. The authors argue that impulsivity is a failure in the serotonergic system. The $5-HT_{1B}$ KO mouse has more impulsive aggression, and the mouse

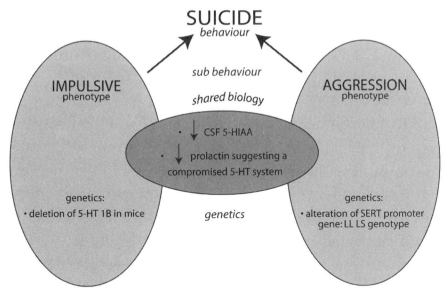

Fig. 1. Schematic diagram illustrating the genetic contributions and shared biological alteration in aggression, impulsivity, and suicide. A serotonergic abnormality, in part genetic, may contribute to reduced serotonergic function and underlie aggressive and impulsive behavior and may be the diathesis for suicidal behavior.

also acquires cocaine self-administration faster and drinks more alcohol, making it an animal model of addiction and motor impulsivity.

It is difficult to extrapolate aggression in animals to suicide as a behavioral dimension in humans. For example, New et al. (2001) found that the G allele of the 5-HT$_{1B}$ polymorphism is associated with a history of suicide attempt in Caucasian patients with a personality disorder. They did not find a relationship between the polymorphism with impulsive aggression. Nishiguchi and colleagues (2001) found no evidence of an association between the 5-HT$_{1B}$ receptor polymorphism and suicide in a Japanese population. We found no association between the 5-HT$_{1B}$ genotypes and aggression, suicide, or major depression (Huang et al., 1999).

What can be argued is that the single deletion of the 5-HT$_{1B}$ receptor gene in the mouse results in the alteration of a complex behavior producing, most notably, increased aggression and impulsivity. The specific events relating the gene deletion to the aggressive behavior is unknown, but the sequence is clearly that the gene deletion precedes the altered behavior. The importance of this temporal sequence is that it proves in a definitive way that the aggressive–impulsive behavior does not produce, lead to, or contribute to the behavior. Rather, selective deletion of the 5-HT$_{1B}$ receptor in the mouse results, by an as yet unknown chemistry and sequence, in increased aggression.

CONCLUSION

Aggression, suicide, and impulsivity are complex behaviors with interrelationships and overlapping biological alterations involving 5-HT and the prefrontal cortex (Fig. 1).

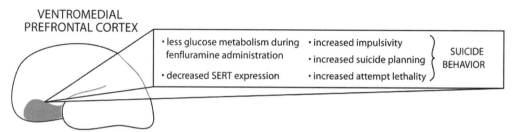

Fig. 2. Anatomical diagram highlighting the ventromedial prefrontal cortex with an accompanying description of alterations in suicide. A serotonergic deficiency in the ventral prefrontal cortex has a permissive effect of allowing suicidal behavior.

Indeed, aggression, suicidal behavior, and impulsivity are all positively correlated with each other (Horesh et al., 1999). Similarly, the rates of aggression and impulsivity are greater in suicide attempters (Mann et al., 1999). The onset of the development of the serotonergic alteration contributing to the behavioral spectrum of aggression–suicide is unknown. Interestingly, Brodsky et al. (2001) has shown a relationship between childhood abuse and higher impulsivity and aggression and more suicide attempts, raising the possibility of the abuse being an environmental risk factor for development of the aggression–impulsivity trait. Alternatively, not all abused children become aggressive. Caspi and colleagues (2002) found a functional polymorphism in the gene encoding monoamine oxidase A that moderates the effect of childhood maltreatment, suggesting an interaction between genes and environment, which may result in disparate behavioral outcomes in adulthood.

Studies in animals, such as the 5-HT$_{1B}$ transgenic mouse, suggest that deletion of a single neurotransmitter receptor can result in alterations in complex behaviors. Whether or not the role of the 5-HT$_{1B}$ receptor in aggression in the mouse is relevant or analogous to aggression in humans is less important than the indication that the biological alteration precedes the behavioral change. The totality of the data, therefore, implicates that 5-HT, in an anatomically specific fashion in the brain, acts to set the level of behavior (Fig. 2). Reduced serotonergic neurotransmission, particularly in the prefrontal cortex, can result, alone or in the presence of a modulating genetic phenotype, in an increased risk for aggressive, impulsive, or suicidal behavior.

ACKNOWLEDGMENTS

This work was supported in part by Public Health Service (PHS) grants AA11293, MH62185, MH40210, the Stanley Foundation, and the Diane Goldberg Foundation. We gratefully acknowledge the artistic contributions of Jasmine Francis.

REFERENCES

Abbar, M., Courtet, P., Bellivier, F., et al. (2001) Suicide attempts and the tryptophan hydroxylase gene. *Mol. Psychiatry* **6**, 268–273.

Arango, V., Underwood, M. D. (1997) Serotonin chemistry in the brain of suicide victims, in *Review of Suicidology*. (Maris, R., Silverman, M., Canetto, S., eds.), Guilford Press, New York, pp. 237–250.

Arango, V., Underwood, M. D., Gubbi, A. V., and Mann, J. J. (1995) Localized alterations in pre- and postsynaptic serotonin binding sites in the ventrolateral prefrontal cortex of suicide victims. *Brain Res.* **688,** 121–133.

Åsberg, M., Träskman, L., and Thorén, P. (1976) 5-HIAA in the cerebrospinal fluid. A biochemical suicide predictor? *Arch. Gen. Psychiatry* **33,** 1193–1197.

Brodsky, B. S., Oquendo, M. A., Ellis, S. P., Haas, G. L., Malone, K. M., and Mann, J. J. (2001) The relationship of childhood abuse to impulsivity and suicidal behavior in adults with major depression. *Am. J. Psychiatry* **158,** 1871–1877.

Brunner, D. and Hen, R. (1997) Insights into the neurobiology of impulsive behavior from serotonin receptor knockout mice. *Ann. NY Acad. Sci.* **836,** 81–105.

Caspi, A., McClay, J., Moffitt, T., et al. (2002) Role of genotype in the cycle of violence in maltreated children. *Science* **297,** 851–854.

Coccaro, E. F., Siever, L. J., Klar, H. M., et al. (1989) Serotonergic studies in patients with affective and personality disorders. Correlates with suicidal and impulsive aggressive behavior. *Arch. Gen. Psychiatry* **46,** 587–599.

Courtet, P., Baud, P., Abbar, M., et al. (2001) Association between violent suicidal behavior and the low activity allele of the serotonin transporter gene. *Mol. Psychiatry* **6,** 338–341.

Cremniter, D., Jamain, S., Kollenbach, K., et al. (1999) CSF 5-HIAA levels are lower in impulsive as compared to nonimpulsive violent suicide attempters and control subjects. *Biol. Psychiatry* **45,** 1572–1579.

Goodwin, F. K. and Post, R. M. (1983) 5-hydroxytryptamine and depression: a model for the interaction of normal variance with pathology. *Br. J. Clin. Pharmacol.* **15,** 393S–405S.

Gorwood, P., Batel, P., Ades, J., Hamon, M., and Boni, C. (2000) Serotonin transporter gene polymorphisms, alcoholism, and suicidal behavior. *Biol. Psychiatry* **48,** 259–264.

Horesh, N., Gothelf, D., Ofek, H., Weizman, T., and Apter, A. (1999) Impulsivity as a correlate of suicidal behavior in adolescent psychiatric inpatients. *Crisis* **20,** 8–14.

Huang, Y., Grailhe, R., Arango, V., Hen, R., and Mann, J. J. (1999) Relationship of psychopathology to the human serotonin$_{1B}$ genotype and receptor binding kinetics in postmortem brain tissue. *Neuropsychopharmacology* **21,** 238–246.

Linnoila, M., Virkkunen, M., Scheinin, M., Nuutila, A., Rimond, R., and Goodwin, F. K. (1983) Low cerebrospinal fluid 5-hydroxyindoleacetic acid concentration differentiates impulsive from non-impulsive violent behavior. *Life Sci.* **33,** 2609–2614.

Mann, J. J., Brent, D. A., and Arango, V. (2001) The neurobiology and genetics of suicide and attempted suicide: A focus on the serotonergic system. *Neuropsychopharmacology* **24,** 467–477.

Mann, J. J., Huang, Y., Underwood, M. D., et al. (2000) A serotonin transporter gene promoter polymorphism (5-HTTLPR) and prefrontal cortical binding in major depression and suicide. *Arch. Gen. Psychiatry* **57,** 729–738.

Mann, J. J. and Malone, K. M. (1997) Cerebrospinal fluid amines and higher lethality suicide attempts in depressed inpatients. *Biol. Psychiatry* **41,** 162–171.

Mann, J. J., Malone, K. M., Sweeney, J. A., et al. (1996) Attempted suicide characteristics and cerebrospinal fluid amine metabolites in depressed inpatients. *Neuropsychopharmacology* **15,** 576–586.

Mann, J. J., Marzuk, P. M., Arango, V., McBride, P. A., Leon, A. C., and Tierney, H. (1989) Neurochemical studies of violent and nonviolent suicide. *Psychopharmacol. Bull.* **25,** 407–413.

Mann, J. J., McBride, P. A., Malone, K. M., DeMeo, M. D., and Keilp, J. G. (1995) Blunted serotonergic responsivity in depressed patients. *Neuropsychopharmacology* **13,** 53–64.

Mann, J. J., Waternaux, C., Haas, G. L., and Malone, K. M. (1999) Towards a clinical model of suicidal behavior in psychiatric patients. *Am. J. Psychiatry* **156,** 181–189.

Manuck, S. B., Flory, J. D., Ferrell, R. E., Dent, K. M., Mann, J. J., and Muldoon, M. F. (1999) Aggression and anger-related traits associated with a polymorphism of the tryptophan hydroxylase gene. *Biol. Psychiatry* **45,** 603–614.

New, A. S., Gelernter, J., Goodman, M., et al. (2001) Suicide, impulsive aggression, and HTR1B genotype. *Biol. Psychiatry* **50,** 62–65.

Nielsen, D. A., Goldman, D., Virkkunen, M., Tokola, R., Rawlings, R., and Linnoila, M. (1994) Suicidality and 5-hydroxyindoleacetic acid concentration associated with a tryptophan hydroxylase polymorphism. *Arch. Gen. Psychiatry* **51,** 34–38.

Nishiguchi, N., Shirakawa, O., Ono, H., et al. (2001) No evidence of an association between 5HT1B receptor gene polymorphism and suicide victims in a Japanese population. *Am. J. Med. Genet.* **105,** 343–345.

Oquendo, M. A. and Mann, J. J. (2000) The biology of impulsivity and suicidality, in *Borderline Personality Disorder.* (Paris, J., ed.), W. B. Saunders, Philadelphia, pp. 11–25.

Oquendo, M. A., Placidi, G. P. A., Malone, K. M., et al. (2002) Positron emission tomography of regional brain metabolic responses to a serotonergic challenge and lethality of suicide attempts in major depression. *Arch. Gen. Psychiatry*, in press.

Parsey, R. V., Oquendo, M. A., Simpson, N., et al. (2001) *In vivo* imaging of receptors in mood disorders: is there agreement with postmortem studies? *Biol. Psychiatry* Abstr., 2001.

Rotondo, A., Schuebel, K., Bergen, A., et al. (1999) Identification of four variants in the tryptophan hydroxylase promoter and association to behavior. *Mol. Psychiatry* **4,** 360–368.

Stanley, B., Molcho, A., Stanley, M., et al. (2000) Association of aggressive behavior with altered serotonergic function in patients who are not suicidal. *Am. J. Psychiatry* **157,** 609–614.

Stein, D. J., Hollander, E., and Liebowitz, M. R. (1993) Neurobiology of impulsivity and the impulse control disorders. *J. Neuropsychiatry Clin. Neurosci.* **5,** 9–17.

Turecki, G., Zhu, Z., Tzenova, J., et al. (2001) TPH and suicidal behavior: a study in suicide completers. *Mol. Psychiatry* **6,** 98–102.

Zalsman, G., Frisch, A., Bromberg, M., et al. (2001) Family-based association study of serotonin transporter promoter in suicidal adolescents: no association with suicidality but possible role in violence traits. *Am. J. Med. Genet.* **105,** 239–245.

5
Behavioral and Neuropharmacological Differentiation of Offensive and Defensive Aggression in Experimental and Seminaturalistic Models

Philip M. Wall, D. Caroline Blanchard, and Robert J. Blanchard

INTRODUCTION

The term "aggression" is used to mean so many different things in ordinary speech that it is impossible to directly relate them all to a single scientific concept. However, building on a remark by Wilson (Wilson, 1975) that aggression reflects resource competition, we have attempted to analyze many instances of human aggression as a response to challenge to an individual's resources or to his or her rights (Blanchard et al., 1984, 1999). The latter reflects that people are intensely social animals and that, among social animals, access to resources is typically determined by status and its privileges, rather than by competition over the specific resource. Challenge to an individual's status represents disputation of its associated rights including, but not limited to, rights to resources.

This concept of "challenge" covers a lot of territory, from the insults that are more or less ritualized forms of interaction among young men in many cultures, through innumerable incidents of normal social life (Averill, 1982; Blanchard et al., 1999), as well as through electoral processes and other legal procedures involved in competition for status and specific rights. Virtually every human culture has been deeply concerned with informal or formal rules concerning challenges to rights and resources and the method of determining a resolution to these challenges (Blanchard et al., 1999). Aggression as conceptualized and used in ordinary psychological language intersects with these situations in several ways. First, the frequent polarization of "good" types or expressions of aggression vs "bad aggression": aggression is typically regarded as good in the context of sports, business, law, military endeavors, and other areas in which competition is assumed and rules of engagement are honored; but aggression is regarded as bad when competition is deemed inappropriate or when the rules are flagrantly broken, whether or not violence occurs. Second, vigorous pursuit of rights or status in almost any realm may be characterized as aggressive, e.g., an aggressive businessman, athlete, scientist, and so forth. In particular, an individual who consistently pushes the rules (or the "envelope") may be deemed aggressive, with a view of whether this is good or bad aggression depending on which side the viewer is on.

These attributions are common, even if no violence is involved. However, real violence, involving physical, fiscal, emotional, and social damage to other individuals frequently does occur in the context of status and/or rights challenge, and it is usually in violation of social norms or legal statutes. Generally, it is this violation rather than the damage itself, that incites approbation. A collegiate wrestler who breaks another's arm in the course of a match is not likely to be regarded as aggressive, unless this involves some nonsanctioned maneuver. There are, moreover, situations in which violence and potential damage to the opponent are intrinsic components of the appropriate response to a challenge. While such situations (e.g., gangs) may be regarded as intrinsically aggressive from an outsider's perspective, an individual in this milieu responding to a status challenge by violent attack on the challenger may not be self-regarded or viewed by his compatriots as aggressive. He (increasingly, she) (Chesney-Lind et al., 1999) is doing what is necessary. Similarly, while all instances of illicit taking of property might be regarded as involving challenge to the victim's control over resources, the law distinguishes between robbery, in which that challenge is exercised directly with an accompanying potential for damage, and burglary, in which no face-to-face challenge (and, usually, no possibility of direct violence) is involved.

These and other complexities in the concept of aggression have led to the notion that aggressive behaviors are situationally and functionally diverse and specific. However, they might equally well be interpreted as indicating that while many instances of aggression have a common origin in response to challenge to status or resources, the situations that fall under this functional rubric cover a wide range of human activities and are subject to a complex nexus of rules. This view of a common origin suggests that all aggression situations involving competition and challenge may potentially be analyzable in terms of some core paradigms.

There are, however, violent situations that these paradigms may not fit. One of these is bodily self-defense. It is difficult to obtain information on this topic as it relates to people, as acts of violence are seldom classified in terms of self-defense vs some other rubric, and (in partial contrast to competition or challenge-based aggression) (e.g., Taylor et al., 1993) satisfactory laboratory paradigms for this are not available. The core feature of self-defense situations is that they involve a primary motivation of fear of bodily damage. This deserves some further clarification. Many challenge–aggression situations involve the possibility of body damage to the attacker, but in the context of motivation related to the challenge. This doubtless reflects an emotion, which is probably anger (Averill, 1978; Fukunaga-Stinson, reported in Blanchard, D.C. et al., 1984), while fear of the opponent acts as a deterrent to such attack. This may be seen in the evaluative process often undertaken by those preparing for a fight, in which they attempt to analyze their opponent's fighting qualities before committing themselves to the engagement (Blanchard et al., 1999). Fighting because you are desperately afraid is something else.

We have tried to tap this human dimension in several ways, first interviewing an array of young adults about their own experiences of this type (reported in Blanchard et al., 1984). Their recollections suggested that important components were high level fear on being attacked by one or several assailants, and second, an inability to escape. In these situations, people typically reported screaming and attacking the opponent, often using methods that were not characteristic of fights in a challenge situation, such as biting. A number of interviewees emphasized an unplanned, reflex-like nature of these attacks, saying "I just went berserk," often considerably damaging their opponents.

A second approach involved scenarios designed to set up situations of serious bodily threat from another individual without any competitive aspect, and asking subjects to select their first response from a list of possible actions (or to supply one, if none of these fit). In such situations, vocal threats and attack were prominent responses. In fact if the threat were clearly serious and the situation and/or attacker very difficult to escape, defensive threat (screaming from women) and attack (men) were the first defensive responses chosen; whereas less serious or more ambiguous or escapable threats elicited different first responses (Blanchard et al., 2001b).

There are several complications associated with differentiating defensive threat–attack from competition–challenge-based aggression, on a human level. If an instance of violence is accepted as involving self-defense, it is almost always viewed as potentially justifiable. Even here there appear to be implicit, and sometimes explicit, rules; for example, that self-defense should not involve unnecessary escalation or damage disproportionate to the danger offered[1]. However, these "rules" may be somewhat flexible, in keeping with an understanding that a person under physical attack might not have the time, the facts, or the clear head necessary to make fine judgments about what is permissible behavior. The greater potential for justification inherent in self-defense as opposed to competitive aggression situations has not gone unnoticed, and claims of self-defense for violent acts against another are common, ranging from criminal pleas to self-justifications for acts of war.

Still more common and confusing to the basic distinction between competitive and self-defense aspects of aggression is the interpretation that any act of hostility or violence can be attributed to "defense" or "protection" of something: "I was defending my (interests; standing; girlfriend; rights; reputation; property, and so forth)." This has a nice ring of defense to it, but often means no more than that the individual was competing with another for some right or resource, or for status, putting it directly back into the rubric of competitive aggression. As an example, while "defending my girlfriend" might mean attempting to thwart an attack on her, it may equally well refer to attacking a rival for the girlfriend's favor; in which case competition for the girlfriend rather than danger to the girlfriend is the issue. "Defending my rights" is an even more insidious example and can cover almost any sort of situation. What it does not mean, however, is self-defense. Such usages, while perfectly understandable in social situations in which aggression in violation of relevant rules is likely to be punished, nonetheless obscure the difference between competitive aggression and self-defense, a difference that, as will be seen, is relatively clear in the experimental animal literature. In terms of a conceptual method of differentiating the two types of aggression objectively, one possibility is to determine what the aggressive individual would do if a wider range of possibilities were offered: the "goal" of defensive threat–attack is to eliminate the danger. Subjects in an earlier scenario (Fukunaga-Stinson, reported in Blanchard, D.C. et al., 1984)

---

[1] See "Self-defense; The 'Lectric Law Library (http://www.lectlaw.com/def/d030.htm)" for a general discussion of rights and rules pertaining to legal views of self-defense.

answered "leave" as their first choice in such situations. In the competition–challenge situation, however, the primary goal is to control the challenger, with ancillary goals of punishment and domination, depending on the situation.

While people show other behaviors, such as rough and tumble play in children that are sometimes labeled "aggressive," these "offensive" and "defensive" aggression modes are likely the two major situations in which human emotion-based aggression occurs. Much has been made of the fact that a lot of human violence is expressed as an operant, i.e., in aid of some goal that is extraneous to the violence itself, robbery, for example. This is hardly surprising, as people can express virtually any voluntary skeletomuscular behavior (and some not typically regarded as voluntary) for extraneous rewards. It should not be overlooked, however, that the goal of competition–challenge-based aggression is to control the behavior of the (challenging) opponent, such that extraneous goals, particularly those achieved through coercion of others, are a natural reward of such aggressive altercations. A personal interview with a habitual wife-beater brought forth the following comment on the outcome of such activity: "It means you win the argument!"

Given this diversity, it might be felt that creation and analysis of animal models of human aggression, in which aggressive behaviors can be manipulated and measured, would be a considerable challenge. In fact, this may be stating the situation backwards; much of this analysis actually stems from animal research, and many of the human studies reported above reflect an attempt to determine if factors outlined in animal experimentation are also relevant to human aggression. Thus, in the present chapter, we attempt to review animal studies relevant to the notion of an important difference between competitive or offensive aggression and defensive aggression. Although the major focus is on situational and behavioral differences between these two distinct forms of aggression, we will also briefly outline some of the potential or accepted pharmacological differences between the two.

ETHOLOGICAL AND EXPERIMENTAL–LABORATORY APPROACHES TO THE STUDY OF ANIMAL AGGRESSION

For aggression, as for many aspects of behavior, there exist two distinct investigative approaches (Blanchard et al., 1989c); it would be charitable, but not always accurate to view them as complementary. The ethological approach typically involves the study of animals in their own natural situations, with an extensive focus on description and sequential analysis of behavior and with varying amounts of theoretical interpretation. The value of the ethological approach is that it permits a relatively full analysis of what animals normally do under particular circumstances, such as the presence of threat stimuli or sexually accessible members of the opposite sex. These situations permit an animal to choose its own behaviors and to change them, depending on, for example, the reactions of conspecific or nonconspecific others.

In contrast, traditional experimental approaches almost always involve animals tested in laboratory situations, with very limited access to some key characteristics of the environments in which their species evolved, such as the occasional or continuous presence of conspecifics and a terrain with features facilitating expression of a wide variety of behaviors. Experimental approaches permit manipulation of important variables in a test situation, affording a superior basis for interpretation of changes in behavior that may follow such manipulation. Because experimental research can involve alterations of features of a test situation to selectively foster particular behaviors, it can ensure that they will be rapidly and precisely elicited. This is a great boon in the study of physiological mechanisms underlying behavior, although the dependent variables measured in such tests may not adequately reflect the behaviors that occur in natural or seminatural situations.

The Differentiation of Attack (Offense) and Defense

Both of the aforementioned traditions have been applied to the study of aggression, as have "ethoexperimental" approaches that attempt to merge the two (Blanchard et al., 1989c). An important series of early studies (Grant et al., 1958, 1963; Grant 1963) that combined elements of ethological and experimental work involved description and analysis of behavioral interactions of dyads of male rats over successive encounters. These studies provided a detailed description of the interactive behaviors between both animals, along with sequential analyses within a session, and analyses of changes over sessions. The results provided a solid grounding for the view that agonistic (i.e., fighting) experience produces changes in both animals of such a dyad, with one, the winner, coming over time to show a particular subset of the behaviors that were initially displayed, while the other, the loser, tends to show a considerably different subset of behaviors. The two behavioral subsets empirically differentiated on the basis of this consistent grouping were attack (characteristic of the consistent winner) and defense.

Although we were at the time abysmally ignorant of the Grant et al. studies, our own work using seminatural colony situations unwittingly took off where theirs had stopped. In large open bin colonies with mixed-sex rat groups, adult males fight, and dominance hierarchies form on the basis of winning and losing fights. When an adult male intruder is placed in such an established colony, only the dominant male attacks it (virtual abolition of attack in the presence of the dominant is one of the characteristics of colony subordinate males). If, as is usual, the intruder is naïve, then a strongly polarized encounter occurs, with the colony dominant attacking, and the intruder showing an exclusively defensive pattern. The major focus of our analyses soon came to be the response of one member of such a dyad to the immediately preceding behavior of the other member of the dyad. The question became: What does the defender do following specific attack behaviors of the other rat, and what does the attacker do in response to particular behaviors of the defender? These analyses (Blanchard et al., 1977), in conjunction with observations that the animals were biting specific targets on the bodies of their opponents, produced a remarkably coherent picture, one that is characteristic of both dominant–subordinate interactions in chronic grouping situations and of residents and intruders; while naïve intruders may not be skilled in these defenses, they do show all of them even during first attacks.

Attack typically involves an initial social investigatory approach, including sniffing of the anogenital region, nose, or body of the opponent, quickly followed by piloerection of the guard hairs of the animal's coat (rat) or by tooth chattering and/or tail rattling (mice). This preliminary sniffing is apparently aimed at obtaining information as to the intruder's gender and sexual development status, in that resident rats do not attack male intruders until the latter are sexually mature. This established, the resident may then deliver a bite without further preliminaries.

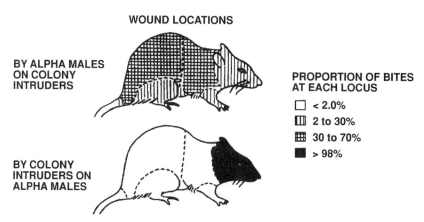

Fig. 1. Proportions of bite wounds located on various body sites following colony resident–intruder tests. Upper figure represents colony intruders, indicating bite targets of dominant or alpha male colony attackers. Lower figure represents colony dominant or alpha males, indicating bite targets of intruders. These bites tend to occur in response to bites by the dominant males.

Back-Attack and Back-Defense

The overwhelming majority of bites by an attacker on a defender were made on the latter's back and flank areas, whereas the few bites of defenders on an attacker were almost entirely on the latter's head (*see* Fig. 1). Moreover, the behaviors of defenders in response to attack, consisting of either flight, an upright stance facing the attacker, or lying on its back, all had in common that they tended to remove the defender's back from the attacker's teeth. Flight, of course, represents a generic removal of self, but the upright and on the back postures enabled a defender to stay in the same place while interposing other parts of its body between the attacker and that animal's apparent target, the defender's back. A typical sequence of such activities, for animals previously unknown to each other, will be presented under the "resident–intruder" model.

That targeting of bites to the back represents an important mechanism in the control of agonistic behavior, which was demonstrated by additional observations indicating that, faced with an upright defense, attackers showed a lateral attack motion crowding against the upright defender, sometimes throwing it off balance. They might also lunge forward and around the defender from this lateral position, providing a second means of access to its back. Similarly, faced with a defender lying on its back, attackers stood over the supine animal and often pushed or pulled at it to roll it over, again potentially exposing the back. To flight, the relevant attack was "chase." In all three cases, flight–chase, upright–lateral attack, on the back–stand over, the result tended to be exposure of the defender's back, which the attacker would then, typically, bite. These elaborate back-attack (for the attacker) and back-defense (for the defender) strategies clearly had some evolutionary basis, in that they could be seen, albeit in a clumsy and only partly successful form, in the attack and defensive behaviors of naïve rats.

Why these behaviors were adaptive was further clarified by studies presenting terminally anesthetized "intruder" rats to colony dominants, in such a way that only selected body parts were available for biting (Blanchard et al., 1977). When the back was available, attackers bit it freely. When only the ventrum was available, it was not bitten at all. Interestingly, another target site emerged in these studies of anesthetized intruders, the head. Notably, unanesthetized intruders were virtually never bitten on the head, whereas this was a prime target on intruders when they were incapable of defending themselves. The difference was in the use of vibrissae, those long and stout hairs embedded in the prominent nose-pads of rats. An unanesthetized defender would erect its vibrissae and direct them forward, to serve as a sort of limbus around the rats face. An attacker biting at the defender's face or head would inevitably encounter these vibrissae, meeting them with its own complement of vibrissae, such that the two sets of hairs would cross, touch, and lock. The lock reflects the extreme sensitivity of vibrissae to tactile stimulation. Fossorial and nocturnal animals like rats use these primarily for locomoting in the dark: when a vibrissa touches something, it halts movement in that direction. That the vibrissae lock was stopping attackers from reaching the head of their opponents was made clear by the results of vibrissae removal: head bites increased dramatically (Blanchard et al., 1978). That defenders can make bites on the attacker's head and/or snout appears to reflect that when biting, the defenders fold back their vibrissae, while the vibrissae of the attacker are seldom erected.

Offensive Attack Vs Defensive Behaviors

The aforementioned relationships, observed in wild and laboratory *Rattus norvegicus* and *R. rattus* in both laboratory and field conditions, and, with some modifications in mice (reviewed in Blanchard, 1997) provide a number of independent differentiators of attack and defense patterns: situational (e.g., which animal is in its home territory); organismic (e.g., previous experience with winning or being defeated; drugs and other biological factors enter here also); behavioral (attack vs defense patterns; in the rat back-attack vs back-defense); and outcome (bites on the back vs bites at the face and/or snout). While a number of different labels have been applied to the two patterns, the most common designation for attack by dominant animals on intruders into their territory was offensive aggression, while the threat and attack components of the defense pattern are usually called defensive threat and attack.

These differentiators have been analyzed in other species as well. Geist (1968) had earlier noted that in sheep and goats, antlers and horns are typically the targets of blows, as well as the weapons that deliver these. Pellis and his associates (1987) confirmed target sites for attack in rats and mice and soon extended target site findings to other rodent species (e.g., hamsters, Pellis et al., 1988; voles, Pellis et al., 1992). Parmigiani and his associates have differentiated offensive and defensive targets for aggression in both male and female mice, noting that for the latter, attack on intruder males is largely defensive, while attack on intruder females is offensive (Ferrari et al., 1996). Leyhausen provided beautiful descriptions of attack and defense patterns in house cats, including targeting of bites and blows (Leyhausen, 1979). While most rodent species do target the back or flanks for offensive attack, other species, such as cats and ungulates, have different specific attack targets. We have suggested (Blanchard et al., 1984) that such targeting is particularly marked in social species. This is because, in animals that live in relatively small groups with high degrees of kinship among group members, death as the result of conspecific attack is likely to reduce the (related) attacker's inclusive fitness.

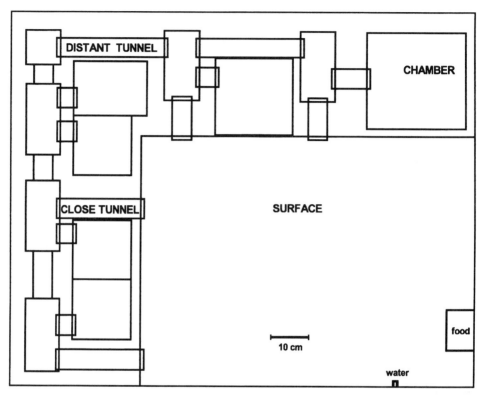

Fig. 2. Schematic diagram of the visible burrow system (VBS). Food and water provided *ad libitum* in the open surface arena.

Thus, a dominant male rat, highly likely to be the father of young animals in the group, may attack and bite juvenile males, driving them from the group and reducing his own competition for females; however, he is better served by simply driving them away, via painful but relatively harmless back bites, than by killing them. In larger groups, such as are characteristic of many herd animals, kinship is a less potent factor. However, the benefits of a larger group in terms of enhanced protection from predators may lead to the same result: attacking and defeating another member of your group may be advantageous, but killing that individual probably is not.

THE OFFENSE–DEFENSE DISTINCTION AND COMMONLY USED AGGRESSION TASKS AND MEASURES

The Colony Model

Colonies with mixed-sex groups of rats or mice provide simulations or, better yet, exemplars, of social (e.g., aggressive, defensive, and sexual–reproductive) behaviors. The combination of a large seminatural habitat with burrows and the presence of females results in a high level of aggression for the dominant male (Adams, 1985; Blanchard et al., 1985b), a particular advantage when dominant–subordinate relationships and chronic social stress are under investigation (*see* Fig. 2). The stress that may be engendered in such colonies induces changes in virtually every aspect of the activity of subordinates, producing a pattern that has been suggested to parallel many of the behavioral changes characteristic of the target symptoms of clinical depression (Blanchard et al., 1990a). This provides a specific rationale for use of colony systems aside from aggression research, *per se* (Blanchard et al., 1995, 1998a, 1996b).

When colony groups have stabilized, the agonistic behaviors of dominant and subordinate males are strongly polarized, as with some resident–intruder paradigms. Although colony models may be used as a variant of the resident–intruder model (and there is a striking relationship between aggression scores in single resident–intruder tests and attainment of dominance in colony situations) (Blanchard et al., 1988), they are not efficient in this context: the (single or paired with female) resident–intruder situation is much less expensive, takes less space, and is easier to maintain. In addition, colony situations may elicit such high levels of agonistic behavior and subordinate stress that subordinate male longevity is dramatically reduced (Blanchard et al., 1985b, 1990b).

The Resident–Intruder Model

The single resident (or resident paired with a female, which is removed just before intruder testing) model has many similarities to colony models, but with some subtle differences as well. For example, while colony dominants and subordinates usually have somewhat equivalent agonistic experience, residents (experienced) and intruders (usually naïve) do not. Naïve residents may also be used, but seldom are, because naïve rats may show highly variable attack tendencies, whereas an experienced victor typically proceeds directly to the attack.

As in colony situations, this involves initial social investigatory approaches, including sniffing of the anogenital region, nose, or body of the intruder, quickly followed by piloerection of the guard hairs of the animal's coat (rat) or by tooth chattering and/or tail rattling (mice). The preliminary sniffing of the resident is apparently aimed at obtaining information as to the intruder's gender and sexual development status, in that resident rats do not attack male intruders until the latter are sexually mature. This established, the resident may then deliver a bite without further preliminaries, leading to the same back-defense and back-attack behaviors characteristic of colony situations. Defensive behaviors of the intruder are followed by attack behaviors that counter the defense. Thus, with the exception of the initial bites, the behavior of the defender leads that of the attacker, resulting in what looks like a "dance" of response and counterresponse. For both animals, this gets better with practice, which is likely why experienced defenders are bitten less often (Lore et al., 1980), particularly when facing an experienced attacker. As one example, experienced defenders avoid flight, the most dangerous defensive strategy within the confines of the small test situation. In such situations, the chasing attacker has virtually unlimited access to the intruder's posterior back and rump.

Resident–intruder tests are also frequently used with mice. Studies of aggression in mouse colonies are much less frequent, as male mice in mixed-sex groups show such high levels of fighting that an unacceptable level of injury and death may result. Mice show a progression of attack and defensive behaviors similar to those of rats, with the exception that the on-the-back defense (and the accompanying on-top or pinning offense) is rare, likely due to the fact that mice show less inhibition of bites to the ventrum of an opponent (Blanchard et al., 1979). A further distinction is often made in mice, between an upright but active (i.e., turning to face or fend off attack) defense and a submissive upright, which involves drooping forepaws and little attempt to counter the actions of the attacker. In rats, the term submission is often applied to the on-the-back defense, but the implication of that designation, that the animal is submitting passively to attack, appears not to be correct. Supine rats use both fore- and hindpaws to push off attackers and often roll to face the attacker, keeping the vulnerable back protected (Blanchard et al., 1977).

Defensive Bites in Colony and Resident-Intruder Tests

Defensive bites by intruders and subordinates do sometimes occur, but are rare, probably because such bites are under tight situational control. They occur almost exclusively during or immediately (approx 1 s) after a bite by the resident on an intruder, when the resident's head and/or snout is close enough to be bitten (Blanchard et al., 1977). This preferential targeting of defensive bites at the head and/or snout of the attacker is also seen when the attacker is a predator (cat) (Blanchard et al., 1980), and these antipredator bites are also facilitated by pain in male rats, although female rats will bite the cat's face and/or snout without shock (Blanchard et al., 1978). As will be seen, defensive bites are the aggression measures in some other tests. These offensive and defensive bite targets and behavior patterns are virtually identical to those seen in wild rats in laboratory or natural settings (Barnett, 1960; Mackintosh, 1981; Takahashi et al., 1982; Blanchard et al., 1985b).

Competition and Frustration Tests

The frustration–aggression hypothesis, formulated in respect to human aggressive behavior and hugely influential over a period of many years (Dollard et al., 1939), views the thwarting of goal-directed behavior as an important antecedent of aggression. It has spawned a number of influential behavioral paradigms. In one of these, rats are initially trained to traverse a narrow tunnel for food. After training, subjects are placed at opposite ends and meet in the runway itself. A "contest" ensues in which a "winner" pushes the "loser" out of the tunnel. Fighting may then occur in the goalbox (Work et al., 1969; Grossman, 1970). In another paradigm, two severely food-deprived animals are housed with an apparatus that limits food access to only one animal (e.g., Zook et al., 1975) and fighting may occur during competition for food access.

Although these procedures have been used to assess drug actions (Eichelman et al., 1973; Miczek et al., 1975), they have a number of deficiencies. Dominance as measured by winning in these tests correlates poorly with dominance measured during spontaneous agonistic encounters (Grossman, 1972; Masur, 1975; Miczek et al., 1975). This fits with data indicating that colony dominant rats do not have a striking advantage in terms of obtaining water or food items under mild deprivation; suggesting that for highly social species, fighting is more closely tied to status than to access to food and water, at least in animals that are only mildly deprived and have little history of previous deprivations. Moreover, these agonistic interactions may wane with repeated testing (Grossman, 1970), again suggesting that the development of dominant–subordinate relationships may be more important than competition over a food item. The close confinement of the subjects inhibits measurement of a full aggressive–defensive behavioral profile, as well as other behavioral changes (e.g., hyperactivity, grooming) useful in determining the specificity of independent variable effects.

Pain-Elicited Aggression

For many years (e.g., Ulrich et al., 1962), pain-elicited aggression or "reflexive fighting" was the primary model for aggression in laboratory rodents. In the reflexive fighting test, two animals are placed in a small chamber (the smaller the better) and repeatedly administered foot shock. After a few trials, one or both animals will respond to shock by assuming an upright posture, displaying fore and hind paw movements that resemble kicking or striking with occasional bites directed toward the opponent's snout. This procedure produces consistent high rates of behavior, which can be brought under stringent experimental control.

However, as analyses of attack and defensive behaviors became available, it became increasingly clear that shock evokes only a narrow range of conspecific agonistic behaviors. These include the upright posture (often assumed by both participants) and biting directed toward the opponent's snout (Blanchard et al., 1978; Blanchard, 1984). The fore- and hindlimb movements (labeled striking and kicking) seen during reflexive fighting can also be reliably elicited in solitary animals that are held in an upright position while being shocked (Blanchard, 1984), and thus appear to reflect reflexive responses to scrambled foot shock. None of the postures consistently associated with offensive attack (lateral attack, chasing, pinning), however, are observed during such tests (Takahashi et al., 1982). Finally, the fact that reflexive fighting is virtually impossible to elicit in larger test chambers, calls the offensive aggressive motivation of the subjects into question.

An alternative "pain-elicited aggression" procedure is to confine a subject in a narrow open-ended tube with its tail secured to an electrode and its head and/or snout near to an inanimate biting target such as a wooden dowel (Azrin et al., 1968). Rats receiving repeated tail shocks during such tube tests will show high frequencies of target biting, and the biting increases with higher levels of shock (Blanchard et al., 1980). Because subjects are tightly confined in the tube, actions (e.g., lateral attack, defensive upright), which might help clarify the offensive or defensive nature of the bites, are impossible. However, when an anesthetized conspecific or potential predator (anesthetized cat) is substituted for the inanimate target, bites are directed at the snout (a defensive target) but never directed at the back (offensive target) (Blanchard et al., 1980). In mice, target biting occurs reliably in the total absence of shock and shows few parallels with either resident–intruder or maternal aggression (Brain et al., 1983). These findings provide little evidence that pain elicits offensive aggression, and the defense components observed in these situations appear to be mixed with a variety of indiscriminately displayed reflexive behaviors to shock.

Defensive Threat–Attack to a Predator

Unlike (offensive) aggression, which is most often aimed at conspecifics, defensive behaviors have evolved on the basis of their adaptive value against attack by both predators and conspecifics. Defensive attack is relatively rare in laboratory rats (Blanchard, 1997), a factor in the development of tests using predators to elicit a full range of defensive behaviors, first for rats and later for mice. When mice are approached and contacted by a predator (a hand-held, terminally anesthetized, rat; mice are common prey for wild rats) in a relatively inescapable situation, they face the oncoming animal and may show sonic vocalizations, followed by biting attack (Blanchard et al., 1997b, 1998a, 2001a). While these mouse bites typically involve the face and/or snout of the predator, it is not known if this is preferential targeting, as this is the part of the rat's body that first contacts the mouse subject. However, rats confronted by a terminally anesthetized cat show a high-level preference for biting the face and/or snout when different body parts are offered under exactly the same conditions. In fact, although tail shock facilitates rat biting at a predator, many sites on a cat are not bitten even with substantial tail shock (Blanchard et al., 1978, 1980). Both their relationship to shock and the fact that these defensive threat (sonic vocalizations: screams) and defensive attack bites are elicited by predators indicate that they reflect defensive rather than offensive aggression, while other behaviors seen in the context of mouse response to a rat predator (Blanchard et al., 1997b, 1998a, 2001a) make it clear that defensive threat–attack occur as part of a defense pattern that is consistently elicited by predators.

NEUROPHARMACOLOGICAL DIFFERENTIATION OF OFFENSIVE AND DEFENSIVE AGGRESSION

Whereas a great number of drugs have been evaluated in tests of offensive aggression, mostly involving resident–intruder models, very few drugs have been used in studies that unequivocally reflect defensive threat and attack. Thus, a wider literature is available involving models that suggest offensive attack. However, many of these, for example those that measure aggression in an unfamiliar area rather than the home cage of a resident or those that attempt to measure drug effects on aggression elicited by prior drug treatment, are difficult to interpret. For these reasons, this section will be confined to drugs for which there is evidence of effects on both offensive and defensive attack and that can be relatively straightforwardly attributable to these behavioral distinctions in both humans and animals.

Serotonin

Serotonin (5-HT) is the most frequently investigated neurotransmitter, in terms of effects on aggression. Most relevant to the differentiation between offensive and defensive attack, "serenic" compounds with principal effects at two 5-HT receptor subtypes (5-HT_{1A} and 5-HT_{1B}) have been shown to consistently reduce offensive attack in rats and mice and have been measured in a variety of relevant tasks and situations (Olivier et al., 1994). When the same compounds were investigated in both offensive and defensive attack tests, however, they failed to alter, or only marginally reduced, the defensive attack (or other defense) component, yet produced profound reductions in offensive attack parameters (Blanchard et al., 1985a; Flannelly et al., 1985; Parmigiani et al., 1989; Ferrari et al., 1996).

Antidepressants, including the selective serotonin reuptake inhibitor (SSRI) fluoxetine and the serotonin reuptake inhibitor (SRI) imipramine, have been reported to not only reduce the aggressive behaviors of resident male rats following acute administration, but also to increase it after chronic administration (Mitchell et al., 1991, 1997). Ferris et al. (1997) also reported reductions in resident aggression in hamsters following acute administration of fluoxetine, while Ho et al. (2001) reported reductions in the heightened aggression displayed by resident female rats during diestrus, after short-term (4 to 5 d) fluoxetine administration. However, chronic fluoxetine reduced, rather than enhanced, aggression in prairie voles (Villalba et al., 1997).

In a mouse defense test battery (MDTB) measuring a wide range of mouse reactions to a rat, fluoxetine and imipramine enhanced defensive threat–attack following acute administration, but decreased these response patterns when given chronically, i.e., daily for 21 d (Griebel et al., 1995a). Thus, in rodents, most studies suggest that the effects of acute fluoxetine and imipramine are opposite for offensive and defensive aggression (reducing and increasing these, respectively), while chronic fluoxetine and imipramine enhance offensive, but reduce defensive, aggression.

In humans, deficiencies in 5-HT metabolites measured in cerebrospinal fluid have been associated with impulsive violence (George et al., 2001; Placidi et al., 2001), but not with more premeditated violence (Linnoila et al., 1983). However, a recent review (Walsh et al., 2001) of SSRI studies indicated a paucity of major effects on aggression in patients prescribed these drugs. It seems reasonable to assume that the aggressive behaviors of these patient groups did not include a high proportion of premeditated violent acts, but it is unclear how the aggressive behaviors measured in each of the studies that were included in this analysis might relate to an offensive vs defensive aggression dichotomy. Insofar as human aggression may involve both of these, the opposite effects on the two, for both acute and chronic drug administration, might be expected to result in something of a washout of overall net effects. The animal findings do, however, render the suggestion that more distinctive effects might be obtained if attempts were made to analyze serotonergic drug effects on human aggression, separately, for offensive and defensive behavioral patterns. Of course to facilitate such an analysis, more precise operational definitions of the two behavioral patterns of aggression in humans will be required.

Benzodiazepines

Classic benzodiazepine agonists have been shown to have extremely consistent effects in the MDTB and in a closely related task involving attack by rats to human handling. Acute chlordiazepoxide, diazepam, or midazolam administration reduced defensive threat and attack behaviors in both rats and mice (Blanchard et al., 1989b; Griebel et al., 1995b), whereas benzodiazepine inverse agonists potentiated defensive threat–attack (Griebel et al., 1995). In mice, acute diazepam has consequently been used in a number of MDTB studies as a positive control, owing to consistently demonstrated effects on defensive threat and attack (Griebel et al., 1995b, 1996; Blanchard et al., 1997; Griebel et al., 2001, 2001a).

In contrast, low dose benzodiazepines appear to enhance offensive attack (Olivier et al., 1991), particularly under circumstances in which an animal is confronted by an intruder, but defensive attack is inhibited. Thus maternal attack on intruders is high when the male intruder is the same size as the female, but much lower when a larger male is used. In the latter situation, chlordiazepoxide substantially enhanced maternal attack (Mos et al., 1987). Olivier et al. suggested that the actions of benzodiazepines on aggression are not direct, but involve central effects on factors that modulate its expression (1991). This interpretation is consonant with the lack of an interaction between chlordiazepoxide and the serenic fluprazine on maternal aggression in rats; fluprazine may be interpreted as directly reducing offense, while chlordiazepoxide enhances offense, perhaps by reducing the inhibitory effects of emotional factors, such as fear (Olivier et al., 1986).

In mice, there appears to be a similar relationship between dose levels and pro-aggression effects of benzodiazepines. For example, acute administration of 5 mg/kg chlordiazepoxide increased social aggression in a neutral arena in individually housed males, while higher doses (10 and 20 mg/kg) consistently reduced social aggression, except for previously defeated animals that were less aggressive initially (Ferrari et al., 1997). Chronic (daily for 21 d) administration of 2 mg/kg chlordiazepoxide also stimulated mouse social aggression (Cutler et al., 1997).

The data on what is definitely a defensive threat–attack response (in the MDTB) and a number of studies involving offense (Olivier et al., 1991) appear to be clear that classic benzodiazepines reduce the former and, at low doses, tend to increase the latter. However, some additional studies while potentially interpretable as in agreement with this view, are not as straightforward. Diazepam reduced biting and also upright defensive postures of male rats reared in isolation toward unfamiliar animals encountered outside their home cages (Wongwitdecha et al., 1996). Socially reared rats, however, did not show either biting or defensive upright attack under these circumstances, suggesting (as does the effect on defensive upright) that the isolate response to the unfamiliar animal was defensive rather than offensive. Diazepam, and the 5-HT_{1A} agonists buspirone and 8-hydroxy-2(di-n-propylamino) tetraline (8-OH-DPAT), reduced maternal rat attack toward male intruders (Ferreira et al., 2000). Such maternal attack patterns, evaluated in terms of bite targets and response to serenics (Parmigiani et al., 1989), were clearly defensive. In female mice, chlordiazepoxide decreased maternal aggression to intruder males only in nonexperienced (and presumably more fearful) females, but increased it in experienced females confronted by a male (Palanza et al., 1996). Notably, the female mice in the study of Ferreira et al. (2000), appear to have been inexperienced, as well. Consonant with findings that (low dose) benzodiazepines reduce defensive attack and enhance offensive attack, chlordiazepoxide consistently produced a switch in female attack strategy (evaluated on the basis of target sites) toward male intruders, from defensive to offensive (Palanza et al., 1996).

These animal experimental findings agree well with an emerging body of experimental literature suggesting that some benzodiazepines enhance aggressive behavior in people (Cherek et al., 1991; Weisman et al., 1998; Ben-Porath et al., 2002). Clinical studies are more variable, perhaps in keeping with the diverse nature of the individuals to whom benzodiazepines are prescribed. However, in addition to studies suggesting that these drugs may decrease aggression in clinical settings (Griffith, 1985), some studies have described enhanced aggression following chronic (Bond et al., 1995; Mathew et al., 2000) or acute (Rodrigo, 1991) benzodiazepine administration. Some such enhancements appear to involve quite serious aggressive behavioral patterns (French, 1989). Discrepancies between findings of experimental studies utilizing computer-based competitions involving simulated delivery of "shock" to other participants, which consistently suggest enhancement of aggression, and the more variable clinical studies may also reflect differential expression of offensive vs defensive forms of aggression in the (typically nonpatient) subjects of the former studies, compared to those in patient populations.

SUMMARY

This chapter has attempted to outline some features of two behavior patterns, offensive-based or competition–challenge-based aggression and defensive threat–attack-based aggression. While these are not the only behaviors that may be subsumed under the rubric of aggression, particularly as that concept is applied to humans, they appear to reflect the major biobehavioral systems underlying serious attack on conspecifics. They may be clearly separated in laboratory animal research paradigms, and such research provides additional discriminanda, such as the targets of offensive vs defensive bites and blows. Pharmacological studies involving serenic and other serotonergic drugs and benzodiazepines, indicate differential involvement of neurotransmitter or neuromodulatory systems in offense and defense. Insofar as pharmacological studies are available and comparable, they suggest similarities between human and nonhuman mammal offensive vs defensive aggression drug effects. However, research on the distinction between offensive and defensive aggression in people is lacking. Barriers to this research, in addition to the inherent difficulty of direct scientific observation of aggression in natural settings, include the admixture of the two in many aggression-provoking settings; as well as the aforementioned social and legal advantages of self-justification for aggression in terms of defense. Nonetheless, the plethora of aggression-linked problems in contemporary society suggest that analysis of potentially differentiable systems of human aggression would be a valuable undertaking.

REFERENCES

Adams, N. (1985) Establishment of dominance in domestic Norway rats: effects of the degree of captivity and social experience. *Anim. Learn. Behav.* **13,** 93–97.

Averill, J. R. (1978) *Anger. Nebr. Symp. Motiv.* **26,** 1–80.

Averill, J. R. (1982) *Anger and Aggression: An Essay on Emotion.* Springer-Verlag, New York.

Azrin, N. H., Rubin, H. B., and Hutchinson, R. R. (1968) Biting attack by rats in response to aversive shock. *J. Exp. Anal. Behav.* **11,** 633–639.

Barnett, S. A. (1960) Social behavior among tame rats and among wild-white hybrids. *Proc. Zool. Soc. (Lond)* **134,** 611–621.

Ben-Porath, D. D. and Taylor, S. P. (2002) The effects of diazepam (valium) and aggressive disposition on human aggression: an experimental investigation. *Addict. Behav.* **27,** 167–177.

Blanchard, D. C. (1997) Stimulus and environmental control of defensive behaviors, in *The Functional Behaviorism of Robert C. Bolles: Learning, Motivation and Cognition.* (Bouton, M. and Fanselow, M., eds.), American Psychological Association, Washington, DC, pp. 283–305.

Blanchard, D. C. and Blanchard, R. J. (1984) Affect and aggression: an animal model applied to human behavior, in *Advances in the Study of Aggression.* (Blanchard, R. J. and Blanchard, D. C., eds.), Academic Press, New York, pp. 1–63.

Blanchard, D. C. and Blanchard, R. J. (1989a) Experimental animal models of aggression: what do they say about human behaviour? in *Human Aggression: Naturalistic Approaches.* (Archer, J. and Browne, K., eds.), Routledge, London, England, pp. 94–121.

Blanchard, D. C. and Blanchard, R. J. (1990a) Behavioral correlates of chronic dominance-subordination relationships of male rats in a seminatural situation. *Neurosci. Biobehav. Rev.* **14,** 455–462.

Blanchard, D. C. and Blanchard, R. J. (1990b) The colony model of aggression and defense, in *Contemporary Issues in Comparative Psychology.* (Dewsbury, D. A., ed.), Sinauer Associates, Sunderland, MA, pp. 410–430.

Blanchard, D. C., Griebel, G., and Blanchard, R. J. (2001a) Mouse defense behaviours: pharmacological and behavioural assays for anxiety and panic. *Neurosci. Biobehav. Rev.* **25,** 205–218.

Blanchard, D. C., Hebert, M. A., and Blanchard, R. J. (1999) Continuity vs (political) correctness: animal models and human aggression, in *Animal Models of Human Psychopathology.* (Huag, M. and Whalen, R., eds.), American Psychological Association, Washington, DC, pp. 297–316.

Blanchard, D. C., Hori, K., Rodgers, R. J., Hendrie, C. A., and Blanchard, R. J. (1989b) Attenuation of defensive threat and attack in wild rats (*Rattus rattus*) by benzodiazepines. *Psychopharmacology* **97**, 392–401.

Blanchard, D. C., Hynd, A. L., Minke, K. A., Minemoto, T., and Blanchard, R. J. (2001b) Human defensive behaviors to threat scenarios show parallels to fear- and anxiety-related defense patterns of non-human mammals. *Neurosci. Biobehav. Rev.* **25**, 761–770.

Blanchard, D. C., Spencer, R. L., Weiss, S. M., Blanchard, R. J., McEwen, B., and Sakai, R. R. (1995) Visible burrow system as a model of chronic social stress: behavioral and neuroendocrine correlates. *Psychoneuroendocrinology* **20**, 117–134.

Blanchard, D. C., Takushi, R., Blanchard, R. J., Flannelly, K. J., and Kemble, E. D. (1985a) Fluprazine hydrochloride does not decrease defensive behaviors of wild and septal syndrome rats. *Physiol. Behav.* **35**, 349–353.

Blanchard, R. J. (1984) Pain and aggression reconsidered. *Prog. Clin. Biol. Res.* **169**, 1–26.

Blanchard, R. J. and Blanchard, D. C. (1977) Aggressive behavior in the rat. *Behav. Biol.* **21**, 197–224.

Blanchard, R. J., Blanchard, D. C., and Flannelly, K. J. (1985b) Social stress, mortality and aggression in colonies and burrowing habitats. *Behav. Proc.* **11**, 209–213.

Blanchard, R. J., Blanchard, D. C., and Hori, K. (1989c) An ethoexperimental approach to the study of defense, in *Ethoexperimental Approaches to the Study of Behavior. NATO Advanced Science Institutes Series. Series D: Behavioural and Social Sciences, Vol. 48.* (Blanchard, R. J., Brain, P. F., Blanchard, D. C., and Parmigiani, S., eds.), Kluwer Academic Publishers, Dordrecht, Netherlands, pp. 114–136.

Blanchard, R. J., Blanchard, D. C., and Takahashi, L. K. (1978) Pain and aggression in the rat. *Behav. Biol.* **23**, 291–305.

Blanchard, R. J., Blanchard, D. C., Takahashi, T., and Kelley, M. (1977) Attack and defense behavior in the albino rat. *Anim. Behav.* **25**, 622–634.

Blanchard, R. J., Griebel, G., Henrie, J. A., and Blanchard, D. C. (1997b) Differentiation of anxiolytic and panicolytic drugs by effects on rat and mouse defense test batteries. *Neurosci. Biobehav. Rev.* **21**, 783–789.

Blanchard, R. J., Hebert, M. A., Ferrari, P., et al. (1998a) Defensive behaviors in wild and laboratory (Swiss) mice: the mouse defense test battery. *Physiol. Behav.* **65**, 201–209.

Blanchard, R. J., Hebert, M. A., Sakai, R. R., et al. (1998b) Chronic social stress: changes in behavioral and physiological indices of emotion. *Aggr. Behav.* **24**, 307–321.

Blanchard, R. J., Hori, K., Tom, P., and Blanchard, D. C. (1988) Social dominance and individual aggressiveness. *Aggr. Behav.* **14**, 195–203.

Blanchard, R. J., Kleinschmidt, C. F., Fukunaga-Stinson, C., and Blanchard, D. C. (1980) Defensive attack behavior in male and female rats. *Anim. Learn. Behav.* **8**, 177–183.

Blanchard, R. J., Nikulina, J. N., Sakai, R. R., McKittrick, C., McEwen, B., and Blanchard, D. C. (1998c) Behavioral and endocrine change following chronic predatory stress. *Physiol. Behav.* **63**, 561–569.

Blanchard, R. J., O'Donnell, V., and Blanchard, D. C. (1979) Attack and defensive behaviors in the albino mouse. *Aggr. Behav.* **5**, 341–352.

Blanchard, R. J., Pank, L., Fellows, D., and Blanchard, D. C. (1985c) Conspecific wounding in free-ranging *R. norvegicus* from stable and unstable populations. *Psychol. Rec.* **35**, 329–335.

Bond, A. J., Curran, H. V., Bruce, M. S., O'Sullivan, G., and Shine, P. (1995) Behavioural aggression in panic disorder after 8 weeks' treatment with alprazolam. *J. Affect. Disord.* **35**, 117–123.

Brain, P. F., Al-Maliki, S., Parmigiani, S., and Hammour, H. A. (1983) Studies on tube restraint-induced attack on a metal target by laboratory mice. *Behav. Proc.* **8**, 277–287.

Cherek, D. R., Spiga, R., Roache, J. D., and Cowan, K. A. (1991) Effects of triazolam on human aggressive, escape and point-maintained responding. *Pharmacol. Biochem. Behav.* **40**, 835–839.

Chesney-Lind, M. and Hagedorn, J. M. (Eds.) (1999) *Female Gangs in America: Essays on Gender and Gangs,* Lakeview Press, Chicago, IL.

Cutler, M. G., Rodgers, R. J., and Jackson, J. E. (1997). Behavioural effects in mice of subchronic chlordiazepoxide, maprotiline, and fluvoxamine. I. Social interactions. *Pharmacol. Biochem. Behav.* **57,** 119–125.

Dollard, J., Doob, L. W., Miller, N. E., Mowrer, O. H., and Sears, R. R. (1939) *Frustration and Aggression.* Yale University Press, New Haven, CT.

Eichelman, B. S., Jr. and Thoa, N. B. (1973) The aggressive monoamines. *Biol. Psychiatry* **6,** 143–164.

Ferrari, P. F., Palanza, P., Rodgers, R. J., Mainardi, M., and Parmigiani, S. (1996) Comparing different forms of male and female aggression in wild and laboratory mice: an ethopharmacological study. *Physiol. Behav.* **60,** 549–553.

Ferrari, P. F., Parmigiani, S., Rodgers, R. J., and Palanza, P. (1997) Differential effects of chlordiazepoxide on aggressive behavior in male mice: the influence of social factors. *Psychopharmacology (Berl)* **134,** 258–265.

Ferreira, A., Picazo, O., Uriarte, N., Pereira, M., and Fernandez-Guasti, A. (2000) Inhibitory effect of buspirone and diazepam, but not of 8-OH-DPAT, on maternal behavior and aggression. *Pharmacol. Biochem. Behav.* **66,** 389–396.

Ferris, C. F., Melloni, R. H., Jr., Koppel, G., Perry, K. W., Fuller, R. W., and Delville, Y. (1997) Vasopressin/serotonin interactions in the anterior hypothalamus control aggressive behavior in golden hamsters. *J. Neurosci.* **17,** 4331–4340.

Flannelly, K. J., Muraoka, M. Y., Blanchard, D. C., and Blanchard, R. J. (1985) Specific anti-aggressive effects of fluprazine hydrochloride. *Psychopharmacology* **87,** 86–89.

French, A. P. (1989) Dangerously aggressive behavior as a side effect of alprazolam. *Am. J. Psychiatry* **146,** 276.

Geist, V. (1968) Horn-like structures as rank symbols, guards and weapons. *Nature* **220,** 813–814.

George, D. T., Umhau, J. C., Phillips, M. J., et al. (2001) Serotonin, testosterone and alcohol in the etiology of domestic violence. *Psychiatry Res.* **104,** 27–37.

Grant, E. C. (1963) An analysis of the social behaviour of the male laboratory rat. *Behaviour* **21,** 260–281.

Grant, E. C. and Chance, M. R. A. (1958) Rank order in caged rats. *Anim. Behav.* **6,** 183–184.

Grant, E. C. and Mackintosh, J. H. (1963) A comparison of the social postures of some common laboratory rodents. *Behaviour* **21,** 246–259.

Griebel, G., Blanchard, D. C., Agnes, R. S., and Blanchard, R. J. (1995a) Differential modulation of antipredator defensive behavior in Swiss-Webster mice following acute or chronic administration of imipramine and fluoxetine. *Psychopharmacology (Berl)* **120,** 57–66.

Griebel, G., Blanchard, D. C., Jung, A., and Blanchard, R. J. (1995b) A model of 'antipredator' defense in Swiss-Webster mice: effects of benzodiazepine receptor ligands with different intrinsic activities. *Behav. Pharmacol.* **6,** 732–745.

Griebel, G., Blanchard, D. C., Rettori, M. C., Guardiola-Lemaitre, B., and Blanchard, R. J. (1996) Preclinical profile of the mixed 5-HT1A/5-HT2A receptor antagonist S 21,357. *Pharmacol. Biochem. Behav.* **54,** 509–516.

Griebel, G., Moindrot, N., Aliaga, C., Simiand, J., and Soubrie, P. (2001a) Characterization of the profile of neurokinin-2 and neurotensin receptor antagonists in the mouse defense test battery. *Neurosci. Biobehav. Rev.* **25,** 619–626.

Griebel, G., Perrault, G., and Soubrie, P. (2001b) Effects of SR48968, a selective non-peptide NK2 receptor antagonist on emotional processes in rodents. *Psychopharmacology (Berl)* **158,** 241–251.

Griffith, J. L. (1985) Treatment of episodic behavioral disorders with rapidly absorbed benzodiazepines. *J. Nerv. Ment. Dis.* **173,** 312–315.

Grossman, S. P. (1970) Avoidance behavior and aggression in rats with transections of the lateral connections of the medial or lateral hypothalamus. *Physiol. Behav.* **5,** 1103–1108.

Grossman, S. P. (1972) Aggression, avoidance, and reaction to novel environments in female rats with ventromedial hypothalamic lesions. *J. Comp. Physiol. Psychol.* **78,** 274–283.

Ho, H. P., Olsson, M., Westberg, L., Melke, J., and Eriksson, E. (2001) The serotonin reuptake inhibitor fluoxetine reduces sex steroid-related aggression in female rats: an animal model of premenstrual irritability? *Neuropsychopharmacology* **24,** 502–510.

Leyhausen, P. (1979) *Cat Behavior.* Garland Press, New York.

Linnoila, M., Virkkunen, M., Scheinin, M., Nuutila, A., Rimon, R., and Goodwin, F. K. (1983) Low cerebrospinal fluid 5-hydroxyindoleacetic acid concentration differentiates impulsive from nonimpulsive violent behavior. *Life Sci.* **33,** 2609–2614.

Lore, R., Nikoletseas, M., and Flannelly, K. (1980) Aggression in rats: does the colony-intruder model require a colony? *Behav. Neural. Biol.* **28,** 243–245.

Mackintosh, J. H. (1981) Behavior of the house mouse. *Symp. Zool. Soc. (Lond)* **47,** 337–365.

Masur, J. (1975) Competitive behavior between rats: some definition problems. *Behav. Biol.* **13,** 533–535.

Mathew, V. M., Dursun, S. M., and Reveley, M. A. (2000) Increased aggressive, violent, and impulsive behaviour in patients during chronic-prolonged benzodiazepine use. *Can. J. Psychiatry* **45,** 89–90.

Miczek, K. A. and Barry, H., III (1975) What does the tube test measure? *Behav. Biol.* **13,** 537–539.

Mitchell, P. J., Fletcher, A., and Redfern, P. H. (1991) Is antidepressant efficacy revealed by drug-induced changes in rat behaviour exhibited during social interaction? *Neurosci. Biobehav. Rev.* **15,** 539–544.

Mitchell, P. J. and Redfern, P. H. (1997) Potentiation of the time-dependent, antidepressant-induced changes in the agonistic behaviour of resident rats by the 5-HT1A receptor antagonist, WAY-100635. *Behav. Pharmacol.* **8,** 585–606.

Mos, J., Olivier, B., and van Oorschot, R. (1987) Maternal aggression towards different sized male opponents: effect of chlordiazepoxide treatment of the mothers and d-amphetamine treatment of the intruders. *Pharmacol. Biochem. Behav.* **26,** 577–584.

Olivier, B., Mos, J., and Miczek, K. A. (1991) Ethopharmacological studies of anxiolytics and aggression. *Eur. Neuropsychopharmacol.* **1,** 97–100.

Olivier, B., Mos, J., Raghoebar, M., de Koning, P., and Mak, M. (1994) *Serenics. Prog. Drug Res.* **42,** 167–308.

Olivier, B., Mos, J., and van Oorschot, R. (1986) Maternal aggression in rats: lack of interaction between chlordiazepoxide and fluprazine. *Psychopharmacology* **88,** 40–43.

Palanza, P., Rodgers, R. J., Ferrari, P. F., and Parmigiani, S. (1996) Effects of chlordiazepoxide on maternal aggression in mice depend on experience of resident and sex of intruder. *Pharmacol. Biochem. Behav.* **54,** 175–182.

Parmigiani, S., Rodgers, R. J., Palanza, P., Mainardi, M., and Brain, P. F. (1989) The inhibitory effects of fluprazine on parental aggression in female mice are dependent upon intruder sex. *Physiol. Behav.* **46,** 455–459.

Pellis, S. M. and Pellis, V. C. (1987) Play-fighting differs from serious fighting in both target and attack and tactics of fighting in the laboratory rat, *Rattus norvegicus. Aggr. Behav.* **13,** 227–242.

Pellis, S. M. and Pellis, V. C. (1988) Identification of the possible origin of the body target that differentiates play-fighting from serious fighting in Syrian Golden hamsters (*Mesocricetus auratus*). *Aggr. Behav.* **14,** 437–449.

Pellis, S. M., Pellis, V. C., and Pierce, J. D. (1992) Disentangling the contribution of the attacker from that of the defenderin the differences in the intraspecific fighting in two species of voles. *Aggr. Behav.* **18,** 425–435.

Placidi, G. P., Oquendo, M. A., Malone, K. M., Huang, Y. Y., Ellis, S. P., and Mann, J. J. (2001) Aggressivity, suicide attempts, and depression: relationship to cerebrospinal fluid monoamine metabolite levels. *Biol. Psychiatry* **50,** 783–791.

Rodrigo, C. R. (1991) Flumazenil reverses paradoxical reaction with midazolam. *Anesth. Prog.* **38,** 65–68.

Takahashi, L. K. and Blanchard, R. J. (1982) Attack and defense in wild norway and black rats. *Behav. Proc.* **7,** 49–62.

Taylor, S. P. and Chermack, S. T. (1993) Alcohol, drugs and human physical aggression. *J. Stud. Alcohol* **11,** 78–88.

Ulrich, R. E. and Azrin, N. H. (1962) Reflexive fighting in response to aversive stimulation. *J. Exp. Anal. Behav.* **5,** 511–520.

Villalba, C., Boyle, P. A., Caliguri, E. J., and De Vries, G. J. (1997) Effects of the selective serotonin reuptake inhibitor fluoxetine on social behaviors in male and female prairie voles (*Microtus ochrogaster*). *Horm. Behav.* **32,** 184–191.

Walsh, M. T. and Dinan, T. G. (2001) Selective serotonin reuptake inhibitors and violence: a review of the available evidence. *Acta Psychiatr. Scand.* **104,** 84–91.

Weisman, A. M., Berman, M. E., and Taylor, S. P. (1998) Effects of clorazepate, diazepam, and oxazepam on a laboratory measurement of aggression in men. *Int. Clin. Psychopharmacol.* **13,** 183–188.

Wilson, E. O. (1975) *Sociobiology.* Harvard University Press, Cambridge, MA.

Wongwitdecha, N. and Marsden, C. A. (1996) Social isolation increases aggressive behaviour and alters the effects of diazepam in the rat social interaction test. *Behav. Brain Res.* **75,** 27–32.

Work, M. S., Grossen, N., and Rogers, H. (1969) Role of habit and androgen level in food-seeking dominance among rats. *J. Comp. Physiol. Psychol.* **69,** 601–607.

Zook, J. M. and Adams, D. B. (1975) Competitive fighting in the rat. *J. Comp. Physiol. Psychol.* **88,** 418–423.

6
Neuroendocrine Stress Responses and Aggression

Jozsef Haller and Menno R. Kruk

INTRODUCTION

The Mechanisms of the Stress Response

Environmental challenges induce the relatively nonspecific activation of two important neuroendocrine mechanisms: the adrenergic–noradrenergic system and the hypothalamus–pituitary–adrenocortical (HPA) axis (Fig. 1). Both systems have a relatively intense basal activity, which shows ultradian, diurnal, and seasonal variations. However, more then a 10-fold increase in their activity is noted after challenges of various kinds. Collectively, these systems constitute the main elements of the neuroendocrine stress response.[1]

The systems that employ adrenaline and norepinephrine (NE) are probably unique in the sense that they form an interrelated network consisting of a central neuronal unit (a few NE-secreting brain centers), a peripheral neural network (the sympathetic system), and a remote endocrine unit (the adrenal medulla). The neural elements of this system use mainly NE as chemical signals, while the endocrine unit secretes mainly adrenaline. Noteworthy, NE reaches the blood via "leaking" peripheral synapses; therefore, plasma NE originates mainly from the neural system. The main stress-related outputs of the HPA axis are the glucocorticoid hormones. These are secreted in the adrenal cortex, which is activated by the adrenocorticotrop hormone (ACTH), which originates from the pituitary. ACTH-secreting pituitary cells are stimulated by the corticotropin-releasing factor, which is secreted by the parvocellular part of the hypothalamic paraventricular nucleus. The latter nucleus is under the modulatory influence of various limbic structures.

Thus, an environmental challenge results in: (*i*) an increased noradrenergic neurotransmission; (*ii*) an increased adrenaline production; and (*iii*) an increased glucocorticoid production. The outputs of these systems (the neurotransmitter NE and the hormones adrenaline and glucocorticoids) have multiple effects at different levels of the organism. The noradrenergic neurons of the brainstem send multiple projections to various components of the central nervous system (CNS), which receive the signal via noradrenergic receptors. The sympathetic system innervates internal organs (the circulatory system,

---

[1]Naturally, the stress response cannot be reduced to the activation of these two systems. The role of other stress-responsive systems in aggression (e.g., gonadal hormones, serotonin, etc.) are discussed in Chapters 3 and 4.

From: *Neurobiology of Aggression: Understanding and Preventing Violence*
Edited by: M. Mattson © Humana Press Inc., Totowa, NJ

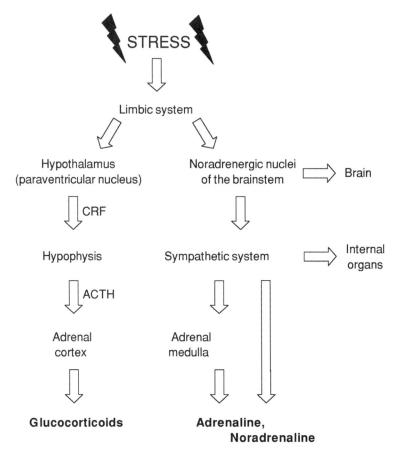

Fig. 1. The activation of the main elements of the stress system: the HPA axis and the central noradrenergic and sympathoadrenal system.

the gastrointestinal tract, etc.) and has a large impact on their functioning. The hormone adrenaline does not penetrate the blood–brain barrier (thus, it has no direct effects on the CNS), but reaches internal organs and affects their functioning. Its effects are often similar or complementary to the effects of the sympathetic system, for which the two systems are generically termed the sympathoadrenal system. Glucocorticoids exert their action both at the periphery and in the CNS via glucocorticoid receptors. The well-known effects of glucocorticoids involve the modulation of genomic mechanisms. These effects develop slowly and are persistent. Recent evidence, however, suggests that glucocorticoids are able to influence target cells by rapid nongenomic mechanisms, which are poorly known at present.

Short- and Long-Term Effects of Stress

The stress response promotes the development of adequate responses to challenges. This goal is reached by multiple mechanisms. These mechanisms range from the stimulation of energy-providing biochemical processes to promoting vigilance and situation-specific behavioral reactions. When the challenge is adequately met, the stress response is terminated, and the neuroendocrine activity returns to basal levels. Challenges that activate the neuroendocrine stress response acutely are part of daily life and do not normally have serious long-term consequences. However, exceptionally strong single stressors (traumatic events) may have long-term effects on the functioning of the organism, even after the disappearance of the stress factor. Such long-lasting effects have been observed in both humans and experimental animals, in early phases of development as well as in adulthood. Persisting or repeated stressors activate the neuroendocrine system chronically. It is generally believed that a chronic stress response is a causal factor of many somatic and behavioral disorders.

Aggression and Stress: An Overview

The neuroendocrine response in laboratory rodents demonstrates that social defeat is the most severe stressor of all (Koolhaas et al., 1997). The social challenge-induced neuroendocrine stress response occurs rapidly, even before the actual aggressive encounter starts (Haller et al., 1995; Schuurman, 1980). Moreover, an anticipatory stress response occurs in environments repeatedly associated with aggression, even if the opponent is not present (Tornatzky and Miczek, 1994). The neuroendocrine stress response affects the mechanisms involved in the behavioral stress response (i.e., aggressive behavior).

The functional properties of the adrenergic–noradrenergic system and HPA axis suggest that these stress-related systems affect aggressive behavior in multiple ways:

1. Aggressive behavior appears to be affected by both increased and decreased basal levels of neuroendocrine activity. Basal activity depends on individual features such as genetic factors, or past experience with stressors, and sets the propensity of individuals to behave aggressively. Moreover, long-term changes in basal levels determine the *patterns* of aggression. Changes in basal activity levels usually induce changes in stress reactivity. Therefore, the effects of changes in basal functions can only be studied in concert with changes in stress reactivity.
2. Ultradian, diurnal, and seasonal changes in the activity of the neuroendocrine stress system induce corresponding oscillations in aggressive behavior. Such oscillations may be important in synchronizing the activity patterns of animals.
3. The acute stress response induced by other factors or developed during the aggressive encounter affects ongoing behavior. Aggression is usually preceded by preparatory behaviors (e.g., the exploration of the opponent and intention signaling), during which the neuroendocrine stress response can fully develop. Thus, aggressiveness is preceded by an acute stress reaction, which is important for the subsequent behavioral reaction.
4. Single traumatic events lead to long-term changes in brain function and stress reactivity, which result in long-term behavioral changes including changes in aggressiveness. If such changes occur at neonatal age, they may last a lifetime, but traumas occurring later in life may also have long-lasting behavioral consequences (Fig. 2).

THE ADRENERGIC–NORADRENERGIC SYSTEM

Preparing the Organism for Aggression: Adrenalin and NE

Aggression leads to a massive activation of the peripheral sympathoadrenal and central noradrenergic systems (Sgoifo et al., 1996; Halász et al., 2002; Korzan et al., 2000; Levine et al., 1990). Some effects of these activated systems are not intimately linked to the control of aggression. However, such effects play an important role in this behavior by adapting the organism in multiple ways to the requirements set by the behavioral response to social challenges.

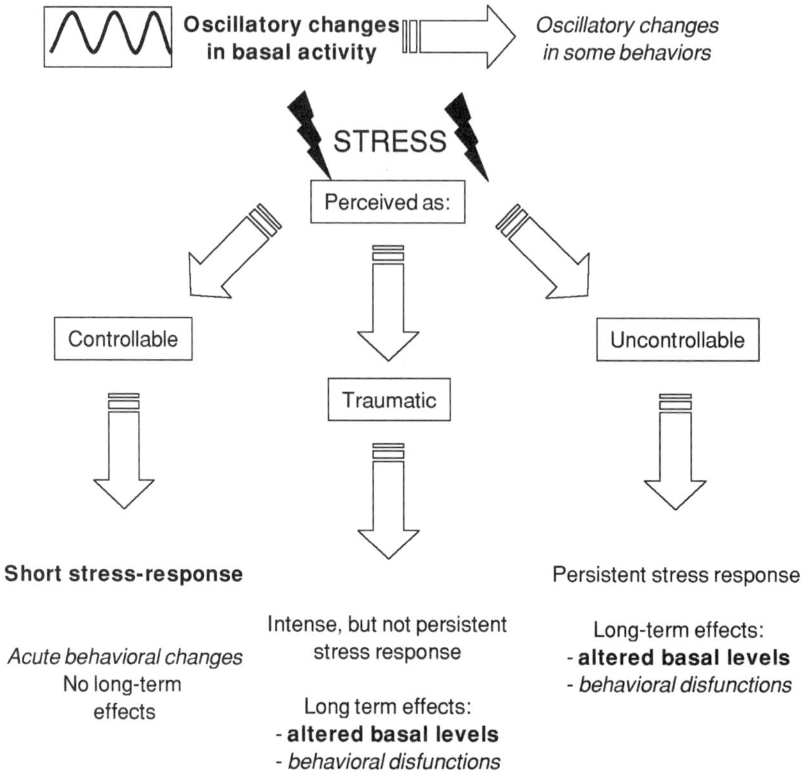

Fig. 2. The effect of the stress-responsive systems on behavior: an overview. CRF, corticotropin-releasing factor; ACTH, adrenocorticotrop hormone.

Energy Metabolism

Aggressive behavior involves strenuous activity, which needs to be fueled by biochemical mechanisms that provide energy for the skeletal muscles. Energetic constraints due to poor food supply, exhaustion as a consequence of recent aggressive encounters, and so forth, restrain the behavioral competence of animals, including aggression (for a review, *see* Haller, 1995b; Haller et al., 1996b). A series of energy-yielding biochemical processes are accelerated during aggression. Such processes include oxygen consumption, glycogenolysis, recycling of lactic acid into glucose, and lipid and amino acid degradation. Energy metabolism is, to a large extent, controlled by the sympathoadrenal system. Thus, this system appears important for meeting the energetic requirements set by aggressive behavior.

Circulation

Oxygen and the substrates for energy metabolism reach their target via the circulation. Not surprisingly, metabolic preparations for fights involve changes in circulatory functions. Heart rate and blood pressure are increased during aggressive interactions in both animals and humans (Holmes and Will, 1985; Sgoifo et al., 1994). In addition, blood flow is redistributed from less essential vascular beds (e.g., mesenteric arteries) to crucial vascular beds (e.g., muscle arteries) (Viken et al., 1991). Such changes in circulatory functions are likely to be governed (at least partly) by the sympathoadrenal and noradrenergic systems.

Attention and Vigilance

Central NE has long been involved in attention and vigilance (Clark et al., 1988). Animal studies implicate NE in a wide range of attention-related behaviors, such as exploratory activity, distractibility, response rate to and discriminability of stimuli, and switching of attention. Although direct experimental proof is lacking at present, it is highly probable that the increase of attention and vigilance by the central noradrenergic system promotes the response to social challenges.

Pain Perception

Attacks are painful for the recipient. Since most aggressive encounters involve reciprocal attacks, both contestants are exposed to painful stimuli. The ability of individuals to maintain aggressive behavior despite the associated pain is strongly promoted by endogenous mechanisms that lower pain sensitivity. It has been shown that opioid mechanisms have an important role in this process (Miczek et al., 1982; Rodgers and Hendrie, 1983). However, it has also been shown that attack-induced endogenous analgesia has a nonopioid component (Miczek and Winslow, 1987; Rodgers and Randall, 1988). Despite the fact that the noradrenergic system plays an important role in pain perception, little attention has been paid so far to the role of the noradrenergic neurotransmission in this phenomenon. Pain sensitivity is potently inhibited by noradrenergic fibers descending into the medulla (Eide and Hole, 1992). Since the noradrenergic system is strongly activated by aggression, it is highly probable that these descending fibers play a role in inhibiting aggression-induced pain.

Olfaction

Aggression is dramatically reduced when olfaction is blocked by various means (Murphy, 1976; Stowers et al., 2002). This phenomenon is probably explained by the fact that olfaction is the main information source for many animals, and aggressive encounters are preceded by intense social exploration that involves mainly sniffing. The noradrenergic system enhances olfaction (Mouly et al., 1995). Since odor-related memory determines individual recognition and the number of attacks delivered to an intruder (Garcia-Brull et al.,1993), the effect of NE on olfaction appears relevant for aggression control in animals.

Memory

The correct identification of social partners is of primordial importance for aggressive behavior. In addition to the opponent, situations and locations associated with dangerous events (e.g., aggression) should also be memorized. It is well known that the central noradrenergic system promotes learning and memory processes. In animals, the memory-enhancing effect of NE often occurs in situations that involve olfaction, which may be less important for humans. However, the noradrenergic system promotes memory in other situations as well, in both animals and humans (Ferry et al., 1999; O'Carroll et al., 1999).

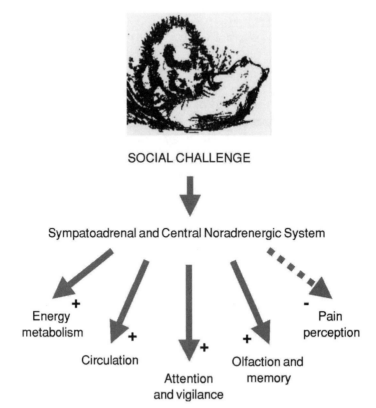

Fig. 3. The effects of the adrenergic–noradrenergic system on processes that are not directly linked to the control of aggression, but enhance the ability of animals to cope with the challenge.

Conclusions

The sympathoadrenal and the central noradrenergic systems enhance energy metabolism, circulation, attention and vigilance, olfaction, and memory, and inhibit pain perception. These effects are indirectly linked to aggression, but are important for enabling the organism to cope with aggressive conflicts (Fig. 3).

The Effects of NE on Aggression Control

NE is a chemical signal used by a limited number of brainstem nuclei. The locus coeruleus plays a pivotal role in controlling aggression. The projections of the locus coeruleus reach almost the whole brain and form synapses at multiple locations. After release, NE binds to a number of different receptors (see below), while the excess is either metabolized by monoamino oxidase A (MAOA) or taken up by glia cells or the noradrenergic cell. The noradrenergic system can be relatively easily manipulated by selectively destroying noradrenergic cells, increasing the availability of precursors, inhibiting its synthesizing enzymes, depleting storage, increasing release, inhibiting MAOA, blocking reuptake, stimulating reuptake, and activating or blocking noradrenergic receptors. These manipulations are frequently used for the amelioration of psychological disorders, including some disorders with aggressive components.

The Effects of Correlated Changes in Basal Levels and Reactivity

Brunner et al. (1993) have identified a large Dutch kindred showing a genetic deficiency of the MAOA enzyme. All affected males in this family showed very characteristic aggressiveness and sometimes violent behavior. Other types of impulsive behavior included arson, attempted rape, and exhibitionism. Subsequent research in MAOA knock-out mice (i.e., mice in which the MAOA gene has been disrupted) confirmed human findings: these mice exhibited increased aggressiveness in adulthood (Cases et al., 1995). The MAOA enzyme is not entirely specific, because it degrades both NE and serotonin. Thus, the behavioral deficiency appears to be linked to an enhanced NE and/or serotonin availability. The role of NE is outlined by the fact that brain serotonin levels normalized in MAOA knock-out mice until adulthood, while brain NE levels were still elevated at this age. In MAOA knock-out pups, the combined increase in brain serotonin and NE levels resulted in trembling, difficulty in righting, and fearfulness. These symptoms disappeared toward adulthood. Thus, the increase in aggressiveness appears to be associated specifically with increased brain NE levels. Collectively, the above data suggest that enhanced noradrenergic neurotransmission increases aggressiveness in both humans and laboratory animals. The involvement of serotonin in this phenomenon cannot be ruled out, but data support the assumption that the main factor was NE.

Brain NE production can be blocked by the selective destruction of noradrenergic neurons. Such manipulations do not lead to a total disappearance of brain noradrenergic activity, but reduce it by 70–90%. Spontaneous aggression (e.g., in a territorial setting) is potently inhibited by such manipulations (Crawley and Contrera, 1976; Ellison, 1976). These data suggest that a chronic decrease in NE neurotransmission (which affects both basal levels and reactivity) potently decreases aggressiveness. Noteworthy, the effect of chronic NE depletion has dissimilar effects on other types of aggressiveness (e.g., shock-induced, and δ-tetrahydrocannabinol-induced aggressive behavior (Mogilnicka et al., 1983; Fujiwara et al., 1984). The reasons of this discrepancy are unknown, but one can hypothesize that NE depletion influenced the reactivity of animals to these manipulations, which induce rather diverse forms of aggressive behavior.

Taken together, the above data suggest that there is a positive correlation between noradrenergic neurotransmission and the natural aggressive response; factors that increase noradrenergic neurotransmission increase aggressiveness, while the reverse is true for factors that decrease noradrenergic neurotransmission.

Oscillatory NE Neurotransmission

Both brain and plasma NE levels show ultradian, diurnal, and seasonal variations in various species including humans (Davidovic and Petrovic, 1981; Tapp et al., 1981). Several lines of evidence suggest that these variations affect other forms of behavior (*see*, e.g., Margules et al., 1972). Unfortunately, no data are available on the role of these oscillations in aggressive behavior. However, the strong effects of basal and stress-induced noradrenergic neurotransmission on aggressiveness suggest that rhythmic oscillations in noradrenergic neurotransmission should be relevant for aggressiveness.

Acute Effects of NE

Aggression induces the activation of the central noradrenergic system. In animals, aggressive interactions increase c-Fos staining in the locus coeruleus, NE turnover of specific brain regions, and locus coeruleus firing rate in various species (Halász et al., 2002; Higley et al., 1992; Korzan et al., 2000; Levine et al., 1990). Agents that stimulate NE release or block re-uptake have a biphasic effect: low doses increase, while large doses decrease aggressiveness (Cai et al., 1993; Hodge and Butcher, 1975). The involvement of NE in the pro-aggressive effects of low doses is suggested by the fact that noradrenergic denervation inhibits aggressiveness induced by the blockade of NE reuptake (Matsumoto et al., 1995). This suggests that a slight increase in the acute NE response promotes aggressiveness. NE increased aggressiveness when injected into hypothalamic regions that are involved in attack control (Barrett et al., 1990). This finding suggests that the acute effects of NE on aggressiveness are at least partly mediated by brain mechanisms that directly control attack.

The aggression-inhibiting effects of noradrenergic "overexcitation" (e.g., by large doses of reuptake blockers) are intriguing. One can hypothesize that the effect of large doses are not specific, i.e., their anti-aggressive effects may be unrelated to NE neurotransmission.

The pro-aggressive effects of acute NE activation is supported by the acute blocking of aggression by agents that decrease the availability of NE (e.g., reuptake stimulators or enzyme inhibitors) (Hodge and Butcher, 1975; Sheard, 1975; Smith, 1977). Taken conjointly, the above data suggest that the acute activation of the brain noradrenergic system promotes aggressiveness.

Traumatic Events and NE

EARLY STAGES OF LIFE

In laboratory animals, early traumatic experiences induce life-long behavioral disturbances and changes in stress reactivity (Tonjes et al., 1986; Workel et al., 2001). Human studies suggest that traumas experienced in early life result in elevated levels of aggressiveness in adulthood (Buka et al., 2001; Reiss, 1993). However, studies on the long-term endocrinological effects of early traumas focus mainly on the HPA axis. In addition, laboratory studies rarely, if ever, tackle the interaction between early stressors and aggressiveness in adulthood. Nevertheless, it has been suggested that early traumas increase reactivity to environmental stimuli and induce hypersensitivity in interpersonal situations, which were related to noradrenergic mechanisms activated by traumas (Figueroa and Silk, 1997). The interaction between early stressors, noradrenergic stress reactivity, and aggressiveness still remains unclear.

TRAUMATIC EVENTS IN ADULTS

Traumatic experiences lead to the development of post-traumatic stress disorder (PTSD) in humans. There are many causes of this disorder, but it appears that the up-regulation of the noradrenergic systems plays an important role. PTSD patients show increased NE excretion rates and exaggerated NE responses to stressful stimuli, especially when these are related to the traumatic event experienced previously. In addition, agents that block noradrenergic neurotransmission are partially effective in the treatment of this disorder (Southwick et al., 1999). Thus, it appears that increased noradrenergic neurotransmission plays an important role in the pathophysiology of PTSD. As shown above, a noradrenergic background similar to that induced by PTSD (i.e., increased noradrenergic neurotransmission) is associated with heightened levels of aggressiveness. In line with these data, PTSD patients have been frequently reported to exhibit high levels of aggressive and violent behavior (Freeman and Roca, 2001; Lasco et al., 1994). Taken together, these data suggest that increased noradrenergic neurotransmission contributes to the aggressiveness of PTSD patients. Unfortunately, no direct evidence for this assumption is available.

NE Receptors and Aggression

NE exerts its actions via noradrenergic receptors. Three major classes of noradrenergic receptors have been identified so far: (*i*) α_1-adrenoceptors are linked to inositol phospholipid hydrolysis; (*ii*) β-adrenoceptors activate the adenyl cyclase via Gs type of G proteins; and (*iii*) α_2-adrenoceptors inhibit the adenyl cyclase via Gi type of G proteins. All three classes are divided further into adrenoceptor subtypes. The interaction between NE and aggression cannot be fully understood without understanding the receptor mechanisms underlying the overall effect of NE. The functions of β- and α_1-receptors can be studied by drugs that specifically block (antagonists) or activate (agonists) these receptors and, in this way, interfere with the normal action of endogenous NE on its receptors. However, agents that affect the α_2-adrenoceptor have more complex effects, because these receptors also have an autoreceptor function.

THE PRO-AGGRESSIVE EFFECTS OF β-ADRENOCEPTORS

The involvement of β-adrenoceptors is the most clear-cut. β-Adrenoceptor antagonists inhibit aggression in a variety of situations, including spontaneous aggression in a territorial setting (Bell and Hobson, 1984; Hegstrand and Eichelman, 1983). Also, the β-antagonist propranolol is one of the few drugs that is able to block aggression elicited by electrical stimulation in the aggressive area of the hypothalamus (Kruk, 1991; Kruk et al., 1998). Moreover, β-adrenoceptor antagonists appear effective in blocking human aggression associated with psychiatric disorders (Lader, 1988). In accordance with the effects of blocking agents, β-adrenoceptor agonists increase aggressiveness (Mogilnicka and Zazula, 1986). Thus, β-adrenoceptors contribute to the general proaggressive effects of NE.

It occurs that this effect is β-adrenoceptor *subtype* specific. Evidence for this phenomenon derives from experiments using NE reuptake inhibitors. These compounds enhance the concentration of endogenous NE by inhibiting the removal of NE by a highly efficient active reuptake process. As shown above, when the release of endogenous NE is enhanced by reuptake inhibitors, aggression increases. This effect was potently inhibited by the β_2-adrenoceptor antagonist ICI-118,551, but not by the β_1-adrenoceptor antagonist prolol tartrate (Matsumoto et al., 1994).

UNCLEAR INVOLVEMENT: THE α_1-ADRENOCEPTOR

The role of α_1 adrenoceptors on aggressive behavior was poorly investigated. Some data suggest that the role of this receptor is modest in mediating the effect of NE on aggressive behavior. Neither aggression induced by inhibiting NE reuptake nor aggression induced by the α_2-agonist clonidine was affected by the α_1-antagonist prazosin (Matsumoto et al., 1991; Fujiwara et al., 1988). This may suggest that α_1-adrenoceptors have little or no effect on aggressive behavior, and the lack of appropriate studies is explained by this circumstance.

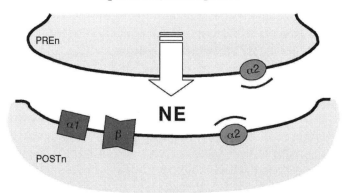

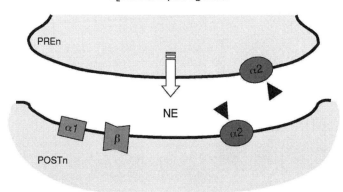

Fig. 4. The mechanism of action of α_2-adrenoceptor ligands. Blockers inhibit α_2-adrenoceptors both pre- and postsynaptically. As a consequence of presynaptic blockade, NE secretion is increased, therefore, α_1- and β-adrenoceptors are activated more strongly. Agonists activate α_2-adrenoceptors, which leads to a decrease in NE release. As a consequence, the activation of α_1- and β-adrenoceptors is decreased. NE, norepinephrine; α_2, pre- and postsynaptic α_2-adrenoceptors; α_1, α_1-adrenoceptors (only postsynaptic); β, β-adrenoceptors (only postsynaptic); PREn, presynaptic neurone terminal; POSTn, postsynaptic neuronal membrane.

COMPLEX EFFECTS: THE α_2-ADRENOCEPTOR

The role of α_2-adrenoceptors in mediating NE effects on aggression are complex, mostly because this receptor has a dual function: (*i*) it conveys the NE message to the postsynaptic cell via postsynaptic α_2-adrenoceptors and (*ii*) it decreases NE release through α_2-adrenoceptors situated presynaptically. Therefore, agonists stimulate postsynaptic α_2-receptors directly, but decrease the postsynaptic activation of β- and α_1-receptors by the decrease of NE release (Fig. 4). Antagonists block noradrenergic neurotransmission via postsynaptic α_2-receptors, but increase the activation of postsynaptic β- and α_1-receptors by increasing NE release. Such complex molecular effects at the level of the receptors result in apparently contradictory behavioral effects. Both α_2-adrenoceptor antagonists (Gentsch et al., 1989; Haller, 1995a) and agonists (Barrett et al., 1990; Buus Lassen, 1978) were shown to increase aggressive behavior in rats.

Results obtained with α_2-adrenoceptor agonists suggest that NE stimulates aggressive behavior via postsynaptic α_2-adrenoceptors, even while the activation of presynaptic α_2-receptors would diminish NE release (i.e., it would decrease the activation of the postsynaptic β-receptors). This implies that NE effects mediated by α_2-adrenoceptors are sufficient to increase aggression even when β-adrenoceptor activation is reduced. Similarly, results obtained with α_2-adrenoceptor antagonists suggest that the activation of postsynaptic β-receptors (by the increase in NE release) is sufficient for the induction of aggressive behavior, even if this effect is not supported by a concurrent activation of α_2-adrenoceptors (which are blocked by α_2-adrenoceptor antagonists) (Fig. 4). Taken together, these data suggest that both α_2- and β-adrenoceptors mediate the aggression-enhancing effects of NE, and either receptor alone is sufficient to produce the aggression-stimulating effect of NE.

CONCLUSIONS

The results reviewed above suggest that: (*i*) the pro-aggressive effects of NE are mediated independently by both α_2- and β-adrenoceptors; (*ii*) among β-adrenoceptor subtypes, β_2-adrenoceptors appear to play the most important role; and (*iii*) α_1-adrenoceptors may play a less important role in aggression, but data are scarce. The anti-aggressive effect of some noradrenergic drugs in clinical trials support these assumptions.

THE ROLE OF GLUCOCORTICOIDS IN CONTROLLING AGGRESSION

The adrenal cortex synthesizes a large variety of steroid hormones, including mineralocorticoids, glucocorticoids, and neuroactive steroids. Among these, glucocorticoids (corticosterone in reptiles, birds, rodents, and some other mammals; cortisol in some mammals like golden hamsters, dogs, primates, and humans) have an extensively studied role in the stress response. Glucocorticoids are distributed in the organism by circulation. They easily penetrate the blood–brain barrier, and variations in cerebrospinal fluid parallel variations in plasma level (Butte et al., 1976; Carroll et al., 1975). The message conveyed by these hormones is received by cells that possess glucocorticoid receptors in the brain. There are two well-known receptors for glucocorticoids: the mineralocorticoid and glucocorticoid receptors (MR and GR), which bind glucocorticoids with high and low affinity, respectively. The MR is a shared mineralocorticoid–glucocorticoid receptor, and cross-reactions are inhibited by local enzymatic processes, which convert the glucocorticoid into inactive products in tissues, where glucocorticoid action on mineralocorticoid receptors in not required. When not bound to the ligand, both the MR and GR form complexes with heat-shock proteins in the cytoplasm. Upon binding glucocorticoids, the receptor complex is dismantled, and the glucocorticoid-bound receptor migrates to the cell nucleus, where it modifies gene transcription. Glucocorticoid receptors can both inhibit and stimulate gene transcription, and the number of affected genes appears relatively large. However, exactly which genes are affected is not yet fully known. The earliest consequences of these genetic processes can be observed after 15–30 min, but most effects are manifested after a significantly longer delay (e.g., hours, days, and even weeks) (for a review, *see* Joëls and de Kloet, 1994; Webster and Cidlowski, 1999). Recent evidence suggests that glucocorticoids also have nongenomic effects. Such nongenomic effects can be caused via several mechanisms, e.g., binding to active membrane proteins (the synaptic voltage-dependent calcium channel), some neurotransmitter

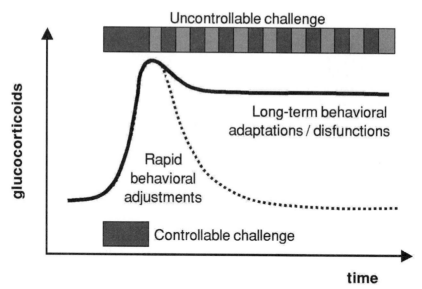

Fig. 5. Short- and long-term effects of stressors. The stressor induces a rapid activation of physiological and behavioral mechanisms that are aimed at controlling the situation. If physiological and behavioral mechanisms are inefficient, both the stress factor and the stress response become persistent. Long-term behavioral adaptations do not aim at escaping the stressor, but at surviving the critical period.

receptors (the κ-opioid and acetylcholine receptors), and specific membrane receptors for glucocorticoids (for a review, see Falkenstein et al., 2000; Haller et al., 1998; Makara and Haller, 2001). By such mechanisms, glucocorticoids affect neuronal functions within seconds and minutes, and cause behavioral effects in the range of 3–7 min.

Glucocorticoids constitute a general "trouble signal" for the organism. Their acute increase signals the occurrence of a challenge, while their chronic increase means that the challenge could not be responded to adequately (Fig. 5). These simple messages are of primordial importance for the organism, because they have a large impact on survival. When the challenge occurs for the first time, the organism should mobilize all its capacities to respond adequately. This requires the activation of various behaviors, including aggression when the challenge is social. If the efforts to meet an acute challenge are not successful, the challenge will develop into a disturbing factor and, ultimately, into a chronic stress state. If that happens, a different kind of adaptation to the conditions set by the chronic stressor is required. The characteristics of adaptation depend on the nature of the stressor represented by factors such as intensity, duration, controllability, and predictability. In general, chronic stressors lead to behavioral depression. This response enables the organism to spare energies and to avoid the risks deriving from active behaviors. Thus, such changes can be considered adaptive in the short run. However, if chronic stressors exceed the normal adaptive capacities of organisms, they lead to behavioral disturbances, including disturbed aggressive behavior.

Apparently, glucocorticoids contribute to the orchestration of behavioral and physiological responses to both acute and chronic stressors. In addition, these hormones are involved in the development of stress-related behavioral disorders. The adrenocortical stress response offers an intriguing question: the question of how one single signaling molecule, such as a glucocorticoid, which circulates throughout the entire organism, can mediate so many distinct, but functionally related responses. That question becomes even more intriguing if one bears in mind that the nature of the behavioral changes caused by corticosteroids are often context-dependent.

The answer to that question lies in the complexity and specific distribution of the mechanisms, which decode that message. The complexity of the decoding mechanisms allow this molecule to affect processes in a time frame of seconds to weeks, to provoke changes on an intensity scale that ranges from a temporary increase in neuronal firing rate to massive cell death, and processes as different as gluconeogenesis and aggression. Moreover, the interaction between the different decoding mechanisms probably allow for the well-known context dependency of the responses.

Glucocorticoid Background and Aggression

It has long been assumed that glucocorticoids play an important role in aggression. Long-lasting increased levels of plasma glucocorticoids were implicated in several psychological and other disorders. Therefore, research on the relation between glucocorticoids and aggression focused on chronic stress and chronic treatment with glucocorticoids. Recent data suggest, however, that glucocorticoid hypofunction also plays important roles in aggressive behavior.

Glucocorticoid Hypofunction: A Role in Pathological Aggression?

Almost two decades ago, Virkunnen et al. (1985) found that habitually violent offenders with antisocial personality show lower levels of plasma glucocorticoids than both healthy subjects and violent offenders without the habitual tendency. Similar results were reported in subjects with other psychological disorders. It has been shown that plasma glucocorticoid levels are inversely correlated with aggressiveness in children with conduct disorder (McBurnett et al., 2000). Subsequent research has confirmed these findings by showing that low basal levels and hyporesponsiveness of plasma glucocorticoids are associated with persistent aggression in humans (including females) (Kariyawasam et al., 2002; Pajer et al., 2001) and various animal species (e.g., dogs and fish) (Hennessy et al., 2001; Pottinger and Carrick, 2001). Studies based on correlation, however, cannot clarify the relationship between the phenomena involved. Recently, we have suggested that glucocorticoid hypofunction is causally linked to abnormal forms of aggressive behavior. This assumption was based on the fact that adrenalectomized rats (supplemented with a subcutaneous glucocorticoid pellet to ensure low and constant levels of plasma glucocorticoids) showed abnormal forms of aggressive behavior when faced with intruders of smaller size. Their aggressive behavior was abnormal in two ways: (*i*) offensive threats were markedly reduced and (*ii*) attacks were preferentially directed toward vulnerable body parts of opponents (Haller et al., 2001). Offensive threats have the role of signaling attack intentions, thus, the decrease in threats make the behavior of adrenalectomized rats less predictable. Targeting attacks toward non-vulnerable body parts is a behavioral mechanism that prevents the delivery of lethal injuries during aggressive encounters. This mechanism is so efficient that the exposure of vulnerable body parts by defeated animals is the usual tactic that deflects the attacks

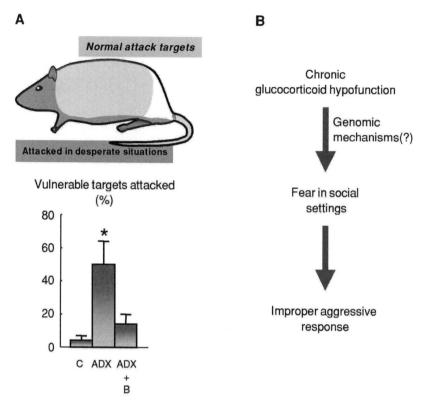

Fig. 6. The effects of chronic glucocorticoid hypofunction on aggressive behavior: a hypothesis. Glucocorticoid hypofunction was associated persistent aggression in psychologically disordered humans and induced abnormal forms of aggression in rats. (**A**) Rats with experimentally induced glucocorticoid hypofunction attack vulnerable body parts of their opponents, which are usually attacked in desperate situations only. Glucocorticoid hypofunction was associated with fear and anxiety in both humans and rats. (**B**) We hypothesize that glucocorticoid hypofunction increases fear in social settings, which leads to the occurrence of inappropriate forms of attack. C, sham operated rats; ADX, adrenalectomized rats supplanted with glucocorticoid pellets that ensured stable low levels of plasma glucocorticoids; ADX+B, adrenalectomized rats supplanted with glucocorticoid pellets that ensured stable low levels of plasma glucocorticoids and acutely treated with glucocorticoids.

of the winner. Whether the reduction of violent attacks results from the direct perception of a "submission signal" emitted by the vulnerable parts of the opponent's body, or from concealing the attack-releasing, or directing stimuli on the back of the opponent, is not known. Attacks directed toward vulnerable body parts are only common in life-threatening or otherwise desperate situations (Blanchard and Blanchard, 1981; Parmigiani et al., 1988). These considerations suggest that the experimentally induced glucocorticoid hypofunction activated attack patterns that were inappropriate under the current circumstances (i.e., in residents fighting with smaller intruders). Apparently, adrenalectomized rats have difficulty in correctly interpreting either the context of the situation or the social signals of the opponent or both. Because "desperate" situations also induce considerable anxiety, these unrestrained violent attacks may be due to an activation of fear-related mechanisms. Accordingly, territorial aggressive behavior of adrenalectomized rats massively activated brain centers involved in fear (central amygdala) and the stress response (the parvocellular part the paraventricular hypothalamic nucleus). No such activation was observed in rats with intact adrenals following normal territorial fighting (Halász et al., 2002). These findings support the assumption that glucocorticoid hypofunction-induced aggression is associated with strong fear reactions. Noteworthy, glucocorticoid hypofunction-associated aggression in humans was also associated with high levels of anxiety (McBurnett et al., 2000) (Fig. 6).

The above-mentioned animal studies suggest that glucocorticoid hypofunction is a causal factor in the development of a deviant aggressive response. The specific destruction of glucocorticoid-producing adrenocortical cells induces behavioral effects similar to those caused by adrenalectomy (Haller et al., unpublished data). This shows that glucocorticoids appear to have a specific role, while the many other consequences of adrenalectomy (e.g., loss of adrenomedular function) are less important.

Studying the effect of glucocorticoid hypofunction on aggressive behavior may develop into an important and exciting new direction of aggression research. Glucocorticoid hypofunction-induced aggressive behavior appears to be pathological because it is linked to (*i*) an endocrinological malfunction and (*ii*) psychological disorders in humans and abnormal forms of aggression in animals. Earlier, aggression research focused mainly on natural aggressiveness (or supposedly functional aggressive behavior in experimental settings). There were a few exceptions only, e.g., the study of attacks elicited by the electrical stimulation of the hypothalamic attack area or brain lesions (Albert and Walsh, 1982; Kruk, 1991). It was shown, for instance, that certain hypothalamic and septal lesions change normal aggressive behavior (which is dependent on testosterone) into defensive behavior (which is not dependent on testosterone). Such research strategies are important, because the mechanisms underlying pathological aggressive behavior are probably different from those that underlie natural aggressive behavior.

More insights into the interactions between glucocorticoid hypofunction and aggressiveness may provide information on how certain forms of pathological aggressiveness develop. Tentatively, the interaction of developmental or traumatic factors blunt the adrenocortical stress response, which in turn conceivably could lead to desperate and inappropriate aggression on a relatively harmless social stressor. Such a speculative mechanism may underlie the so-called "cycle of violence" in genetically or socially vulnerable persons.

The importance of this concept for our understanding of pathological aggression in humans is far from completely established. For instance, neither the prevalence nor the causes of the glucocorticoid hypofunction are known in humans. The endocrinological abnormality may be due to genetic anomalies, developmental problems (e.g., early adverse effects), endocrinological diseases, long-term effects of traumatic stressors, or an interaction of several of these factors. For instance, patients with PTSD are often reported to have low basal levels of cortisol (see below). Noteworthy, such patients also show heightened levels of aggressiveness. It is also unclear how the glucocorticoid insufficiency translates into a behavioral disorder. The genomic effects of glucocorticoids are probably involved, because a short-term inhibition of glucocorticoid synthesis results in behavioral changes that are different from those of a chronic glucocorticoid hypofunction (Mikics and Haller, manuscript in preparation).

Inhibited Aggressiveness in Glucocorticoid Hyperfunction

Chronic stressors increase the plasma levels of glucocorticoid hormones, and this results in the inhibition of aggressive behavior (Leshner et al., 1980; Politch and Leshner, 1977). The role of glucocorticoids in this response is suggested by the fact that chronic administration of glucocorticoids also decreases aggressiveness. The decrease in aggressiveness is associated with an increase in submissive behaviors, and it has been hypothesized that the decrease in aggressiveness is secondary to this behavioral change. The involvement of other behaviors (e.g., submissiveness) suggests that the inhibition of aggression constitutes a particular case of the general depressive effect of chronic stress (Price et al., 1994).

Plasma levels of glucocorticoids are elevated when the stress factor is uncontrollable and permanent or recurring. The hormonal change indicates the inability of the organism to respond adequately the challenge, i.e., to escape, control, or predict the stressor. Under such conditions, the lowering of aggressiveness appears functional, because aggressiveness involves substantial energy loss and the risk of injury, which may lower the chance of surviving the critical situation. The chronic increase in plasma glucocorticoids may constitute the physiological mechanism by which unnecessary engagement in aggressive acts is inhibited.

Rhythmic Variations in Glucocorticoid Secretion Lead to Rhythmic Variations in Aggression

Glucocorticoid secretion is not constant but shows large ultradian, diurnal, and seasonal variations in all the species studied, including humans (Stupfel and Pavely, 1990). The amplitude of such variations is rather large. For example, the amplitude of ultradian variations may be as high as 500 nmol/L in rats, and peak values are at least twice as high as values measured during the nadir. The relevance of these changes for aggression control has rarely been investigated. There are a few studies, however, which suggest that the oscillations in glucocorticoid levels are paralleled by oscillations in the propensity to behave aggressively. In rats, the increasing phase of both ultradian and diurnal glucocorticoid variations is correlated with an increase in aggressiveness (Haller et al., 2000a, 2000b). Similarly, the aggressiveness of children showed a positive correlation with diurnal changes in a plasma glucocorticoids under clinical conditions (Dettling et al., 1999). In rats, the relationship appears causal, because the inhibition of glucocorticoid increase or the blockade of glucocorticoid receptors in relevant phases prevent the increase in aggressiveness. These phenomena can be considered particular cases of the acute effects of glucocorticoids on aggressive behavior (see below) and suggest that the aggressive response to a social challenge depends on the phase of the rhythmic changes in glucocorticoid secretion. Therefore, an organism may respond more aggressively to a challenge if it occurs in the increasing phase of glucocorticoid oscillations.

The relevance of the interaction between variations in glucocorticoid production and aggressiveness is difficult to assess at present. The functional relevance may be that diurnal increases of circulating corticosteroids reflect impending physiological, psychological, or social challenges. Increasing glucocorticoid levels that facilitate aggressiveness would then help removing rivals that could prevent access to resources, which could relieve such an impending problem. Studies in humans suggest that the relation between aggression and corticosteroid response may have relevance for the pathophysiology of the mechanisms underlying disordered social relations and violence. In psychologically disordered subjects, persistent aggressiveness was associated with both decreased basal levels and decreased diurnal variability of plasma glucocorticoids (see the previous section). This may suggest that the "flattening" of glucocorticoid variations may contribute to the occurrence of glucocorticoid-associated aggressiveness. Ultradian variations in glucocorticoid secretion may have a role in intermittent explosive behavior. This disorder is characterized by intermittently occurring outbursts of aggressiveness. One can hypothesize that these outbursts occur when the challenge coincides with the peak of (possibly exaggerated) ultradian glucocorticoid variations. Presently, there is no valid evidence to substantiate these assumptions.

Acute Effects of Glucocorticoids: Promoting Aggressiveness

The aggressive encounter *per se* induces the rapid activation of the HPA axis; moreover, this occurs even if the opponent can be perceived without any possibility for physical contact (Haller et al., 1995; Schuurman, 1980). In addition, the electrical stimulation of hypothalamic centers involved in attack induce the rapid activation of the HPA axis (Kruk et al., 1998). Since attacks usually occur several minutes after the sight of the opponent, the aggressive encounter is performed under the acute effect of glucocorticoid hormones, which rapidly increase aggressiveness if injected to animals previous to the aggressive encounter. The consequences of acute glucocorticoid treatments include the shortening of attack latencies and the increase of the frequency and duration of agonistic interactions (Brain and Haug, 1992; Haller et al., 1997). Glucocorticoids exert these effects even when they are locally injected into hypothalamic sites that control aggressive behavior (Hayden-Hixon and Ferris, 1991), i.e., the effect appears centrally mediated.

Taken together, the above data suggest the existence of a positive feedback loop between brain mechanisms involved in attack and the stress response (Fig. 7). According to this model, the sight of the opponent elicits a mild HPA axis activation, which stimulates attack behavior. The latter, when started, increases the activation of the stress response further. We have shown earlier, that the intensity of the attack-induced stress response depends on the behavior of the opponent, since counterattacking opponents induced a significantly larger stress response than submissive ones (Haller et al., 1995). This finding fits well with the proposed model, because resistance to attacks by opponents maintains aggressive behavior at high levels, which stimulates HPA axis activation.

The rapidity with which acute glucocorticoid injections stimulate attack, raises questions regarding the mechanism of action. In the available studies, the pro-aggressive effect of glucocorticoids was expressed within a time frame of 15 min. This suggests that the mechanism was nongenomic. In a recent study, we have applied corticosterone 1 to 2 min before the aggressive encounter and monitored behavior only for 5 min (Mikics et al., manuscript in preparation). The pro-aggressive effect of corticosterone was significant even within this time frame. In addition, this effect was resistant to the protein synthesis inhibitor cycloheximide. Taken together, these data strongly suggest that corticosterone stimulates aggressive behavior via nongenomic mechanisms. This may be the reason why rapid ultradian oscillations in glucocorticoid synthesis induce rapid oscillations in the propensity of rats to respond a challenge aggressively (see above). Rapid on–off functions of genomic mechanisms appear unlikely, but nongenomic receptors may

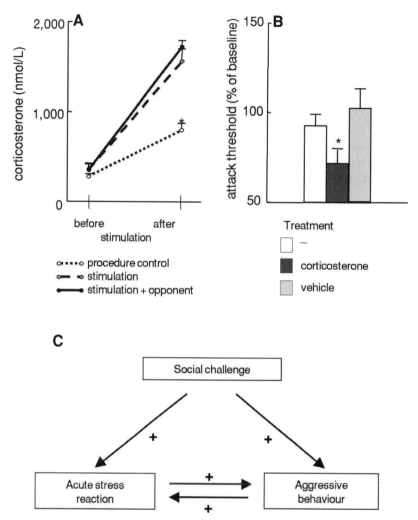

Fig. 7. Positive feed-forward relationships between mechanisms that control the acute stress reaction and aggressive behavior. (**A**) The electrical stimulation of the hypothalamic attack area induces a strong activation of the HPA axis, irrespective to the presence of an opponent (i.e., the presence of fights) (Kruk and Haller, manuscript in preparation). (**B**) Glucocorticoids in turn decrease the electric thresholds for the induction of attacks (i.e., they increase the sensitivity of neurons involved in the elicitation of attacks)(Kruk and Haller, manuscript in preparation). Similar results were obtained in tests of naturally occurring aggressive behavior. (**C**) These data suggest that brain mechanisms involved in the stress response and aggressive behavior reciprocally support each other.

well cause such oscillations. The nongenomic effects of glucocorticoids constitute a novel mechanism of action that was disregarded previously.

The fast positive feedback between the mechanisms controlling the adrenocortical stress response and the mechanisms controlling aggression may explain why aggression is often precipitated by stressors and why aggression is difficult to control once it has started. In a natural setting, such a mechanism is probably advantageous by bringing a fight over resources to a conclusion. However, a dysregulation of the underlying regulation mechanisms will probably result in a dysfunctional responses to a social challenge.

The Effect of Traumatic Events

As shown above, traumatic events induce long-term changes in stress physiology and behavior. Such traumas (e.g., the experience of war, assault, or natural disaster) lead to the development of PTSD in humans. Behavioral consequences of PTSD include a marked increase in aggressive behavior. The long-term hormonal changes include a last-ing disturbance of the diurnal glucocorticoid secretion rhythm (Aardal-Eriksson et al., 2001) and a general lowering of basal cortisol levels for months and even years (Goenjian et al., 1996; Mason et al., 2001). In one study, basal levels of cortisol correlated negatively with PTSD symptoms, i.e., hormonal levels were the lowest in those individuals that showed the strongest symptoms (Anisman et al., 2001). Thus, PTSD involves a long-term reduction in basal HPA axis function. As shown above, such a hormonal profile is associated with persistent aggression in several psychological disorders and induces abnormal aggression patterns in animals. Taken together, these data suggest that the aggressiveness of PTSD patients may be related to the hormonal change elicited by traumas. This assumption, however, needs to be substantiated by future research.

CONCLUSIONS

The Neuroendocrine Stress Response and Aggressiveness

Success in aggressive behavior confers advantages to individuals that include access to territory, females, food, and shelter. However, aggressiveness also involves risks, e.g., the risk of injury and substantial energy loss. In nature, aggressive behavior also increases the risk of predation. Injuries, beyond the possibility of infections, may hamper the ability to feed, mate, etc. Energy lost in aggressive behavior may disadvantage animals in a subsequent danger. Such negative consequences of aggressive interactions affect mainly the defeated individual, but the winner is also at risk. A behavior as critical as aggressiveness needs a fine-tuned control, which implies that controlling mechanisms are complex. Probably, this is the reason why the role of the neuroendocrine stress system in aggression cannot be described in simple terms.

The studies reviewed here allow the following general conclusions:

1. Normal levels and forms of aggression require a balanced functioning of the brain noradrenergic system and HPA axis. If the noradrenergic system is chronically down-regulated, aggressiveness is diminished. When the basal levels of glucocorticoids are chronically lowered, abnormal forms of aggressiveness occur. Thus, appropriate responses to a social challenge require a precisely tuned functioning of these two systems.
2. The acute social challenge-induced activation of the two systems promotes aggressive behavior. A social challenge activates the sympathoadrenal system, the brain noradrenergic system, and the HPA axis. The activation usually takes place before the fights actually start. The sympathoadrenal system supports fights indirectly, via inducing appropriate somatic changes. The noradrenergic system and glucocorticoids appear to support brain mechanisms involved in aggressive behavior more directly. Indirect evidence and our unpublished results suggest that the acute effect of glucocorticoids involves a nongenomic mechanism. The reciprocal stimulatory effects of the acute stress response and aggressiveness suggest the existence of a positive feedback loop between the neuroendocrine and the behavioral systems. According to this hypothesis, the neuroendocrine stress response is activated by the social challenge *per*

se, the activation of the neuroendocrine stress system promotes the development of an aggressive response, which further augments the stress response. The cycle is probably stopped or inhibited by the submissive behavior of the opponent or possibly by an overriding effect of losing.

3. Chronic increases in brain NE and glucocorticoids have opposite effects on aggressiveness. While a chronic increase in brain NE increases aggressiveness, a chronic increase in plasma glucocorticoids inhibits it. Noteworthy, the pro-aggressive effect of the chronic increase of brain noradrenergic function was not demonstrated in situations involving chronic stress, but in humans and animals with a genetical MAOA deficiency.

4. Both the noradrenergic system and the HPA axis show considerable ultradian, diurnal, and seasonal variations. The effects of these two systems on aggressiveness suggests that these variations are relevant for aggression control. However, the area has been insufficiently studied. Some studies show that glucocorticoid oscillations affect aggression, and these appear to be related to the acute effects of glucocorticoids (i.e., the increasing phase of glucocorticoid oscillations stimulate aggressiveness). One can hypothesize that the behavioral response given to a social challenge depends on the relationship between the timing of the challenge and the phase of glucocorticoid secretion. Such a relationship may be valid also for the noradrenergic system, but evidence is lacking.

5. Single traumatic events have long-term consequences for both the functioning of the neuroendocrine stress system and aggressive behavior. The changes in the neuroendocrine systems appear to promote aggressiveness, since traumas up-regulate the noradrenergic system and decrease plasma glucocorticoids. The former consequence of PTSD (the increase in central noradrenergic activity) may increase aggressiveness, while the latter (the reduction in plasma glucocorticoids) may render it abnormal. Aggressiveness is a general problem with PTSD patients, and their specific neuroendocrine background may cause or at least amplify these problems. However, there is insufficient evidence to support this assumption. In addition, no animal studies have been performed in this respect, despite the fact that there are at least three laboratory models available for the study of the long-term consequences of traumatic events (Adamec and Shallow, 1993; Koolhaas et al., 1997; van Dijken et al., 1992).

Multiple Interactions

Obviously, the adrenergic–noradrenergic systems and the HPA axis interact in controlling aggressiveness. These interactions are, however, poorly understood. Owing to the lack of information, one may only speculate on some possible links. In the case of PTSD, for instance, the changes in the NE system (up-regulation) and HPA axis (down-regulation) appear to support each others' behavioral effects. Agents that suppress NE neurotransmission appear efficient in treating this disorder. This may either mean that glucocorticoid down-regulation is insufficient to induce pathological forms of aggressiveness in the case of this disorder or such treatments may normalize glucocorticoid regulation. This consideration may be relevant also for the prospective treatment of glucocorticoid-induced pathological forms of aggressiveness in the case of other disorders.

In contrast to PTSD, the chronic elevation of noradrenergic neurotransmission and glucocorticoid secretion have conflicting consequences. This may either mean that the effects of glucocorticoid hyperfunction are "stronger" than the effects of NE up-regulation or that the joint activation of the two systems leads to other types of behavioral dysfunctions. The up-regulation of the noradrenergic system alone may lead to an increased aggressiveness, while the co-activation of the two systems may induce anxiety and depression, with negative consequences for aggressiveness.

Both the noradrenergic system and HPA axis interact with other systems relevant for aggression control, e.g., gonadal steroids and the serotonergic and vasopressinergic systems. These interactions may be relevant for aggression control, but this aspect is poorly understood.

Pathological Aggression and the Neuroendocrine Stress Response

Labeling aggression with the term pathological is a delicate social issue that is prone to social abuse in itself. Killing people *per se* cannot be considered pathological. In legal terms, aggressive self-protection or the protection of the life of other people (e.g., by the police) is allowed and even institutionalized. Laws ban the use of aggression, but law-breaking cannot actually be considered pathological. In addition, social competitiveness may involve some sort of aggressiveness as well. Usually, aggression is viewed as a medical problem only if it obviously relates to pathological conditions like brain injury, Alzheimer's disease, psychological disorders (depression, schizophrenia, antisocial personality, conduct disorder), or drug abuse. Presently, there is substantial evidence suggesting that the malfunction of the neuroendocrine stress systems is among the underlying causes of such abnormal forms of aggressiveness (Feldmann, 2001; Gerra et al., 2001; Russo-Neustadt et al., 1998).

Relatively new evidence suggests that the malfunction of the neuroendocrine stress system may cause the occurrence of abnormal aggression. MAOA deficiency and glucocorticoid hypofunction are good examples for this relationship. This suggests that the malfunction of the neuroendocrine stress system can be considered a pathological state that has important consequences for aggressive behavior. Much more research is required to elucidate: (*i*) the frequency of such neuroendocrine malfunctions in the human population; (*ii*) their correlation with aggressiveness in general and with abnormal forms of aggressiveness in particular; (*iii*) the neural processes that connect these neuroendocrine malfunctions to brain mechanisms that control aggressiveness; and (*iv*) the possibilities of counteracting the neuroendocrine malfunction or its adverse effects on aggressive behavior.

REFERENCES

Aardal-Eriksson, E., Eriksson, T. E., and Thorell, L. H. (2001) Salivary cortisol, posttraumatic stress symptoms, and general health in the acute phase and during 9-month follow-up. *Biol. Psychiatry* **50**, 986–993.

Adamec, R. E. and Shallow, T. (1993) Lasting effects on rodent anxiety of a single exposure to a cat. *Physiol. Behav.* **54**, 101–109.

Albert, D. J. and Walsh, M. L. (1982) The inhibitory modulation of agonistic behaviour in the rat brain: a review. *Neurosci. Biobehav. Rev.* **6**, 125–143.

Anisman, H., Griffiths, J., Matheson, K., Ravindran, A. V., and Merali, Z. (2001) Posttraumatic stress symptoms and salivary cortisol levels. *Am. J. Psychiatry* **158**, 1509–1511.

Barrett, J. A., Edinger, H., and Siegel, A. (1990) Intrahypothalamic injections of norepinephrine facilitate feline affective aggression via alpha 2-adrenoceptors. *Brain Res.* **525**, 285–293.

Bell, R. and Hobson, H. (1984) Effects of (-)pindolol and SDZ 216-525 on social and agonistic behavior in mice. *Pharmacol. Biochem. Behav.* **46**, 873–880.

Blanchard, R. J. and Blanchard, D. C. (1981) The organization and modeling of animal aggression, in *The Biology of Aggression.* (Brain, P. F. and Benton, D., eds.), Sijthoff et Noordhoff, Alphen aan den Rijn (The Netherlands), pp. 529–563.

Brain, P. F. and Haug, M. (1992) Hormonal and neurochemical correlates of various forms of animal "aggression." *Psychoneuroendocrinology* **17,** 537–551.

Brunner, H. G., Nelen, M., Breakefield, X. O., Ropers, H. H., van Oost, B. A. (1993) Abnormal behavior associated with a point mutation in the structural gene for monoamine oxidase A. *Science* **262,** 578–580.

Buka, S. L., Stichick, T. L., Birdthistle, I., and Earls, F. J. (2001) Youth exposure to violence: prevalence, risks, and consequences. *Am. J. Orthopsychiatry* **71,** 298–310.

Butte, J. C., Kakihana, R., and Noble, E. P. (1976) Circadian rhythm of corticosterone levels in rat brain. *J. Endocrinol.* **68,** 235–239.

Buus Lassen, J. (1978) Piperoxane reduces the effects of clonidine on aggression in mice and on noradrenaline dependent hypermotility in rats. *Eur. J. Pharmacol.* **47,** 45–49.

Cai, B., Matsumoto, K., Ohta, H., and Watanabe, H. (1993) Biphasic effects of typical antidepressants and mianserin, an atypical antidepressant, on aggressive behavior in socially isolated mice. *Pharmacol. Biochem. Behav.* **44,** 519–525.

Carroll, B. J., Heath, B., and Jarrett, D. B. (1975) Corticosteroids in brain tissue. *Endocrinology* **97,** 290–300.

Cases, O., Seif, I., Grimsby, J., et al. (1995) Aggressive behavior and altered amounts of brain serotonin and norepinephrine in mice lacking MAO_A. *Science* **268,** 1763–1766.

Clark, C. R., Geffen, G. M., and Geffen, L. B. (1988) Catecholamines and attention. I Animal and clinical studies. *Neurosci. Biobehav. Rev.* **11,** 341–352.

Crawley, J. N. and Contrera, J. F. (1976) Intraventricular 6-hydroxy-dopamine lowers isolation-induced fighting behaviour in male mice. *Pharmacol. Biochem. Behav.* **4,** 381–384.

Davidovic, V. and Petrovic, V. M. (1981) Diurnal variations in the catecholamine content in rat tissues. Effects of exogenous NA. *Arch. Int. Physiol. Biochim.* **89,** 457–460.

Dettling, A. C., Gunnar, M. R., and Donzella, B. (1999) Cortisol levels of young children in full-day childcare centers: relations with age and temperament. *Psychoneuroendocrinology* **24,** 519–536.

Eide, K. and Hole, K. (1992) Interactions between substance P and norepinephrine in the regulation of nociception in mouse spinal cord. *Pharmacol. Toxicol.* **70,** 397–401.

Ellison, G. (1976) Monoamine neurotoxins: selective and delayed effects on behavior in colonies of laboratory rats. *Brain Res.* **103,** 81–92.

Falkenstein, E., Tillmann, H.-C., Christ, M., Feuring, M., and Wehling, M. (2000) Multiple actions of steroid hormones—a focus on rapid, non-genomic effects. *Pharmacol. Rev.* **52,** 513–555.

Feldmann, T. B. (2001) Bipolar disorder and violence. *Psychiatr. Q.* **72,** 119–129.

Ferry, B., Roozendaal, B., and McGaugh, J. L. (1999) Role of norepinephrine in mediating stress hormone regulation of long-term memory storage: a critical involvement of the amygdala. *Biol. Psychiatry* **46,** 1140–1152.

Figueroa, E. and Silk, K. R. (1997) Biological implications of childhood sexual abuse in borderline personality disorder. *J. Personal Disord.* **11,** 71–92.

Freeman, T. W. and Roca, V. (2001) Gun use, attitudes toward violence, and aggression among combat veterans with chronic posttraumatic stress disorder. *J. Nerv. Ment. Dis.* **189,** 317–320.

Fujiwara, M., Kataoka, Y., Hori, Y., and Ueki, S. (1984) Irritable aggression induced by delta 9-tetrahydrocannabinol in rats pretreated with 6-hydroxydopamine. *Pharmacol. Biochem. Behav.* **20,** 457–462.

Fujiwara, Y., Takeda, T., Kazahaya, Y., Otsuki, S., and Sandyk, R. (1988) Inhibitory effects of carbamazepine on clonidine-induced aggressive behavior in mice. *Int. J. Neurosci.* **42,** 77–84.

Garcia-Brull, P. D., Nunez, J., and Nunez, A. (1993) The effects of scents on the territorial and aggressive behaviour of laboratory rats. *Behav. Processes* **29,** 25–36.

Gentsch, C., Lichteiner, M., and Feer, H. (1989) Behavioral effects of yohimbine and chlordiazepoxide: dependence on the rat's previous familiarisation with the test conditions. *Neuropsychibiology* **22,** 101–107.

Gerra, G., Zaimovic, A., Raggi, M. A., et al. (2001) Aggressive responding of male heroin addicts under methadone treatment: psychometric and neuroendocrine correlates. *Drug Alcohol Depend.* **65**, 85–95.

Goenjian, A. K., Yehuda, R., Pynoos, R. S., et al. (1996) Basal cortisol, dexamethasone suppression of cortisol, and MHPG in adolescents after the 1988 earthquake in Armenia. *Am. J. Psychiatry* **153**, 929–934.

Halász, J., Liposits, Z., Kruk, M. R., and Haller, J. (2002) Neural background of glucocorticoid dysfunction-induced abnormal aggression in rats: involvement of fear- and stress-related structures. *Eur. J. Neurosci.* **15**, 561–569.

Haller, J., Albert, I., and Makara, G. B. (1997) Acute behavioral effects of corticosterone lack specificity but show marked context dependency. *J. Neuroendocrinol.* **9**, 515–518.

Haller, J., Barna, I., and Baranyi, M. (1995) Hormonal and metabolic responses during psychosocial stimulation in aggressive and nonaggressive rats. *Psychoneuroendocrinology* **20**, 65–74.

Haller, J., Halász, J., Mikics, É., Kruk, M. R., and Makara, G. B. (2000b) Ultradian corticosterone rhythm and the propensity to behave aggressively in male rats. *J. Neuroendocrinol.* **12**, 937–941.

Haller, J., Kiem, D. T., and Makara, G. B. (1996a) The physiology of social conflict in rats: what is particularly stressfull? *Behav. Neurosci.* **110**, 353–359.

Haller, J., Van De Schraaf, J., and Kruk, M. R. (2001) Deviant forms of aggression in glucocorticoid hyporeactive rats: a model for 'pathological' aggression? *J. Neuroendocrinol.* **13**, 102–107.

Haller, J. (1995a) Alpha-2 adrenoceptor blockade and the response to intruder aggression in Long-Evans rats. *Physiol. Behav.* **58**, 101–106.

Haller, J. (1995b) Biochemical backgrounds for cost-benefit interrelations in aggression. *Neurosci. Biobehav. Rev.* **19**, 599–604.

Haller, J., Makara, G. B., and Kruk, M. R. (1998) Catecholaminergic involvement in the control of aggression: hormones, the peripheral sympathetic, and central noradrenergic systems. *Neurosci. Biobehav. Rev.* **22**, 85–97.

Haller, J., Miklósi, Á., Csányi, V., and Makara, G. B. (1996b) Behavioral tactics control the energy costs of aggression. The example of Macropodus opercularis. *Aggress. Behav.* **22**, 437–446.

Haller, J., Millar, S., van de Schraaf, J., Kloet, R. E., and Kruk, M. (2000a) The Active phase-related incrrease in corticosterone and aggression. *J. Neuroendocrinol.* **12**, 431–436.

Hayden-Hixon, D. M. and Ferris, C. F. (1991) Steroid specific regulation of agonistic responding in the anterior hypothalamus of male hamsters. *Physiol. Behav.* **50**, 793–799.

Hegstrand, L. R. and Eichelman, B. (1983) Increased shock-induced fighting with supersensitive beta-adrenergic receptors. *Pharmacol. Biochem. Behav.* **19**, 313–320.

Hennessy, M. B., Voith, V. L., Mazzei, S. J., Buttram, J., Miller, D. D., and Linden, F. (2001) Behavior and cortisol levels of dogs in a public animal shelter, and an exploration of the ability of these measures to predict problem behavior after adoption. *Appl. Anim. Behav. Sci.* **73**, 217–233.

Higley, J. D., Mehlman, P. T., Taub, D. M., et al. (1992) Cerebrospinal fluid monoamine and adrenal correlates of aggression in free-ranging rhesus monkeys. *Arch. Gen. Psychiatry* **49**, 436–441.

Hodge, G. K. and Butcher, L. L. (1975) Catecholamine correlates of isolation-induced aggression in mice. *Eur. J. Pharmacol.* **31**, 81–93.

Holmes, D. S. and Will, M. J. (1985) Expression of interpersonal aggression by angered and nonangered persons with the type A and type B behavior patterns. *J. Pers. Soc. Psychol.* **48**, 723–727.

Joëls, M. and de Kloet, E. R. (1994) Mineralocorticoid and glucocorticoid receptors in the brain. Implications for ion permeability and transmitter systems. *Progr. Neurobiol.* **43**, 1–36.

Kariyawasam, S. H., Zaw, F., and Handley, S. L. (2002) Reduced salivary cortisol in children with comorbid Attention deficit hyperactivity disorder and Oppositional defiant disorder. *Neuroendocrinol. Lett.* **23**, 45–48.

Koolhaas, J. M., De Boer, S. F., De Rutter, A. J., Meerlo, P., and Sgoifo, A. (1997) Social stress in rats and mice. *Acta Physiol. Scand. Suppl.* **640**, 69–72.

Koolhaas, J. M., Meerlo, P., De Boer, S. F., Strubbe, J. H., and Bohus, B. (1997) The temporal dynamics of the stress response. *Neurosci. Biobehav. Rev.* **21**, 775–782.

Korzan, W. J., Summers, T. R., and Summers, C. H. (2000) Monoaminergic activities of limbic regions are elevated during aggression: influence of sympathetic social signaling. *Brain Res.* **870**, 170–178.

Kruk, M. R., Westphal, K. G., Van Erp, A. M., et al. (1998) The hypothalamus: cross-roads of endocrine and behavioural regulation in grooming and aggression. *Neurosci. Biobehav. Rev.* **23**, 163–177.

Kruk, M. R. (1991) Ethology and pharmacology of hypothalamic aggression in the rat. *Neurosci. Biobehav. Rev.* **15**, 527–538.

Lader, M. (1988) Beta-adrenoceptor antagonists in neuropsychiatry: an update. *J. Clin. Psychiatry* **49**, 213–223.

Lasko, N. B., Gurvits, T. V., Kuhne, A. A., Orr, S. P., and Pitman, R. K. (1994) Aggression and its correlates in Vietnam veterans with and without chronic posttraumatic stress disorder. *Compr. Psychiatry* **35**, 373–381.

Leshner, A. I., Korn, S. J., Mixon, J. F., Rosenthal, C., and Besser, A. K. (1980) Effects of corticosterone on submissiveness in mice: some temporal and theoretical considerations. *Physiol. Behav.* **24**, 283–288.

Levine, E. S., Litto, W. J., and Jacobs, B. L. (1990) Activity of cat locus coeruleus noradrenergic neurons during the defense reaction. *Brain Res.* **531**, 189–195.

Makara, G. B. and Haller, J. (2001) Non-genomic effects of glucocorticoids in the neural system. Evidence, mechanisms and implications. *Prog. Neurobiol.* **65**, 367–390.

Margules, D. L., Lewis, M. J., Dragovich, J. A., and Margules, A. S. (1972) Hypothalamic norepinephrine: circadian rhythms and the control of feeding behavior. *Science* **178**, 640–643.

Mason, J. W., Wang, S., Yehuda, R., Riney, S., Charney, D. S., and Southwick, S. M. (2001) Psychogenic lowering of urinary cortisol levels linked to increased emotional numbing and a shame-depressive syndrome in combat-related posttraumatic stress disorder. *Psychosom. Med.* **63**, 387–401.

Matsumoto, K., Cai, B., Satoh, T., Ohta, H., and Watanabe, H. (1991) Desipramine enhances isolation-induced aggressive behavior in mice. *Pharmacol. Biochem. Behav.* **39**, 167–170.

Matsumoto, K., Ojima, K., Ohta, H., and Watanabe, H. (1994) Beta2- but not beta 1-adrenoceptors are involved in desipramine enhancement of aggressive behavior in long-term isolated mice. *Pharmacol. Biochem. Behav.* **49**, 13–18.

Matsumoto, K., Ojima, K., and Watanabe, H. (1995) Noradrenergic denervation attenuates desipramine enhancement of aggressive behavior in isolated mice. *Pharmacol. Biochem. Behav.* **50**, 481–484.

McBurnett, K., Lahey, B. B., Rathouz, P. J., and Loeber, R. (2000) Low salivary cortisol and persistent aggression in boys referred for disruptive behavior. *Arch. Gen. Psychiatry* **57**, 38–43.

Miczek, K. A., Thompson, M. L., and Shuster, L. (1982) Opioid-like analgesia in defeated mice. *Science* **215**, 1520–1522.

Miczek, K. A. and Winslow, J. T. (1987) Analgesia and decrement in operant performance in socially defeated mice: selective cross-tolerance to morphine and antagonism by naltrexone. *Psychopharmacology* **92**, 444–451.

Mogilnicka, E., Dooley, D. J., Boissard, C. G., and Delini-Stula, A. (1983) Facilitation of shock-induced fighting in the rat after DSP-4, a selective noradrenergic neurotoxin. *Pharmacol. Biochem. Behav.* **18**, 625–628.

Mogilnicka, E. and Zazula, M. (1986) Interaction between beta-adrenoceptor agonists and alpha 1-adrenergic system. A behavioral study with the clonidine-induced aggression test. *Pol. J. Pharmacol. Pharm.* **38,** 529–534.

Mouly, A. M., Elaagouby, A., and Ravel, N. (1995) A study of the effects of noradrenaline in the rat olfactory bulb using evoked field potential response. *Brain Res.* **681,** 47–57.

Murphy, M. R. (1976) Olfactory stimulation and olfactory bulb removal: effects on territorial aggression in male Syrian golden hamsters. *Brain Res.* **113,** 95–110.

O'Carroll, R. E., Drysdale, E., Cahill, L., Shajahan, P., and Ebmeier, K. P. (1999) Stimulation of the noradrenergic system enhances and blockade reduces memory for emotional material in man. *Psychol. Med.* **29,** 1083–1088.

Pajer, K., Gardner, W., Rubin, R. T., Perel, J., and Neal, S. (2001) Decreased cortisol levels in adolescent girls with conduct disorder. *Arch. Gen. Psychiatry* **58,** 297–302.

Parmigiani, S., Brain, P. F., Mainardi, D., and Brunoni, V. (1988) Different patterns of biting attack employed by lactating female mice (*Mus domesticus*) in encounters with male and female conspecific intruders. *J. Comp. Psychol.* **102,** 287–293.

Politch, J. A. and Leshner, A. I. (1977) Relationship between plasma corticosterone levels and levels of aggressiveness in mice. *Physiol. Behav.* **19,** 775–780.

Pottinger, T. G. and Carrick, T. R. (2001) Stress responsiveness affects dominant-subordinate relationships in rainbow trout. *Horm. Behav.* **40,** 419–427.

Price, J., Sloman, L., Gardner, R., Jr., Gilbert, P., and Rohde, P. (1994) The social competition hypothesis of depression. *Br. J. Psychiatry* **164,** 309–315.

Reiss, D. (1993) The long reach of violence and aggression. *Psychiatry* **56,** 163–165.

Rodgers, R. J. and Hendrie, C. A. (1983) Social conflict activates status-dependent endogenous analgesic or hyperalgesic mechanisms in male mice: effects of naloxone on nociception and behaviour. *Physiol. Behav.* **30,** 775–780.

Rodgers, R. J. and Randall, J. I. (1988) Blockade of non-opioid analgesia in intruder mice by selective neuronal and non-neuronal benzodiazepine recognition site ligands. *Psychopharmacology* **96,** 45–54.

Russo-Neustadt, A., Zomorodian, T. J., and Cotman, C. W. (1998) Preserved cerebellar tyrosine hydroxylase-immunoreactive neuronal fibers in a behaviorally aggressive subgroup of Alzheimer's disease patients. *Neuroscience* **87,** 55–61.

Schuurman, T. (1980) Hormonal correlates of agonistic behavior in adult male rats. *Prog. Brain Res.* **53,** 415–420.

Sgoifo, A., de Boer, S. F., Haller, J., and Koolhaas, J. M. (1996) Individual differences in plasma catecholamine and corticosterone stress responses of wild-type rats: relationship with aggression. *Physiol. Behav.* **60,** 1403–1407.

Sgoifo, A., Stilli, D., Aimi, B., Parmigiani, S., Manghi, M., and Musso, E. (1994) Behavioral and electrocardiographic responses to social stress in male rats. *Physiol. Behav.* **55,** 209–216.

Sheard, M. H. (1975) Lithium in the treatment of aggression. *J. Nerv. Ment. Dis.* **160,** 108–118.

Smith, D. F. (1977) Behavior of rats given lithium salts. A review. *Pharmakopsychiatr. Neuropsychopharmakol.* **10,** 79–88.

Southwick, S. M., Bremner, J. D., Rasmusson, A., Morgan, C. A., III, Arnsten, A., and Charney, D. S. (1999) Role of norepinephrine in the pathophysiology and treatment of posttraumatic stress disorder. *Biol. Psychiatry* **46,** 1192–1204.

Stowers, L., Holy, T. E., Meister, M., Dulac, C., and Koentges, G. (2002) Loss of sex discrimination and male-male aggression in mice deficient for TRP2. *Science* **295,** 1493–1500.

Stupfel, M. and Pavely, A. (1990) Ultradian, circahoral and circadian structures in endothermic vertebrates and humans. *Comp. Biochem. Physiol. A.* **96,** 1–11.

Tapp, W. N., Levin, B. E., and Natelson, B. H. (1981) Ultradian rhythm of plasma norepinephrine in rats. *Endocrinology* **109,** 1781–1783.

Tonjes, R., Hecht, K., Brautzsch, M., Lucius, R., and Dorner G. (1986) Behavioural changes in adult rats produced by early postnatal maternal deprivation and treatment with choline chloride. *Exp. Clin. Endocrinol.* **88,** 151–157.

Tornatzky, W. and Miczek, K. A. (1994) Behavioral and autonomic responses to intermittent social stress: differential protection by clonidine and metoprolol. *Psychopharmacology* **116,** 346–356.

van Dijken, H. H., Mos, J., van der Heyden, J. A., and Tilders, F. J. (1992) Characterization of stress-induced long-term behavioural changes in rats: evidence in favor of anxiety. *Physiol. Behav.* **52,** 945–951.

Viken, R. J., Johnson, A. K., and Knutson, J. F. (1991) Blood pressure, heart rate, and regional resistance in behavioral defense. *Physiol. Behav.* **50,** 1097–1101.

Virkkunen, M. (1985) Urinary free cortisol secretion in habitually violent offenders. *Acta Psychiatr. Scand.* **72,** 40–44.

Webster, J. C. and Cidlowsky, J. A. (1999) Mechanisms of glucocorticoid-receptor-mediated repression of gene expression. *Trends Endocrinol. Metab.* **10,** 396–402.

Workel, J. O., Oitzl, M. S., Fluttert, M., Lesscher, H., Karssen, A., and de Kloet, E. R. (2001) Differential and age-dependent effects of maternal deprivation on the hypothalamic-pituitary-adrenal axis of brown norway rats from youth to senescence. *J. Neuroendocrinol.* **13,** 569–580.

7
Y Chromosome and Antisocial Behavior

Pierre L. Roubertoux and Michèle Carlier

ARE MALES MORE AGGRESSIVE THAN FEMALES?

The possible implication of the Y chromosome in antisocial behavior was suggested by gender differences reported extensively by epidemiology and by the special interest shared by neurobiology and ethology for the initiation of aggressive behavior by males. Very soon, the Y chromosome was recognized as being specific to males. As it was, also, highly conserved across the species, the higher incidence of aggressive behavior in males was considered as evidence for the Y chromosome playing a role in violence, aggression, and antisocial behavior in general.

A detailed examination of epidemiological data makes the picture even more complex and blurs the idea of any simple link between aggression and the Y chromosome in humans and also in other species. The initiation of violent acts does not appear to be the attribute of men only. Although crime rates have always been higher for men, the sex ratio was halved between 1960 and 1970, indicating that the proportion of female offenders, compared to males, was gradually increasing in England. The proportion of women committing antisocial acts varies according to the population group and appears to be higher in Black rather than Caucasian populations (Rutter, 1996). Bohman et al. (1981) observed a higher family-related risk for petty crime with girls compared to boys, although the general prevalence was the same. For these reasons, other hypotheses have challenged the argument that aggression is determined by the Y chromosome. Changes in the criminal sex ratio may be the result of changes in women's social status over the years. Differences in the social status of men and women are also seen to vary between communities.

The situation is even more complex in animal studies. No fewer than 50 out of the 5000 papers dealing with aggression published over the last 4 yr dealt with aggression in females, suggesting that the general belief of aggression as an attribute of maleness was probably the consequence of our ignorance rather than a reflection of the situation. Interestingly, selective breeding for reproductive organ weights (Islam et al., 1976) or agonistic behavior (Hood and Cairns, 1988) in the other sex littermate, for either males or females, produced a correlation, suggesting genetic bases for these phenotypes common to males and females. The hypothesis of Y chromosome involvement in aggression was therefore weakened. Last but not least, advances in molecular genetics have

From: *Neurobiology of Aggression: Understanding and Preventing Violence*
Edited by: M. Mattson © Humana Press Inc., Totowa, NJ

provided opportunities to describe behavioral phenotypes associated with targeted genes. A wide number of genes implicated in aggression have been discovered (Maxson, 1996, 1998). All have been mapped on autosomes, which therefore precludes any possibility of the Y chromosome playing a key role in aggression.

Gradually, the Y chromosome has shifted from being the sole "cause" of violent behavior to being one of a number of contributing factors. Several questions then arise. The first refers to genes carried by the Y chromosome. Are their functions compatible with their putative role in aggression? The second question is the identification of candidate loci involved in aggression on the Y. Classic linkage methods are not appropriate here, because the largest part of this chromosome does not segregate. The third question is to define aggression. Aggression occurs in a wide range of situations, involving different cognitive and emotional processes on different physiological pathways. Does the Y chromosome make an equal contribution to the different forms of aggression? Is the contribution situation-dependent? The fourth question addresses the relationships between the candidate genes, carried by the Y chromosome, and the autosomal genes, which were previously identified, either by overexpression (trangenesis) or invalidation (gene targeting). Do Y genes and the autosomal genes have an additive or interactive effect on aggression? What neurochemical mechanisms operate to produce these effects?

STRUCTURE OF THE Y CHROMOSOME

For a long time, geneticists considered that no recombining event occurred within the Y chromosome. As a corollary, they concluded that the whole chromosome was transmitted from father to son. That was until Koller and Darlington (1934) and then Haldane (1936) postulated one recombining event between the chromosomes of each homologous pair on the basis of cytological observations. Their hypothesis was that the sex chromosomes probably followed this rule. They then hypothesized that one part of the Y and X chromosomes would have to recombine, and as a consequence, there would be an homologous chromosomal region on the two sex chromosomes. Burgoyne (1982) proposed a number of theoretical features for this region in humans. Later, Cooke et al. (1985), Simmler et al. (1985), and Petit et al. (1987) provided experimental evidence for this region and identified its molecular characteristics. Keitges et al. (1985) also observed an homologous region of the X and Y chromosomes that were exchanged during male meiosis in male mice. This region is located on the short arm of the Y in humans and on the telomeric part of the long arm in mice. At this point, the major characteristic of the Y chromosome emerges. This chromosome is comprised of three parts. The largest part, which is labeled the specific part or nonpairing-region (Y^{NPR}), is transmitted solely from father to son. Another part, the pseudoautosomal or pairing region (Y^{PR}), is exchanged with the corresponding region of the X chromosome. Y^{PR} is carried on the X, and after male meiosis, is transmitted from father to daughter with the other part of the X chromosome. Although the region is small (2.6 Mbp) it is the site of frequent recombining events, which occur 10 times more frequently than in other regions. A second pairing region (Y^{PR}) was reported. It is smaller (310–330 kbp) and is located at the end of the long arm (Rappold, 1993).

The Y^{NPR} contains very few functional genes; most of the region is occupied by heterochromatin comprised of noncoding sequences. Some could be of potential interest for aggression: interleukin receptor 3, acetyl serotonin N-methyl transferase, steroid sulfatase, testis-determining factor, and several genes regulating spermatogenesis. The Y^{PR} is located close to the gene for the testis-determining factor at a distance less than 5 kbp. As the boundaries between Y^{NPR} and Y^{PR} have been defined in terms of probability, the testis-determining factor might, exceptionally, be transmitted to a daughter with Y^{PR}.

While Y^{NPR} homologous genes are found in both men and mice, Y^{PR} differ. MIC2 coding for the 12E7 antigen of blood groups mapped on human Y^{PR} (Ellis and Goodfellow, 1989) was not found on the mouse Y^{PR}. However, the steroid sulfatase gene (*Sts*), which is the only functional gene known on the mouse Y^{PR}, is in the specific region of the X in humans, as has been confirmed by X inactivation of *Sts* and by the lack of evidence for an expressed copy on the Y chromosome, whereas mouse *Sts* is not inactivated on the X chromosome, and a copy is expressed on the Y (Migeon et al., 1982; Craig and Tolley, 1986).

The question, then is, how to prove that the Y chromosome genes are involved in antisocial behavior?

DEMONSTRATING THE IMPLICATION OF THE Y CHROMOSOME

Any demonstration of genes carried by the Y chromosome being involved in aggression comes up against the problem of the presence of a large nonsegregating region on the chromosome. As no recombining event occurs within the Y^{NPR}, only association techniques can be used, in both humans and mice. Several genes are carried by Y^{NPR}, and identification of the candidate is a perilous exercise, but with the high frequency of recombination, usual linkage detection can be performed between the phenotype and genes located on Y^{PR}.

Y Chromosome and Human Male Phenotypes

Few strategies are available to test for associations between this chromosomal region and phenotypes, but two methods have been widely used. The first consists of comparing persons with an abnormal number of Y chromosomes. Modification of phenotypic characteristics is observed with an additional or missing Y chromosome, but this does not make it possible to detect the gene(s) carried by the Y chromosome, which might be implicated in the phenotype.

Goldman et al. (1996) emphasized the power of association method for discovering candidate genes implicated in impulsive behaviors. His group extended the approach to the Y^{NPR} (Kittles et al., 1999). To the best of our knowledge, it has only been applied once. The method uses the large number of polymorphisms described on the Y chromosome: the short sequences of length polymorphism (SSLP) (often referred to as microsatellites), restriction fragments of length polymorphisms, and several unique sequences of variants. The tools of molecular genetics can be used to detect these polymorphisms, and with a large population, individual differences appear for these DNA variants, making it possible to associate one variant with one phenotype. A pattern of variants or haplotypes can be obtained for each individual, and a cladogram associating the haplotype configurations can be constructed using the parsimony single-step mutation model.

Implication of the Regions of the Y Chromosome in Murine Phenotypes

Inbred strains or selected lines of mice can be used for performing Mendelian or neo-Mendelian crosses. We first consider inbred strains, in which the situation is simpler, because the animals have identical genotypes and because each locus carries identical allelic forms. The first step consists of choosing two highly contrasting strains for attack behavior. To illustrate the point, the labels chosen here are for two strains, A and B, displaying frequent and rare aggressive bouts, respectively. The second step is to develop the two reciprocal F_1s (A×B F_1 and B×A F_1) by appropriate mating of A and B. In the crosses, the strain of the mother is given first. A statistical difference between the reciprocal F_1s has often led researchers to suspect a Y chromosome effect. When A×B F_1 males do not attack and B×A F_1 do attack, this may be due to the Y chromosome from B, the nonviolent strain, in the first F_1, and to the Y chromosome from A, which attack frequently, in the second F_1. The two F_1s share identical autosomes with one allele from the mother, while the other comes from the father, which led some authors to conclude that the reciprocal difference was due to the Y chromosome, but other genetic factors may contribute to the difference. Reciprocal crosses also differ: (*i*) by the X chromosome (from A and B in A×B F_1 and B×A F_1, respectively); (*ii*) by mitochondrial DNA transmitted exclusively by the mother; and (*iii*) by genomic imprinting. Environmental factors also contribute to differences between reciprocal crosses. The animals did not have the same uterine and postnatal maternal environments (A and B, respectively) or the same ova cytoplasm. To demonstrate any Y effect, therefore, these confounding factors have to be disentangled. The two F_1s may not differ, in which case it could be misleading to conclude that any of these factors has an effect, since other sources of variation can act in opposite directions canceling out the effect (Carlier et al., 1992). Moreover, reciprocal crosses cannot be used to distinguish the respective contributions of Y^{NPR} from Y^{PR}.

Backcrossing of F_1 males with A and B females can eliminate most confounding factors. Males born from crosses of A females with A×B F_1 males and males born from crosses of A females with and B×A F_1 males differ only in the origin of the Y chromosome (B and A, respectively). It is also possible to compare males born from the following two crosses: B females crossed with A×B F_1 males vs B females cross with B×A F_1 males. In these crosses, the pairing and nonpairing regions differ, and it is impossible to know which part of the Y chromosome is linked to the phenotype (Carlier et al., 1992).

A direct demonstration of an effect of the Y^{NPAR} would require the development of strains congenic for this region and then a comparison of congenic and parental strains. The congenic strain A.BY^{NPAR} is bred through successive backcrosses of males carrying the Y^{NPAR} with A females. The symmetrical design is used for developing the second congenic strain, B.AY^{NPAR}. At each generation, the congenic receiver loses 50% of the alleles from the congenic donor. As the mothers are from the same inbred strain in each generation of backcrosses, all the maternal factors are constant. The Y pairing region is also from the mother strain (as are all the autosomes) due to the crossing over of X and Y chromosomes at male meiosis.

To demonstrate an effect of the pairing region is even more complicated. One option is to breed groups with the same Y^{PAR} and different Y^{NPAR}, which is the case for males bred by crossing A females with congenic B.AY^{NPAR} males, and males from crossing A females with B males. This is also the case for males bred from B females and congenic males A.BY^{NPAR} and for males bred from B females with A males. A Y^{PAR} effect can be suspected when no difference is observed within the first two groups and within the second two groups, but when differences are observed when comparing the first two groups with the second two groups (Roubertoux et al., 1994a). A full demonstration, however, would require all other factors to have been ruled out. Another possibility is to map qualitative trait locus (QTL) with markers located on the Y^{PAR}.

When selected lines are available, a similar design can be used with these strains (Van Oortmersen and Sluyter, 1994), but the interpretation is more difficult because of the uncontrolled autosomal background in selected lines.

Y CHROMOSOME AND ANTISOCIAL BEHAVIOR

In Humans

According to Witkin et al. (1976) "few issues in behavior genetics have received more public and scientific attention than that given to the possible role of an extra Y chromosome in human aggression." This interest was probably initiated by Jacob et al. (1965) who found seven males with an extra Y chromosome (XYY karyotype) in a sample of 197 males placed in a penal institution for highly antisocial males. The frequency of the extra Y chromosome was higher than in populations outside penal institutions where it reached around 1 out of 1000 males. Similar differences were observed in other criminal populations, suggesting a possible link between greater "maleness" and aggression (Bénézech, 1975; Denno, 1996). If males carrying two Y chromosomes were more aggressive than normal males, the Y chromosome could be considered as encompassing genes related to aggression, and, under those conditions, men would be expected to be more aggressive than women. In most cases, epidemiological data support this assumption, although the figures are more complex as mentioned earlier. The XYY chromosome syndrome was then investigated more thoroughly, and serious doubts were raised on the alleged causal link between the Y and the incidence of crime (*see* Carey, 1994, for a review) as most of the studies were conducted with selected groups (institutionalized or tall men) and with only a small number of cases in each study. Very often there was no control group for comparison (XY men). Witkin et al. (1976) published the most extensive study dealing with a population-based sample of tall men resident in the municipality of Copenhagen (selecting the tallest from the total group of 28,884 men) and obtained the karyotype of 4139 men (90.8% of the starting group). Each person was visited at home and filled out a questionnaire. A variety of records were available including height, convictions for criminal offenses, intellectual level (army selection test and educational attainment), and the parents' social class at the time of birth. Of the 4139 men for whom sex chromosome determinations were made, Witkin and his group identified 12 XYY and 16 XXY men. The rate of criminal convictions was not statistically higher in XYY than in XXY males (5 out of 12 vs 3 out of 16). The XYY group had a higher criminal conviction rate than the normal control group (4096 XY men), and the XXY did not differ for the normal control group. What is more important in understanding the relationship between the Y chromosome and antisocial behavior is that the XYY and the XXY groups showed no difference in violent acts perpetrated against other persons. In fact, the five XYY males convicted had not been charged with assault or any act of aggression against people. After adjustment for background variables, the main explanations for the differences in antisocial behavior observed between the two groups of abnormal karyotypes appeared to be lower intelligence, educational index, and parental socioeconomic status (SES) in the XYY group. There was, however, a higher crime rate in the XYY group; no significant difference in the crime rate was seen between the XXY and the normal XY groups. Two major points emerge from these data. First, XYY men did not commit more antisocial acts than XXYs. Second, the higher rate of criminal conviction observed in XYYs may simply be the result of lower intelligence, as less intelligent men are less adept at escaping arrest.

Witkin et al. (1976) concluded "No evidence has been found that men with either of these sex chromosome complements are especially aggressive." This is also our opinion. However, some of these men could present neuropsychiatric weaknesses (Bénézech, 1975).

A more recent study screened a population of 34,380 infants at birth between 1967 and 1979. Rates of criminal convictions were examined in 17 XYY men, 17 XXY men, and 60 controls (XY men). XYY males showed a significantly higher frequency of antisocial behavior and of criminal convictions than the controls. Again, the higher rate of criminal convictions in the XYY male group was mediated mainly through lower intelligence, and most of the offenses were property offences rather than aggression against persons (Götz et al., 1999).

These studies show that an extra Y chromosome may have an impact on the liability to antisocial behavior, but no more. The rejection of the argument of a propensity to crime explains why defendants in murder cases usually do not succeed when alleging that their behavior at the time of the murder was not controlled by their free will (Denno, 1996).

Goldman's group (Kittles et al., 1999) tested for an association between the Y^{NPR} haplotypes and personality variables associated with alcohol dependence (antisocial personality disorder, novelty seeking, harm avoidance). They worked with a Finnish sample. Two main results emerged. First, that Y chromosome variability may account for around 7% of the total variance of alcoholism and 15% of the genetic variance of alcoholism, and, second, it proved impossible to show a Y chromosome effect on personality variables independent of alcoholism.

Aggression in Mice

Measuring Aggression in Mice

Aggression in mice appears as one of the easiest phenotypes to observe and define compared to learning or other cognitive performances. In fact, initiation of attack behavior depends upon a large variety of factors, including maternal uterine and postnatal environments, the physiological state of the individual tested, previous experience, and many others (Michard and Carlier, 1983; Roubertoux et al., 1999). The occurrence of aggressive acts also varies under different testing conditions in mice (Roubertoux et al., 1999), indicating that there is no general factor for aggression. This might suggest that different measurements of attack behavior have different genetic bases. This hypothesis fits with epidemiological data as reported by Goldsmith and Gottesman (1995) from twin studies. Their results show concordance declines from juveniles to adults for both monozygotic and dizygotic twins, and differences between the two kinds of twins is higher in adult than in juvenile crimes. Moreover, adult and juvenile crimes are not the same and have different etiologies. With mice, the question becomes, "Is there a component of the aggressive behavior pattern associated with either the pairing or nonpairing part of the Y chromosome?" or "Does the link between one of the two parts of the Y chromosome and aggression depend on the conditions under which the aggression was observed?"

Different Conditions Provide Different Measurements of Aggression

In a previous study, we selected four conditions used in research on mice overexpressing a gene or carrying targeted genes. In these studies, the resident intruder test or the neutral area test are generally chosen. The male being tested has either been isolated or is nonisolated, and his opponent can be from the same strain as the tested male or a different one. We tested aggression across 11 strains of laboratory mice and used different male mice for each test.

The conditions were as follows:

1. Standard opponent test: nonisolated test males, and neutral area for testing aggression (Carlier and Roubertoux, 1986; Roubertoux and Carlier, 1988). Each test was a dyadic encounter with an A/JOrl (A/J) male as the standard opponent from a group male cage. This strain was chosen for its low scores on aggression.
2. Standard opponent test: non-naive test males and neutral cage. Rearing and testing conditions were identical except for one slight modification. Tested males were exposed to the standard opponent prior to being tested.
3. Standard opponent test: isolated test males, one resident and one intruder.
4. Homogeneous set-pair test: isolated test males and resident intruder. We used the same procedure with one exception, the opponent and the tested male belonged to the same strain within a pair. The nonisolated opponent was from a cage housing four males of the same strain.

No correlation was found for the four tests, but there was high reliability of the measurements and, as a consequence, the factor analysis did not reveal a general factor for initiating attack behavior (Roubertoux et al., 1999). Although some strains (NZB/BlNJ or Cast/Ei) attacked under all conditions, other strains only attacked when the males had been previously isolated (CBA/H or DBA/2J) or when placed with specific opponents (Roubertoux et al., 1999), while other strains again never attack. Moreover, the aggression scores under these four conditions showed different correlations with measurements of brain neurotransmitter and gonadal hormones, in all 11 strains (Tordjmann et al., unpublished). The fact that different aggression measurements have different neurochemical correlates could suggest that the genetic bases for these measurements are different. We conducted investigations to determine whether the two parts of the Y chromosome were linked with one of these measurement of aggression.

Implication of the Y Chromosome in Aggression in Mice

Two experiments on mice showed a link for two different types of attack behavior with the Y chromosome. One link was obtained with the Y^{NPR} (Maxson et al., 1979; Selmanoff et al., 1976), and the other with the Y^{PR} (Roubertoux et al., 1994a, 1994b; Van Oortmeersen and Sluyter, 1994). The two experiments started by observing that the reciprocal F_1s from an aggressive and nonaggressive strain showed significant differences in initiating attack behavior.

Nonpairing Region of the Y Chromosome

Male DBA/1bg (D1) mice attacked more frequently than C57BL/10bg (B10). By developing the two reciprocal F_1s (D1.B10 F_1 and B10.D1 F_1), observations showed that the male mice from the first cross attacked less than those from the second cross. Given that the mice from the second cross had received the Y chromosome from D1, which was the most aggressive strain, Selmanoff et al. (1975) concluded that the high frequency of attack behavior in this F_1 was due to the presence of the chromosome. As other genetic and maternal environment factors may be responsible for the difference observed in the reciprocal F_1s, the team developed reciprocal congenic strains for the Y chromosome. The authors used the cross transfer procedure of the Y chromosome described above. Comparing the two congenic strains, they confirmed that there was a higher occurrence of attack behavior in the strain carrying the Y chromosome from D1 (Maxson et al., 1979). The segregation of the two parts of the Y chromosome had not been confirmed when Maxson and collaborators published their results, and they therefore concluded that the Y was a factor in aggression in mice. This was in fact the first experimental evidence for a link between the sole Y^{NPR} and aggression. Although, Maxson's group tested his mice in a neutral area, the tested males were isolated, and the opponent and tested male belonged to the same strain. Our studies have confirmed the implication of the Y^{NPR} under the same conditions, but excluded its involvement under different experimental conditions (Guillot et al., 1995). We tested a quartet of congenic and parental strains (D1 and B10) with nonisolated males, using a standard opponent from the A/J strain in a neutral area, according to Carlier and Roubertoux (1986). Differences between both parental and congenic strains disappeared. Considering the results together, they showed that the Y^{NPR} is involved in one form of aggression characterized by marked reactivity to isolation and sensitivity to the sensorial cues emitted by the opponents (Roubertoux et al., 1998).

Pairing Region of the Y Chromosome

The Carlier and Roubertoux group started by observing spontaneous aggression in NZB/BlNJ (N) males. These males begin to fight other males at the age of 40 d and with ferocious bouts (up to 30 bites/min) by the age of 2 mo. The aggressiveness of N males is always high, although its expression may be reduced slightly, depending on the strain of the opponent (François et al., 1990). N males were compared to CBA/H (H) males, which did not initiate attack behavior under the conditions briefly described above. The males were not isolated, but reared with a female mouse, generally a littermate, and had never experienced the presence of another male; they were weaned at 30 d, before aggressive behavior patterns emerged, and the father was removed before delivery. Aggression was measured in a neutral area, with an A/J male as the standard opponent. One difference observed between reciprocal crosses was that N×H F_1s attacked less than H×N F_1s (Roubertoux and Carlier, 1988), which could suggest involvement of the Y chromosome, but to prove the involvement of Y^{NPR}, backcrosses had to be conducted over 20 generations to successfully obtain the cross transfer of the Y^{NPR} and to eliminate passenger Y^{NPR} genes from the donor; this did not modify the initiation of attack behavior in the congenic strains. These contradictory results were analyzed, and it was concluded that the determining factor was not the Y^{NPR} but the Y^{PR}. The reciprocal F_1s had received the whole Y chromosome, whereas the congenics carried only the Y^{PR} region. Other nonstandard crosses confirmed the result (Roubertoux et al., 1994a). Van Oortmeersen and Sluyter (1994) obtained similar conclusions with selected lines.

Other evidence supports the hypothesis of Y^{PR} involvement in the initiation of attack behavior using the nonisolated male procedure with a standard opponent. A wide genome scan with an F_2 derived from N and C57BL/6by (nonaggressive) strains and with advanced intercrossed lines confirmed a link with SSLP located on the Y^{PR} (Roubertoux et al., unpublished). Observations of initiation of attack behavior in 11 strains of mice, measured under condition 1, showed a strong correlation with an enzymatic marker located on the Y^{PR} (Le Roy et al., 1999).

Y^{PR} involvement seems to be only found in experimental conditions with a nonisolated male and an A/J opponent tested in a neutral area. No link with the SSLP located on the Y^{PR} was found with a wide genome scan performed with another N.B6 F_2 population using the resident intruder procedure (Roubertoux et al., unpublished). No correlation with an enzymatic marker from the Y^{PR} appeared across the 11 strains of mice when aggression was measured with either isolated males, the resident intruder strategy, or with an opponent the same strain.

The next question is to identify the genes located on the Y^{PR} and Y^{NPR}, which may be involved in aggression and to decipher the molecular pathways from genes to behavior patterns.

CANDIDATE GENES ON THE Y CHROMOSOME OF THE MOUSE

Sound knowledge acquired on the physiological mechanisms underlying attack behavior provides the opportunity for assessing candidate genes linked to the two parts of the Y chromosome. Recent studies have focused on some of the few functional genes that may contribute to aggression in mammals and that are carried by the Y^{NPR}. These may be factors initiating attack behavior as observed under the specific conditions described above, i.e., isolated test males placed in a neutral area with an opponent from the same strain.

Y^{NPR} Genes

The *Sry* gene (the sex-determining region of the Y) (Koopman et al., 1991) plays a key role in differentiating the primordial gonads into the testes. A simple way to relate Y^{NPR} to aggression has been to assume that *Sry*, one of the four genes known for testis differentiation signaling in mice (MOUSE GENOME DATABASE [MGD] 1999), is also implicated in reproductive organ weights and testosterone concentration. The hypothesis speculated that testosterone concentration was a consequence of the size of the testes and that it could be used to predict initiation of attack behavior. Very little evidence can be found to support the different stages in this argument. The link between testosterone concentration and aggression appears to be quite complex (Carlier et al., 1990). No testosterone receptors are present in the brain, and this main androgen needs to be transformed before it acts on the central nervous system. It seems that testosterone needs to be present for attack behavior to be initiated, but no correlation has been observed between different testosterone concentrations and the frequency of initiating attack behavior. Animals with absence of testosterone sensitivity caused by a mutation do not initiate attack behavior. Any absence of aggression after castration and any absence of testosterone can be restored by testosterone injection. In intact animals, no significant correlation could be established between testosterone concentrations and the frequency of attack behavior or the intensity of the aggressive bouts (Carlier et al., 1990). This would suggest that testosterone is essential for aggression, but that the concentration does not regulate either the frequency or intensity of the behavior in intact animals.

The association between weight of the testes and Y^{NPAR} is a subject of controversy. Using a quartet of congenic strains for the Y^{NPAR}, we demonstrated that this region made a small contribution to the differences between parental strains. In fact, the QTL contributing to most of the phenotypic variance were not on the Y^{NPAR}, but on the autosomes (Le Roy et al., 2001). Other hypotheses were suggested on the possible implication of the Y^{NPAR} genes in aggression. *Sry* is also expressed in the brain (Lahr et al., 1995). Maxson (1997) pointed out that the corresponding SRY protein interacts with many genes, including non-Y genes. SRY acts as an activating factor initiating cascades that ultimately contribute to regulating genes coding for neuropeptides. It has been shown that SRY can also compete with transcription factors coded by *Sox 1, 2,* and *3*, although, as has recently been suggested, the molecular mechanisms involved in this competing need to be reanalyzed. *Sox 1, 2,* and *3* are expressed in the brain, these competing mechanisms should therefore be explored as a potential source of information on the physiological pathways between Y^{NPR} genes and aggression.

As suggested by Maxson (1997), "a gene of the Y chromosome may code for a different isoform of a protein than that coded for by its homologue elsewhere in the genome." By tracking the different effects of these gene products light could be shed on the contribution of the genes of the Y^{NPR} on aggression. *Smc* and *Rps4* generate different isoforms of proteins depending on their location on X and Y chromosome.

Candidate Genes on the Y^{PR} in Mice

An easier approach would be to investigate candidate genes on the Y^{PR} as the region encompasses very few genes, and the (*Sts*) gene is the only functional one. Steroid sulfatase enzyme STS (E.C.3.1.6.2.) is an ubiquitous arylsulfatase, which binds to reticulum membranes. It is present in mammals where it catalyzes hydrolysis of several 3β-hydroxysteroid sulfates and, specifically, the steroid sulfates that are precursors of estrogens. Yen et al. (1987) and Salido et al. (1996) have cloned and sequenced human and mouse *Sts* genes, respectively.

As seen above, we suggested an implication of the Y^{PR} on the basis of results obtained with different crosses from the N and H parental strains (Roubertoux et al., 1994a, 1994b). We eliminated the hypothesis of an implication of the *Sts* gene in aggression, as no correlation appeared between initiation of aggression behavior and STS enzymatic activity. This conclusion proved to be incorrect as protein concentration is not assessed by measuring the enzymatic activity. To measure STS concentrations, we produced monospecific polyclonal antibodies against murine STS. We checked for STS expression in the brain and observed that the most aggressive mice had the highest brain STS concentrations. Furthermore, STS concentration increased at sexual maturation (Mortaud et al., 1996). We generalized the result with 11 inbred strains of mice. A high correlation ($0.89, p < 0.01$) was found between the STS concentration (expressed as pmol STS/mg protein) and initiation of attack behavior as measured in a standard opponent test in a neutral area with nonisolated males. Measurement of STS showed higher

STS levels in aggressive mice (Le Roy et al., 2000). As a correlation between strains approximates a genetic correlation, we could then assume a causal genetic link between STS concentration and aggression as measured under these conditions. Nicolas et al. (2001) confirmed this result, observing that the inhibition of STS depleted aggression in mice. As STS was expressed in the brain, and as STS and aggression were correlated, the question was to find the physiological pathways followed by STS to induce aggressive behavior.

Relationships between steroids and brain function had to be revisited after Beaulieu (1997) discovered a steroid biosynthesis pathway in the brain. This pathway is used to synthesize neurosteroids from cholesterol (Akwa et al., 1991). 3β-Hydroxy-5pregnen-20-one (pregnenolone) (Δ5P) and 3β-hydroxy-5androsten-17-one (dehydro-epiandrosterone) (DHEA) steroids are present in the central nervous system in mammals as nonconjugated steroids, sulfates, and fat ester acids (Corpéchot et al., 1981). The study by Nicolas et al. (2001) showed that aggression was depleted by an inhibitor of STS and restored by administering DHEA sulfate (DHEA-S), confirming the implication of STS metabolic pathways in aggression.

The recent discovery of free or sulfated steroids, with opposite effects on neurotransmitters, could help gain an understanding of the relationships between Sts gene and aggressive behavior. The gene carried by Y^{PR} interacted with non-Y^{PR} genes as indicated by the genetic analyses showing the implication of this region in aggression (Roubertoux et al., 1994a), suggesting that the contribution of STS to the initiation of attack behavior may be the consequence of cascades rather than a direct effect of STS.

Although the metabolic mechanisms linked to STS activity occur in glial cells, the products of this activity are directed toward the neurons. Steroid biosynthesis, where STS plays a key role, regulates postsynaptic action, either by facilitation or inhibition (Beaulieu, 1997; Mortaud et al., 1996). The alternate action varies, depending on whether the steroids are free or sulfated. Studies of structure–activity demonstrated that several progesterone metabolites, including the 3Δ-hydroxy-steroids, bind with the γ-amino butyric acid (GABA)$_A$/benzodiazepine receptor complex sites via cellular mechanisms, which have been partially identified (Hauser et al., 1995). With the opposite action to free steroids, pregnelone sulfate (Δ5P-S), DHEA-S, and several 3β-hydroxypregnane act as antagonists of the GABA$_A$ receptor. Besides, Δ5P-S also has a facilitating effect on the N-methyl-D-aspartate (NMDA) receptor (Mathis et al., 1994). Both GABA and NMDA have been implicated in aggression as shown, respectively, by Majewska (1992), using pharmacological strategies, and by Chen et al. (1994) via gene targeting. Other transmitter systems involved in aggression are modulated by steroid metabolites. In a review, Mortaud and Degrelle (1996) reported serotonin (5-HT) receptor inhibition by estradiol and opiate inhibition by progesterone. STS is involved in estradiol metabolism (via DHEA-S) and in progesterone metabolism (via Δ5P-S). Once again, the effect of 5-HT (5-HT1$_B$ receptor) (Saudou et al., 1994) and opioids (preproenkephalines) (König et al., 1996) on aggression has been shown using gene targeting methods.

The role of Sts in aggression may solve several paradoxes in aggression research.

Assuming that the Y chromosome is involved in aggression, different genetic mechanisms could be suggested in a bid to explain the fact that female mice can display attack behavior, but the mechanisms seem to be common to males and females. Cairns selected lines to initiate attack behavior in females, thereby precluding any Y chromosome effect.

But males from the same lines displayed aggressive behavior, also precluding any possible effect of the Y chromosome (Hood and Cairns, 1988). This reasoning is only valid when narrowing the Y chromosome down to Y^{NPR}. As males and females share the Y^{PR}, the *Sts* gene may be one of the common mechanisms implicated in aggression.

Mice with an androgen system deficiency did not attack, although the concentration of testosterone was not proportional to the tendency of intact male mice to initiate attacks, as seen before. This suggests that androgens are necessary, but not sufficient, to induce aggression and that neurosteroids may be considered as potentializing the effects of testosterone, *inter alia*.

CAVEAT AND PROSPECTS

Owing to the difficulty in investigating the Y chromosome in humans, we had to conduct animal studies. In the murine species, more than 15 genes or chromosomal regions are known to be linked to aggression. While some putative knock-outs have been shown to be artifacts (Le Roy et al., 2000), most results still stand, offering eloquent illustrations of the complexity of aggression correlates. Aggressive behavior in mice is strongly determined by environmental conditions, including prenatal (Van Oortmeersen et al., 1994) and postnatal environments (Carlier et al., 1991). No correlation can be established between measurements of aggression made under different conditions, which makes it difficult to draw general conclusions on genetic correlates within the mouse species, and therefore, any prospect of developing mouse models of human aggression may be seen as Utopian.

In defense of mouse models, we could present lyrical embryological, neuronal, developmental, and behavioral similarities between mice and humans, but we are not naïve. A male mouse attacking another male in its cage is not a model for aggression in rough urban districts or for petty crime, but animals can be used to investigate the biological mechanisms at work in aggression as the biochemical mechanisms are highly conserved in the two species. The laws of biochemistry and the physiological outcomes are similar in all the different species. An enzymatic defect has the same biochemical and neuronal consequences in a mouse as in a human. Interactions between hormones, the immune system, and neurotransmitters follow invariant structures in vertebrates, showing similar tendencies in behavior patterns. There are murine models for neurobiological pathways underlying behavior patterns, which makes rodents useful for understanding pathophysiological processes. Geneticists are aware that genomic organization is conserved across the mammalian species. Comparative maps for mammalian genomes (DeBry and Seldin, 1996) and particularly for the mouse and human (Nadeau et al., 1991) have been seen as a means of mapping new human genes based on existing homologies between the species. The discovery of a linkage in the mouse would indicate a candidate region for an homologous human gene. This set of considerations suggests that results obtained with aggression in mouse genetics could have an impact on our understanding of neurogenetics and aggression in our species.

In the near future, the investigation of genetic correlates in a wide set of mice, including exotic strains, should prove to be useful in detecting genes, either linked or directly involved in aggression in mice. Several of these genes share strong homologies with human genes and could provide the grounds for hypotheses on the genetic basis of human aggression. Human Y chromosome genes could be cloned to develop transgenic mice and test the impact of those genes on aggression. Success will no doubt come from a top-down strategy, investigating the neurobiological correlates of human aggression with direct searches of genes with wide genome scans and by developing animal models of these mechanisms.

REFERENCES

Akwa, Y., Young, J., Kabbadj, K., et al. (1991) Neurosteroids: biosynthesis, metabolism and function of pregnenolone and dehydroepiandrostenone in the brain. *J. Steroid Biochem. Mol. Biol.* **40,** 71–81.

Beaulieu, E. E. (1997) Neurosteroids of the nervous system, by the nervous system, for the nervous system. *Recent Prog. Horm. Res.* **52,** 1–32.

Benézech, M. (1975) *Aberration du Chromosome Y en Pathologie Médico-Légale.* Masson, Paris.

Bohman, M., Sigvardsson, S., and Cloninger, C. R. (1981) Maternal inheritance of alcohol abuse. Cross-fostering analysis of adopted women. *Arch. Gen. Psychiatry* **38,** 965–969.

Burgoyne, P. S. (1982) Genetic homology and crossing over in the X and Y chromosomes of mammals. *Hum. Genet.* **61,** 85–90.

Carey, G. (1994) Genetics and violence, in *Understanding and Preventing Violence: Biobehavioral Influences, Vol. 2.* (Reiss, A. J., Miczek K. A., and Roth J. A., eds.), National Academy, Washington, DC, pp. 1–58.

Carlier, M., Nosten-Bertrand, M., and Michard-Vanhée, C. (1992) Separating genetic effects from maternal environmental effects, in *Techniques for the Genetic Analysis of Brain and Behavior.* (Goldowitz, D., Wahlsten, D., and Wimer, R. E., eds.), Elsevier Science, Amsterdam, pp. 111–126.

Carlier, M. and Roubertoux, P. L. (1986) Differences between CBA/H and NZB mice on intermale aggression. I. Comparison between parental strains and reciprocal F1s, in *Genetic Approaches to Behavioral Phenotypes.* (Medioni, J. and Vaysse, G., eds.), Privat, Toulouse, France, pp. 47–57.

Carlier, M., Roubertoux, P. L., Kottler, M. L., and Degrelle, H. (1990) Y chromosome and aggression in strains of laboratory mice. *Behav. Genet.* **20,** 137–156.

Carlier, M., Roubertoux, P. L., and Pastoret, C. (1991) The Y-chromosome effect on intermale aggression in mice depends on the maternal environment. *Genetics* **129,** 231–236.

Chen, C., Rainnie, D. G., Greene, R. W., and Tonegawa, S. (1994) Abnormal fear responses and aggressive behavior in mutant mice deficient for a calcium-calmodulin kinase-II. *Science* **266,** 291–294.

Cooke, H. J., Brown, W. R. A., and Rappold, G. A. (1985) Hypervariable telomeric sequences from the human sex chromosome are pseudoautosomal. *Nature* **317,** 687–692.

Corpéchot, C., Robel, P., Lachapelle, F., et al. (1981) Characterization and measurement of dehydroepiandrostenone sulfate in the brain. *Proc. Natl. Acad. Sci. USA* **78,** 4704–4707.

Craig, I. W. and Tolly, E. (1986) Steroid sulfatase and the conservation of mammalian X chromosome. *Trends Genet.* **2,** 201–211.

DeBry, R. and Seldin, M. F. (1996) Human/mouse homology relationships. *Genomics* **33,** 337–351.

Denno, D. (1996) Legal implications of genetics and crime research, in *Genetics of Criminal and Antisocial Behaviour.* Ciba Foundation Symposium 194, Chichester, pp. 248–264.

Ellis, E. and Goodfellow, P. N. (1989) The mammalian pseudoautosomal region. *Trends Genet.* **5,** 406–410.

François, M. H., Nosten-Bertrand, M., Roubertoux, P. L., Kottler, M. L., and Degrelle, H. (1990) Opponent strain effect on eliciting attacks in NZB mice: physiological correlates. *Physiol. Behav.* **47,** 1181–1185.

Goldman, D., Lappalainer, J., and Ozaki, N. (1996) Direct analysis of candidate genes in impulsive behaviours, in *Genetics of Criminal and Antisocial Behaviours.* Ciba Foundation Symposium 194, Chichester, pp. 139–154.

Goldsmith, H. H. and Gottesman, I. I. (1995) Heritable variability and variable heritability in developmental psychopathology, in *Frontiers of Developmental Psychopathology.* (Lenzenweger, M. F. and Haugaard, J. J., eds.), Oxford University Press, New York, pp. 1–19.

Götz, M. J., Johnstone, E. C., and Ratcliff, S. G. (1999) Criminality and antisocial behaviour in unselected men with sex chromosome abnormalities. *Psychol. Med.* **29**, 953–962.

Guillot, P. V., Carlier, M., and Roubertoux, P. L. (1995) The Y chromosome effect on intermale aggression in mice depends on the test situation and on the recorded variables. *Behav. Genet.* **25**, 51–59.

Haldane, J. B. S. (1936) A search for incomplete sex-linkage in man. *Ann. Eugen.* **7**, 28–57.

Hauser, C. A., Chesnoy-Marchais, D., Robel, P., and Baulieu, E. E. (1995) Modulation of recombinant alpha 6 beta 2 gamma 2 GABAA receptors by neuroactive steroids. *Eur. J. Pharmacol.* **289**, 249–257.

Hood, K. E. and Cairns, R. B. (1988) A developmental-genetic analysis of aggressive behavior in mice. II. Cross-sex inheritance. *Behav. Genet.* **18**, 605–619.

Islam, A. B. M., Hill, W. G., and Land, R. B. (1976) Ovulation rate of lines of mice selected for testis weight. *Genet. Res.* **27**, 23–32.

Jacobs, P. A., Brunton, M., Melville, M. M., Brittain, R. P., and McClemont, W. F. (1965) Aggressive behavior, mental subnormality and the XYY male. *Nature* **208**, 1351–1352.

Keitges, M., Rivest, M., Siniscalco, M., and Gartler, S. M. (1985) X-linkage of steroid sulphatase in the mouse is evidence for a functional Y-linked allele. *Nature* **315**, 226–227.

Kittles, R. A., Long, J. C., Bergen, A. W., et al. (1999) Cladistic association analysis of Y chromosome effects on alcohol dependence related personality traits. *Proc. Natl. Acad. Sci. USA* **96**, 4204–4209.

Koller, P. C. and Darlington, C. D. (1934) The genetical and mechanical properties of sex chromosomes 1 *Rattus norvegicus. J. Genet.* **29**, 159–173.

König, M., Zimmer, A. M., Steiner, H., et al. (1996) Pain response and aggression in mice deficient in pre-proenkephalin. *Nature* **383**, 535–538.

Koopman, P., Gubbay, J., Vivian, N., Goodfellow, P., and Lovell-Badge, R. (1991) Male development of chromosomally female mice transgenic for Sry. *Nature* **351**, 117–121.

Lahr, G., Maxson, S. C., Mayer, A., Just, W., Pilgrim, C., and Reisert, I. (1995) Transcription of the Y chromosomal gene, Sry, in adult mouse brain. *Mol. Brain Res.* **33**, 179–182.

Le Roy, I., Mortaud, S., Tordjman, S., et al. (1999) Correlation between expression of the steroid sulfatase gene, mapped on the pairing region of the Y-chromosome, and initiation of attack behavior in mice. *Behav. Genet.* **29**, 131–136.

Le Roy, I., Pothion, S., Mortaud, S., et al. (2000) Loss of aggression, after transfer onto a C57BL/6J background, in mice carrying a targeted disruption of the neuronal nitric oxide synthase gene. *Behav. Genet.* **30**, 367–373.

Le Roy, I., Tordjman, S., Degrelle, H., Migliore-Samour, D., and Roubertoux, P L. (2001) Genetic architecture of reproductive organs weights. *Genetics* **54**, 110–121.

Majewska, M. D. (1992) Neurosteroids: endogenous bimodal modulators of the GABA$_A$ receptor. Mechanism of action and physiological significance. *Prog. Neurobiol.* **38**, 379–395.

Mathis, C., Paul, S. M., and Crawley, J. N. (1994) The neurosteroid pregnenolone sulfate blocks NMDA antagonist-induced deficits in a passive avoidance memory task. *Psychopharmacology* **116**, 201–206.

Maxson, S. C. (1996) Searching for candidate genes with effects on an agonistic behavior, offense, in mice. *Behav. Genet.* **26**, 471–476.

Maxson, S. C. (1997) Sex determination in genetic mechanisms for mammalian brain and behavior. *Biomed. Rev.* **7**, 85–90.

Maxson, S. C. (1998) Homologous genes, aggression and animal models. *Dev. Neuropsychol.* **14,** 143–156.

Maxson, S. C., Ginsburg, B., and Trattner, A. (1979) Interaction of Y-chromosomal gene(s) in the development of intermale aggression in mice. *Behav. Genet.* **9,** 219–226.

Michard, Ch. and Carlier, M. (1983) Les conduites d'agression intraspécifiques chez la souris domestique, différences individuelles et analyses génétiques. *Biol. Behav.* **10,** 123–146.

Migeon, B. R., Shapiro, L. J., Norum, R. A., Mohandas, T., Axelman, J., and Dabora, R. L. (1982) Differential expression of steroid sulphatase locus on active and inactive human X chromosome. *Nature* **299,** 838–840.

Mortaud, S. and Degrelle, S. (1996) Steroid control of higher brain function and behavior. *Behav. Genet.* **26,** 367–372.

Mortaud, S., Donzes-Darcel, E., Roubertoux, P. L., and Degrelle, H. (1996) Murine steroid sulfatase gene expression in the brain during post natal development and adulthood. *Neurosci. Lett.* **215,** 145–148.

Nadeau, J. H., Davisson, M. T., Doolittle, D. P., et al. (1991) Comparative map for mice and humans. *Mamm. Genome* **1,** S461–S515.

Nicolas, L. B., Pinoteau, W., Papot, S., Routier, S., Guillaumet, G., and Mortaud, S. (2001) Aggressive behavior induced by the steroid sulfatase inhibitor COUMATE and by DHEAS in CBA/H mice. *Brain Res.* **20,** 216–222.

Petit, Ch., De La Chapelle, A., Levilliers, J., Castillo, S., Noël, B., and Weissenbach, J. (1987) An abnormal terminal X-Y interchange accounts for most but not all cases of human XX maleness. *Cell* **49,** 595–602.

Rappold, G. A. (1993) The pseudoautosomal regions of the human sex chromosomes. *Hum. Genet.* **92,** 15–24.

Roubertoux, P. L. and Carlier, M. (1988) Differences between CBA/H and NZB on intermale aggression II—maternal effects. *Behav. Genet.* **18,** 175–184.

Roubertoux, P. L., Carlier, M., Degrelle, H., Hass Dupertuis, M. C., Phillips, J., and Moutier, R. (1994a) Co-segregation of intermale aggression with the pseudoautosomal region of the Y chromosome in mice. *Genetics* **135,** 225–230.

Roubertoux, P. L., Degrelle, H., Maxson, S. C., Phillips, J., Tordjman, S., and Dupertuis-Haas, M.-C. (1994b) Polymorphism for the alleles of the microsomal steroid sulfatase gene (Sts) in the pseudoautosomal region of the heterosomes of laboratory mice. *C. R. Acad. Sci. Paris* **317,** 523–527.

Roubertoux, P. L., Le Roy, I., Mortaud, S., Perez-Diaz, F., and Tordjman, S. (1999). Measuring aggression in the mouse, in *Handbook of Molecular-Genetic Techniques for Brain and Behavior Research.* (Crusio, W. E. and Gerlai, R., eds.), Amsterdam, Elsevier, pp. 696–709.

Roubertoux, P. L., Mortaud, S., Tordjman, S., Le Roy, I., and Degrelle, H. (1998) Behaviorgenetic analysis and aggression: the mouse as a prototype, in *Advances in Psychological Science: Biological and Cognitive Aspects.* (Sabourin, M., Craik, F., and Robert, M., eds.), Psychology Press, Hove, UK, pp. 3–29.

Rutter, M. (1996) Introduction: concepts of antisocial behavior, of cause, and of genetic influences, in *Genetics of Criminal and Antisocial Behaviour.* Ciba Foundation Symposium 194, Chichester, UK, pp. 1–20.

Salido, E. C., Li X., M., Yen, P., Martin, N., Mohandas, T. K., and Shapiro, L. J. (1996) Cloning and expression of the mouse pseudoautosomal steroid sulfatase gene (Sts). *Nat. Genet.* **13,** 83–86.

Saudou, F., Amara, D. A., Dierich, A., et al. (1994) Enhanced aggressive behavior in mice lacking 5-HT$_{1B}$ receptor. *Science* **265,** 1875–1878.

Selmanoff, M. K., Jumonville, J. E., Maxson, S. C., and Ginsburg, B. E. (1975) Evidence for a Y chromosome contribution to an aggressive phenotype in inbred mice. *Nature* **253,** 529–530.

Selmanoff, M. K., Maxson, S. C., and Ginsburg, B. (1976) Chromosomal determinants of intermale aggressive behavior in inbred mice. *Behav. Genet.* **6,** 53–69.

Simmler, M. C., Rouyer, F., Vergnaud, G., et al. (1985) Pseudoautosomal DNA sequences in the pairing region of the human sex chromosomes. *Nature* **317,** 692–697.

Van Oortmerssen, G. A. and Sluyter, F. (1994) Studies in wild house mice: V. Aggression in selection lines for attack latency and their congenics for the Y chromosome. *Behav. Genet.* **24,** 73–78.

Witkin, H. A., Mednick, S. A., Schulsinger, F., et al. (1976) Criminality in XYY and XXY men. *Science* **193,** 547–555.

Yen, P. H., Allen, E., Marsh, B., et al. (1987) Cloning and expression of steroid sulfatase cDNA and the frequent occurence of deletions in STS deficiency: implications for X-Y interchange. *Cell* **49,** 443–454.

8
Aggression in Psychiatric Disorders

Shari R. Kohn and Gregory M. Asnis

INTRODUCTION

Despite the numerous studies that have investigated the link between psychiatric disorders and aggression, the association remains controversial. The prevailing stereotype throughout the 1980s and early 1990s was that patients with psychiatric disorders were more likely to be aggressive, despite the scientific evidence concluding otherwise. As a result of more recent epidemiological studies and reviews, it is now accepted by the scientific community that individuals with psychiatric disorders do participate in aggressive behaviors more often than those without psychiatric disorders (e.g., Otto, 1992), with the caveat that the majority of patients with psychiatric disorders are not aggressive (Barlow et al., 2000), and aggression is not common to all diagnoses.

Adding to the controversy, studies of aggression in psychiatric patients have been problematic. For example, investigations are often difficult to compare and draw conclusions from, as results depend on the definition of aggression applied, intervals used for assessment, and criteria for measuring violence (McFall et al., 1999), none of which are standardized. In addition, results differ depending upon the disorder being investigated, and many studies to date have involved small sample sizes. Further adding to the controversy, patient advocate groups have campaigned against associating aggression and psychiatric disorders, predicting negative repercussions, including further stigmatization and discrimination. It is therefore important to understand the link between psychiatric disorders and aggression before rendering any conclusions.

EVIDENCE SUPPORTING THE ASSOCIATION BETWEEN AGGRESSION AND PSYCHIATRIC DISORDERS

The association between aggression and psychiatric disorders is well supported, whether investigating emergency room patients, psychiatric inpatients, psychiatric outpatients, or the general population. For instance, about 20% of patients who present in psychiatric emergency rooms have a history of aggressive behavior (Steadman et al., 1994). Aggression is also a significant problem in inpatient psychiatric facilities. In fact, between 8% and 22% of psychiatric inpatients are reported to have been aggressive in the past 2 wk (Craig, 1982; Tardiff and Sweillam, 1980). Furthermore, one of the primary reasons for admission to inpatient facilities is aggression, with estimates of a 25% increase in inpatient hospitalization due to aggressive behavior expected by the year 2010 (Goldsmith et al.,

1993). Patients discharged from psychiatric hospitals have also been found to be significantly more aggressive than community controls (Rabkin, 1979). Similarly, psychiatric outpatients are more aggressive than community residents (Link et al., 1992), and nonclinical community residents with psychiatric diagnoses are also more aggressive than those without such disorders (e.g., Swanson et al., 1990). Although the association between aggression and psychiatric disorders is significant, it is important to point out that the association is relatively small (Swanson et al., 1990), which may be due to the variability among aggression in different disorders or may indicate that other factors (e.g., substance use) are involved. However, given the relatively large number of patients with psychiatric disorders in the U.S., the total number of aggressive acts committed by these individuals is a cause for concern (Torrey, 1994).

AGGRESSION AND SPECIFIC PSYCHIATRIC DISORDERS

The Diagnostic and Statistical Manual of Mental Disorders, fourth edition (DSM-IV) identifies axis I and axis II disorders. Axis I disorders consist of clinical disorders, including psychotic disorders (e.g., schizophrenia), conduct disorder, major depressive disorder, anxiety disorders (e.g., post-traumatic stress disorder and obsessive–compulsive disorder), and cognitive disorders (e.g., dementia). Axis II includes personality disorders (e.g., borderline personality disorder). Studies of aggression in patients with these more common psychiatric disorders have been conducted and will be reviewed in this chapter.

Schizophrenia

Schizophrenia is a chronic psychotic disorder with symptoms that include, but are not limited to, delusions, hallucinations, disorganized speech, and grossly disorganized or catatonic behavior. Prevalence rates of schizophrenia in the community are estimated to range from 0.2% to 2% (APA, 1994). While some studies have associated schizophrenia with an increased risk of aggression (Sheridan et al., 1990), others have concluded that a diagnosis of schizophrenia is a protective factor, actually lowering the risk of aggressive behaviors (Harris and Rice, 1997). Although the results of risk assessment studies are unclear, we do know that aggression among patients with schizophrenia has been identified in community (Brennan et al., 2000) and inpatient settings (Tam et al., 1996). For example, a study of first admission patients with schizophrenia revealed that two-thirds of these patients (approx 91 patients) were aggressive toward others, but their behaviors were not severe and rarely led to legal prosecution (Steinert et al., 1999a). Further support for this association has come from numerous twin studies (e.g., Coid et al., 1993), evaluations of criminal records (Lindqvist and Allebeck, 1990), and a study of homicides (Eronen et al., 1996). It is, however, important to keep in mind that the majority of patients diagnosed with schizophrenia are not violent (Tardiff et al., 1997).

Correlates and Risk Factors

Steinert et al. (1999a) concluded that the incidence of aggression among patients with schizophrenia was partially the result of inadequate treatment, as incidents of aggression declined when patients with schizophrenia were treated with neuroleptics. Similarly, Barlow et al. (2000) found that during the first 46 h of hospitalization patients are most susceptible to aggressive behavior, but their risk for aggression decreases quickly after that time, most likely due to their quick response to medication and ward structure. Medication noncompliance, which is exacerbated by alcohol abuse, has been found to be related to aggressive behaviors (Swartz et al., 1998). In their prospective study, Arango et al. (1999) found involuntary admission, aggression the week prior to admission, and longer hospital stays to be predictive of aggression in patients with schizophrenia.

One of the most robust findings regarding factors related to aggression in patients with schizophrenia is that of alcohol use. Patients with schizophrenia and co-morbid substance use disorders, especially alcohol, have more than a two-fold risk of committing violent crimes than patients with schizophrenia alone (e.g., Lindqvist and Allebeck, 1990; Swanson et al., 1990). In addition, men with schizophrenia who used alcohol have been found to be more than 25 times more likely to commit violent crimes than those without a psychiatric diagnosis (Rasanen et al., 1998). Interestingly, men with schizophrenia who did not use alcohol were only 3.6 times more likely to be aggressive than men without a psychiatric diagnosis.

To date, the role of symptomatology remains unclear. Yesavage (1983) and Tardiff and Sewillam (1982) found positive symptomatology to be related to aggression, especially persecutory delusions (Nestor et al., 1995), while others have found hostility and anxiety, but not positive symptomatology to be related to aggressive behavior (Palmstierna et al., 1989; Roy et al., 1987). Addressing methodological limitations found in previous studies, Cheung et al. (1997) found aggression in patients with schizophrenia to be related to greater overall psychopathology, and both positive and negative symptoms.

Investigating neurobiological factors, Lachman et al. (1998) concluded that low activity catechol O-methyltransferase genotype is associated with aggression, which replicated earlier findings (cf. Strous et al., 1997). Other studies, on the other hand, failed to find support for neurobiological factors. For instance, given the well-documented association between serotonin (5-HT), a neurotransmitter that has been associated with aggressive and impulsive behavior (Coccaro, 1995; Coccaro et al., 1989), Nolan et al. (2000) investigated polymorphisms that affect 5-HT transmission, but failed to find an association between either the 5-HT transporter (5-HTT) or monoamine oxidase A (MAOA) (an important metabolizer of 5-HT) polymorphisms and aggression in males or females with schizophrenia. Investigating the role of serum cholesterol implicated in the role of aggressive behaviors by numerous epidemiological, animal, and clinical studies, Steinert and colleagues (1999) failed to find support for the role of serum cholesterol in aggression and psychiatric disorders, but recommend that a conclusion not be drawn based on their results, as their sample size was relatively small.

Major Depressive Disorder (MDD)

MDD, estimated to affect 10–25% of women and 5–12% of men in the community (APA, 1994), is characterized by depressed mood, anhedonia, difficulty sleeping, loss of appetite, energy, and motivation. The association between MDD and aggression has been well supported by studies investigating males, females, adolescents, and adults, in both community and clinical populations. For instance, in a study of 42 women and 23 men reporting symptoms of depression, recruited from newspaper advertisements, Bjork et al. (1997) found that for women self-report scores of depressive symptomatology correlated strongly with an objective measure of aggression, namely the Point Subtraction Aggression Paradigm (PSAP). The PSAP measures a person's aggressive response to the periodic loss of a monetary reinforcer attributed to the other participant's behavior. These authors further investigated the association and concluded that the negative affect symptoms, as opposed to somatic symptoms, accounted for the significant association. In another investigation, Feldbau-Kohn et al. (1998) found that depressive symptomatology predicted aggression in treatment-seeking partner aggressive men.

Aggressive behavior has been increasing among adolescents. In 1996, 25% of violent crimes were committed by juvenile offenders (Federal Interagency Forum on Child and Family Statistics, 1998). The risk for aggressive behavior in adolescents appears to be increased by the presence of psychopathology, especially depression (Kazdin et al., 1983). In addition, adolescents with depression may be at risk for future aggression (Schubiner et al., 1993).

Correlates and Risk Factors

One factor implicated in the association between depression and aggression is 5-HT, which has been associated with depression (Meltzer and Lowy, 1987) and aggression (Coccaro, 1995; Coccaro et al., 1989). van Praag (1998) hypothesized that in 5-HT-related depressions, dysregulation of aggression and anxiety is the primary symptom, and mood lowering is a "derivative phenomenon." Support for this hypothesis can be found in some patients whose depressive episodes are precipitated by increased aggressive and anxious behaviors (van Praag, 1998), as well as in other patients whose aggression and anxiety are the first symptoms to respond to antidepressant medications (Katz et al., 1991).

Serum cholesterol has also been implicated in the association between depression and aggression. The cholesterol/5-HT hypothesis of aggression states that low dietary cholesterol intake leads to depressed serotonergic activity, which is associated with both depression and aggression (Kaplan et al., 1997). For example, Hillbrand et al. (2000) looked at 25 patients committed to an inpatient unit after being found guilty by reasons of insanity. In these violent patients, low levels of serum cholesterol were found to be related to negative mood states, consistent with the cholesterol/5-HT hypothesis, and with other studies finding an excess of negative mood states in low cholesterol individuals with depression (Schwartz and Ketterer, 1997) and suicidal ideation (Hillbrand and Spitz, 1999).

The link between depression and aggression may also be the result of common traits, such as irritability (e.g., Knox et al., 2000). For instance, the frustration–aggression hypothesis (Berkowitz, 1989) states that negative affect can induce anger, increasing the likelihood of aggression. Cognitive distortions, not atypical for patients with depression, may also play a role in the association. For example, a patient with depression may be more likely to negatively misinterpret a situation or to focus on the negative aspects of a situation, increasing the likelihood of an aggressive response. Similarly, depressed boys have been found to be less able to control their anger, and the more hopeless they are, the more likely they are to express their anger aggressively (Kashani et al., 1995, 1997). Comorbidity with other psychiatric disorders may also be involved in the association. In one study of depressed adolescents (58 females, 16 males) referred for treatment, Knox et al. (2000) found levels of aggression to be associated with higher comorbidity for adolescents with comorbid oppositional defiant disorder or conduct disorder. Although Knox et al. (2000) did not find gender differences, Tardiff et al. (1997) reported that the increase in aggression among adolescents may be occurring among females. Finally, exposure to high levels of violence has been implicated as a cause of both depression and aggression and could possibly explain the link for some patients (Gorman-Smith and Tolan, 1998).

Bipolar Disorder

Bipolar disorder is a mood disorder characterized by the occurrence of one or more manic or mixed (criteria is met for both a major depressive episode and a manic episode nearly everyday for 1 wk) episodes (bipolar I) or one or more hypomanic and depressive episodes (bipolar II). Prevalence rates have been estimated to range between 0.4% and 1.6% for bipolar I disorder, and 0.5% for bipolar II (APA, 1994).

Although only a few studies have investigated aggression in patients with bipolar disorder, aggression is a well-recognized aspect of this disorder, especially for patients who experience mixed episodes. For instance, patients with mixed episodes have significantly higher hostility scores than patients with other psychiatric disorders (Swann et al., 1993) and have been found to display higher rates of aggression than patients with schizophrenia (Barlow et al., 2000). In the Zurich longitudinal study, Wicki and Angst (1991) found that children with hypomanic episodes had more disciplinary difficulties at school, and more thefts than other children. High comorbidity between pediatric bipolar disorder and disruptive behavior disorders (e.g., oppositional defiant disorder and conduct disorder) has also been reported (e.g., Kovaks and Pollack, 1995; Lewinsohn et al., 1995). In their review, Spencer et al. (2001) reported that children with co-morbid bipolar disorder and conduct disorder participated in attacking and threatening behavior toward family members, teachers, adults, and other children and concluded that, although these behaviors were consistent with a diagnosis of conduct disorder, they may also be due to the behavioral disinhibition that characterizes bipolar disorder.

Correlates and Risk Factors

Higher levels of catecholaminergic activity and hypothalamic–pituitary–adrenocortical axis function have been implicated in the association between bipolar disorder and aggression (Swann et al., 1994). It appears that patients in mixed states have more severe hyperarousal than patients with other psychiatric disorders, which could result in aggression, impulsivity, and anxiety (Swann, 1999). The aggression does not appear to be due to serotonergic dysregulation, as bipolar patients in mixed states do not differ from other psychiatric patients with respect to serotonergic functioning. Therefore, treatments that reduce arousal by increasing γ-amino butyric acid (GABA) or that decrease excitatory amino acids or nonadernergic transmission, may be more effective than serotonergic treatments, often efficacious with other aggressive behaviors (Swann, 1999).

Post-traumatic Stress Disorder (PTSD)

PTSD develops following exposure to extreme traumatic stressors, involving a personal experience of actual or threatened injury or death, or threat to one's personal integrity, or witnessing events that cause death, injury, or threat to the personal integrity of another person, or learning about actual or threatened unexpected or violent death, or injury of a family member or close associate (APA, 1994). Lifetime rates for community samples are reported to range from 8% to 12% (Davidson et al., 1991; Kessler et al.,

1995). The association between PTSD and aggression is well supported. For example, Beckham et al. (1997) found that 57% of Vietnam veterans participating in their study reported high rates of interpersonal aggression during the past year, significantly higher than their control group. Begic and Jokic-Begic (2001) also reported significantly more aggressive acts committed by veterans with PTSD than veterans without PTSD. However, there is a high comorbidity between PTSD and alcohol abuse (Breslau, 2001), which is often associated with interpersonal aggression (e.g., Murdoch et al., 1990). Therefore, it can be difficult to tease apart aggression resulting from PSTD and aggression resulting from substance abuse. For example, in the National Vietnam Veterans Readjustment Study (NVVRS), a community-based study, Kulka et al. (1990) found that Vietnam veterans with PTSD reported an average of 13 acts of aggression in the past year, compared to three acts reported by Vietnam veterans without PTSD. However, Vietnam veterans with PTSD also had more substance abuse than non-PTSD Vietnam veterans. Similarly, in another study, out of 20 patients reporting physical aggression, 60% reported being under the influence of alcohol at the time (Begic and Jokic-Begic, 2001).

Another factor making it difficult to understand the association between aggression and PTSD is combat experience, which is also associated with aggression. For instance, while Kulka and colleagues (1990) found Vietnam veterans with PTSD to be more aggressive than Vietnam veterans without PTSD, Vietnam veterans with PTSD also had more combat experience.

Addressing methodological limitations of the NVVRS study, McFall et al. (1999) investigated psychiatric inpatients with PTSD and found that these patients reported more interpersonal aggression prior to hospitalization and were classified as highly violent more often than psychiatric inpatients without PTSD. McFall et al. (1999) compared their data to the findings from the NVVRS study and found that inpatients with PTSD were more aggressive and were more likely to fall into the highly violent category than patients with PTSD in the community. Unlike the NVVRS study, McFall and colleagues did control for substance abuse and combat exposure, and still found PTSD to significantly contribute to the aggressive behavior among inpatients. In fact, PTSD contributed more to the prediction of aggression than did substance abuse. These findings are consistent with previous studies (e.g., Swanson et al., 1990), which suggest that mental illness increases the risk of aggressive behavior and that substance use appears to amplify this risk.

Correlates and Risk Factors

Beckham and colleagues (1997) investigated the role of psychological and demographic variables in the aggressive behavior of Vietnam veterans with PTSD and found PTSD symptomatology and lower socioeconomic status to be significantly related to the level of interpersonal aggression reported by subjects. In addition, although 45% of subjects endorsed having experienced childhood physical abuse, such abuse was not related to interpersonal aggression in these subjects. However, early violence has been shown to increase the likelihood of PTSD resulting from later traumas (e.g., Bremner et al., 1993) and, therefore, may be indirectly related to aggression in patients with PTSD.

Silva et al. (2000) developed a typology of psychopathological causes of PTSD-related aggression based on a number of case studies. These authors hypothesized that flashbacks, which are a symptom of PTSD, were related to aggression. According to these authors, during the flashback there can be a loss of reality that can lead to aggression, as a result of the associated anxiety, anger, and impairment in reality testing associated with the phenomenon. Furthermore, the authors conceptualize PTSD-related aggression to be a reaction to a perceived external threat and, therefore, also hypothesized that mood and affect lability, often common to patients with PTSD, specifically the intense feelings of anger, were associated with interpersonal aggression. Finally, the authors also contend that "combat addiction violence," defined as a desire to "reexperience thoughts, feelings, and actions related to previous combat experiences," may also contribute to PTSD-related aggression. For a more detailed description of this typology, see Silva et al. (2000).

Obsessive-Compulsive Disorder (OCD)

OCD is characterized by recurrent obsessions or compulsions, recognized to be unreasonable by the patient, that are time-consuming and/or cause marked distress or significant impairment. Lifetime prevalence rates of OCD are estimated to be 2.5% (APA, 1994). Aggression is not usually associated with OCD, but it is associated with OCD subtypes and disorders that are frequently comorbid with OCD. For example, according to Hollander (1999), patients who are hoarders, a subtype of OCD, frequently react with aggression when hoarded objects are removed. OCD spectrum disorders, including autism, often involve impulsivity and aggression (Hollander, 1999).

Correlates and Risk Factors

As with other psychiatric disorders, there appears to be evidence for the role of 5-HT in OCD (Greist et al., 1995; Zohar and Insel, 1987). As aggression does not appear to be higher in patients with OCD when compared to healthy controls and is lower when compared to patients diagnosed with a depressive disorder, the serotonergic presynaptic abnormalities found in patients with OCD may be related to dimensions other than aggression (Marazziti et al., 2001). These authors suggest that that serotonin may play a permissive rather than central role in aggression in OCD.

Conduct Disorder (CD)

CD occurs before age 18 and includes a pattern of behavior in which the rights of others or societal norms or rules are violated. Prevalence rates range from 2% to 10% (APA, 1994). CD, almost by definition, is associated with aggression. In fact, many children diagnosed with CD are often hospitalized as a result of their aggressive behavior (Day et al., 1998). Day et al. (1998) investigated the records of 99 adolescents (43 girls, 56 boys) hospitalized for at least 30 d, and identified 278 incidents of physical aggression during the data collection period, 80% of which were committed by boys. For these aggressive boys, CD was the only diagnosis predictive of aggressive acts.

Correlates and Risk Factors

Similar to previous findings of aggression and psychiatric disorders, 5-HT dysregulation is also found in aggressive children with CD. Kruesi et al. (1990) found whole blood 5-HT to be lower in children with disruptive behavior disorders, including CD, and also predicted aggressive behavior over a 2-yr follow-up period (Krusi et al., 1992). Interestingly, Unis et al. (1997) found an increase in 5-HT levels in adolescents with childhood-onset CD, as compared to adolescents with adolescent-onset CD. Since childhood-onset CD has been found to be a risk factor for a later diagnosis of antisocial personality disorder (Robbins, 1966) and higher levels of aggression (Loeber, 1982), it appears that serotonergic dysregulation may play a larger role in childhood-onset CD, higher levels of aggression, and future aggression.

In addition to neurobiological explanations, cognitive explanations for the association between CD and aggression have also been offered. For instance, children with CD have deficient problem-solving skills, mostly related to their appraisal and interpretation of social interactions. For example, aggressive children are more likely to interpret ambiguous acts of others as intentionally hostile (Dodge and Crick, 1990), thus increasing their likelihood of responding inappropriately and often aggressively (Dodge, 1985). Therefore, interventions often include social skills training and teaching these children to accurately interpret the statements and behaviors of others.

Demographic variables are also associated with children with CD, including having parents who are martially discordant, involved in criminality, use inconsistent discipline, and are punitive. According to Loeber and Dishion (1983) the strongest predictors of antisocial and criminal behavior in children with CD are harsh and inconsistent discipline and poor supervision. Alcoholism is common in families of children with CD, as is multiple separations from caregivers (Reid and Patterson, 1989; Marquiz, 1992; Schlotte, 1992). Although the presence of aggressive behaviors often characterizes children with disruptive behavior disorders, children who demonstrate a pattern of unremitting physical fighting are at the greatest risk for adult antisocial and criminal behavior.

Dementia

Disorders falling under the rubric of dementia involve the development of multiple cognitive deficits due to a general medical condition, the persisting effects of a substance, or to multiple etiologies. The prevalence of dementia is estimated to range from 2% to 4% for the population over age 65 yr old (APA, 1994). Dementia has been associated with aggression in a number of investigations. For instance, of the 541 community-residing patients with dementia investigated by Lyketsos and colleagues, 79 were aggressive during the 2-wk study period (Lyketsos et al., 1999). This 15% prevalence rate was consistent with most clinical studies (15–20%) (e.g., Ryden, 1988) and a little less than the 1-yr prevalence rate reported in a population-based investigation (18%) (Burns et al., 1990). Gormley and colleagues (1998) investigated the relation between Alzheimer's disease (a subtype of dementia) and aggression and found 31 out of 70 patients to be aggressive in the 3-d period prior to assessment, which is a high rate given the short evaluation period. Although there appears to be evidence for an association between dementia and aggression, it is possible that this association may be an artifact of a third variable, specifically depression, which is often associated with dementia (Lyketsos et al., 1997) and aggression. In one particular study, after controlling for depression, aggression was no longer associated with dementia (Lyketsos et al., 1999), leading the authors to conclude that the relation between aggression and dementia was actually a function of the association between depression and aggression. Similarly, Menon et al. (2001) concluded from their study of 1101 newly admitted nursing home residents with dementia, that aggression by such patients was related to depression. In contrast, Rapoport et al. (2001) found aggression to be related to dementia even after controlling for depression, apathy, and impairment in activities of daily living. To date, the association remains unclear.

Correlates and Risk Factors

Risk factors and correlates of aggression in patients with dementia include moderate to severe depression, male gender, greater impairment in activities of daily living, greater severity of dementia, hallucinations and delusions, and sleep disturbance. Similar to other psychiatric disorders, 5-HT dysfunction has been implicated in the association between dementia and aggression. As discussed, 5-HT dysfunction is associated with depression (Richelson et al., 1991) and aggression (Coccaro et al., 1989). As it has also been found to be related to Alzehimer's disease, it is hypothesized that 5-HT dysfunction may be responsible for the association between dementia and aggression via depression (Menon et al., 2001). Partial support for this hypothesis comes from an investigation that found an association between 5-HT dysfunction and aggression in patients with Alzehimer's disease (Procter et al., 1992). Therefore, it appears that some patients with dementia are aggressive, and 5-HT dysfunction may play a significant role in their aggressive behaviors. Further study will be necessary to determine whether or not depression is necessary for the manifestation of these aggressive behaviors.

Borderline Personality Disorder (BPD)

BPD involves a pervasive pattern of instability of interpersonal relationships, self-image, and affect, as well as marked impulsivity, which is present in a variety of contexts. Prevalence rates of BPD are estimated to be about 2% for the general population, 10% for psychiatric outpatients, and 20% for psychiatric inpatients (APA, 1994). Impulsive aggression is often found among patients with BPD. For example, Dougherty and colleagues (1999) compared 14 hospitalized women diagnosed with BPD to 17 community controls on the PSAP (see earlier description), and found women with BPD to be three times more likely to exhibit an aggressive response than women in the control group. Men who abuse their wives have also been found to have higher rates of a diagnosis similar to BPD, referred to as borderline personality organizational disorder, than men who do not abuse their wives (Eronen et al., 1996). However, as BPD is highly co-morbid with other axis I and axis II disorders (Hudziak et al., 1997), to determine if BPD alone was responsible for the aggression, Dougherty and colleagues (1999) controlled for depressive symptomatology and no longer found an association between BPD and aggression. In contrast, Soloff et al. (2000) found that patients with BPD and patients with BPD and co-morbid depression had higher scores on measures of impulsive aggression than patients with a diagnosis of depression alone, with no significant differences found between those with BPD and those with co-morbid BPD and depression. Therefore, it is still unclear whether depression is responsible for the association between BPD and aggression, or whether BPD exacerbates, or is exacerbated by the association between depression and aggression, or is independently responsible.

Correlates and Risk Factors

Similar to other psychiatric disorders, dysfunctional serotonergic activity has been found to be inversely related to aggressive and suicidal behavior in patients with BPD (Goodman and New, 2000). Therefore, treatment studies of aggressive patients with personality disorders have utilized selective serotonin reuptake inhibitors (SSRIs). For example, Coccaro and Kavoussi (1997) treated 40 patients with personality disorders (13 patients met criteria for BPD) with fluoxetine, an SSRI, and found positive results for the majority of subjects. Studies such as this one, as well as other SSRI trials, including citalopram (Kallionemi and Syvalahti, 1993) and sertraline (Kavoussi et al., 1994), have also supported the role of 5-HT in aggressive behavior in patients with personality disorders. Clozapine, an atypical antipsychotic, has also been shown to be beneficial in decreasing aggression in patients with BPD and comorbid psychosis. Although clozapine has antagonist effects at the postsynaptic 5-HT receptors, it does affect 5-HT receptor subtypes, including autoreceptors (Chengappa et al., 1999). Therefore, clozapine's mechanism of action, as it relates to aggression in patients with BPD, may be related to dysfunctional serotonergic activity, similar to the study described earlier (Coccaro and Kavoussi, 1997).

Serum cholesterol is also related to BPD. In their study of 42 patients with personality disorders, New et al. (1999) found that patients with BPD had significantly lower cholesterol levels than non-BPD controls. However, serum cholesterol levels were not found to be inversely related to aggression levels, possibly due to the relatively small sample size.

Impulsive aggression, characteristic of patients with BPD, often leads to clinically important negative outcomes including psychiatric hospitalization, motor vehicle accidents (Hollander, 1999), suicide (9% of patients with BPD die by suicide) (Stone, 1989), and forensic involvements. Therefore, future studies investigating the link between BPD and aggression are extremely important.

CONCLUSIONS

Although continued efforts are needed to understand the correlates and risk factors, the association between aggression and psychiatric disorders is well supported. While aggression is not common to all psychiatric disorders, it does occur more in some psychiatric diagnoses than others and in patients who are substance abusers (Lindquist and Allebeck, 1990), noncompliant with treatment (Tardiff et al., 1997), have a history of violent behavior (Klassen and O'Connor, 1990), comorbid personality disorders (Caton et al., 1994), and/or co-morbid psychopathy (Nolan et al., 1999). Furthermore, although results are mixed, sociodemographic variables do not appear to be associated with aggression in psychiatric patients, with the exception of low socioeconomic status and younger age (Craig 1982; Fulwiler et al., 1997; Link et al., 1992; Swett and Mills, 1997). For instance, studies have reported that men are more likely to be aggressive (Sheridan et al., 1990), others have found women to be at an increased risk (Kho et al., 1998), and a number of studies (e.g., Tardiff et al., 1997; Wessley et al., 1994) report no differences.

In order to better understand the link between psychiatric disorders and aggression, further investigations of correlates and risk factors will be necessary. Investigators will need to make a concerted effort to develop a gold standard for the definition of aggression, selection of evaluation tools, appropriate intervals for assessment, and criteria for measuring violence, thereby rendering comparable results across studies. Investigations of risk factors for specific disorders are also necessary, as risk factors appear to function differently for each disorder. While the continued study of aggression in psychiatric populations is important in order to better understand the association, develop effective treatments, and protect patients and those in contact with them, caution must be taken to avoid generalizations and further stigmatization.

REFERENCES

American Psychological Association (APA) (1994) *Diagnostic and Statistical Manual of Mental Disorders, 4th ed.* Washington, DC.

Arango, C., Barba, A. C., Gonzalez-Salvador, T., and Ordonez, A. C. (1999) Violence in inpatients with schizophrenia: a prospective study. *Schizophr. Bull.* **25,** 493–503.

Barlow, K., Grenyer, B., and Ilkiw-Lavalle, O. (2000) Prevalence and precipitants of aggression in psychiatric inpatient units. *Aust. NZ J. Psychiatry* **34,** 967–974.

Begic, D. and Jokic-Begic, N. (2001) Aggressive behavior in combat veterans with post traumatic stress disorder. *Mil. Med.* **166,** 671–676.

Beckham, J. C., Feldman, M. E., Kirby, A. C., Hertzberg, M. A., and Moore, S. D. (1997) Interpersonal violence and its correlates in Vietnam veterans with chronic post traumatic stress disorder. *J. Clin. Psychol.* **53,** 859–869.

Berkowitz, L. (1989) Frustration-aggression hypothesis: examination and reformation. *Psychol. Bull.* **106,** 59–73.

Bjork, J. M., Dougherty, D. M., and Moeller, F. G. (1997) A positive correlation between self-ratings of depression and laboratory measured aggression. *Psychiatry Res.* **69,** 33–38.

Bremner, J. D., Southwick, S. M., Johnson, D. R., Yehuda, R., and Charney, D. S. (1993) Childhood physical abuse and combat related post traumatic stress disorder in Vietnam veterans. *Am. J. Psychiatry* **150,** 235–239.

Brennan, P. A., Mednick, S. A., and Hodgins, S. (2000) Major mental disorders and criminal violence in a Danish birth cohort. *Arch. Gen. Psychiatry* **57,** 494–500.

Breslau, N. (2001) Outcomes of posttraumatic stress disorder. *J. Clin. Psychiatry* **62,** 55–59.

Burns, A., Jacoby, R., and Levy, R. (1990) Psychiatric phenomena in alzheimer's disease, IV: disorders of behavior. *Br. J. Psychiatry* **157,** 86–94.

Caton, C. L., Shrout, P. E., Eagle, P. F., Opler, L. A., and Felix, A. (1994) Correlates of codisorders in homeless and never homeless indigent men. *Psychol. Med.* **24,** 681–688.

Chengappa, K. N., Ebeling, T., Kang, J. S., Levine, J., and Parepally, H. (1999) Clozapine reduces severe self-mutilation and aggression in psychotic patients with borderline personality disorder. *J. Clin. Psychiatry* **60,** 477–484.

Cheung, P., Schweitzer, I., Crowley, K., and Tuckwell, V. (1997) Aggressive behaviour in schizophrenia: the role of psychopathology. *Aust. NZ J. Psychiatry* **31,** 62–27.

Coccaro, E. F., Siever, L. J., Maurer, G., et al. (1989) Serotonergic studies in patients with affective and personality disorders. *Arch. Gen. Psychiatry* **46,** 587–599.

Coccaro, E. F. (1995) The biology of aggression. *Sci. Am.* **XX,** 38–47.

Coccaro, R. and Kavoussi, R. (1997) Fluoxetine and impulsive behavior in personality-disordered subjects. *Arch. Gen. Psychiatry* **54,** 1081–1088.

Craig, T. J. (1982) An epidemiologic study of problems associated with violence among psychiatric inpatients. *Am. J. Psychiatry* **139,** 1262–1266.

Davidson, J. R. T., Hughes, D., Balzer, D., and George, L. K. (1991) Posttraumatic stress disorder in the community: an epidemiological study. *Psychol. Med.* **21,** 713–722.

Day, H. D., Franklin, J. M., and Marshall, D. D. (1998) Predictors of aggression in hospitalized adolescents. *J. Psychol.* **132,** 427–434.

Dodge, K. A. (1985) Attributional bias in aggressive children, in *Advances in Cognitive-Behavioural Research and Therapy.* (Kendall, P. C., ed.), Academic Press, Orlando, FL, pp. 73–110.

Dodge, K. A. and Crick, N. R. (1990) Social information processing bases of aggressive behavior in children. *Pers. Soc. Psychol. Bull.* **16,** 8–22.

Dougherty, D. M., Bjork, J. M., Huckabee, H. C., Moeller, F. G., and Swann, A. C. (1999) Laboratory measures of aggression and impulsivity in women with borderline personality disorder. *Psychiatry Res.* **22,** 315–326.

Eronen, M., Hakola, P., and Tiihonen, J. (1996) Mental disorders and homicidal behavior in Finland. *Arch. Gen. Psychiatry* **53,** 497–501.

Federal Interagency Forum on Child and Family Statistics. (1998) America's children: key indicators of well-being. Federal Interagency Forum on Child and Family Statistics. US Government Printing Office, Washington, DC.

Feldbau-Kohn, S. R., Heyman, R. E., and O'Leary, K. D. (1998) Major depressive disorder and depressive symptomatology as predictors of husband-to-wife physical aggression. *Violence Vict.* **13**, 347–360.

Fulwiler, C., Grossman, H., Forbes, C., and Ruthazer R. (1997) Early-onset substance abuse and community violence by outpatients with chronic mental illness. *Psychiatr. Serv.* **48**, 1181–1185.

Goldsmith, H. F., Manderscheid, R. W., Henderson, M. J., and Adele, J. S. (1993) Projections of inpatient admissions to specialty mental health organizations: 1990 to 2010. *Hosp. Community Psychiatry* **44**, 478–483.

Goodman, M. and New, A. (2000) Impulsive aggression borderline personality disorder. *Curr. Psychiatry Rep.* **2**, 56–61.

Gorman-Smith, D. and Tolan, P. (1998) The role of exposure to community violence and developmental problems among inner-city youth. *Dev. Psychopathol.* **10**, 101–116.

Gormley, N., Rizwan, M. R., and Lovestone, S. (1998) Clinical predictors of aggressive behavior in Alzheimer's disease. *Int. J. Geriatr. Psychiatry* **13**, 109–115.

Greist, J. H., Jefferson, J. W., Kobak, K. H., Katzenglick, D. J., and Serlin, R. C. (1995) Efficacy and tolerability of serotonin transport inhibitors in obsessive compulsive disorder. *Arch. Gen. Psychiatry* **52**, 53–60.

Harris, G. T. and Rice, M. E. (1997) Risk appraisal and management of violent behaviour. *Psychiatr. Serv.* **48**, 1168–1176.

Hillbrand, M. and Spitz, R. T. (1999) Cholesterol and aggression. *Aggress. Violent Behav.* **4**, 359–370.

Hillbrand, M., Waite, B. M., Miller, D. S., Spitz, R. T., and Lingswiler, V. M. (2000) Serum cholesterol concentrations and mood states in violence psychiatric patients: an experience sampling study. *J. Behav. Med.* **23**, 519–529.

Hollander, E. (1999) Managing aggressive behavior in patienst with obsessive-compulsive disorder and borderline personality disorder. *J. Clin. Psychiatry* **60**, 38–44.

Hudziak, J. J., Boffeli, T. J., Battaglia, M. M., Stranger, C., and Guze, S. B. (1997) Clinical studies of the relation of borderline personality disorder to Briquet's syndrome (hysteria), somatization disorder, antisocial personality disorder, and substance abuse disorders. *Am. J. Psychiatry* **152**, 1598–1606.

Kallionemi, H. and Syvalahti, E. (1993) Citaloroam, a specific inhibitor of serotonin reuptake in treatment of psychotic borderline patients. *Nord. J. Psychiatry* **47**, 79–84.

Kaplan, J. R., Manusck, S. B., Fontenot, M. B., Muldoon, M. F., Shivley, C. A., and Mann, J. (1997) The cholesterol-serum hypothesis: interrelationships among dietary lipids, central serotonergic activity and social behavior in monkeys, in *Lipids Health and Behavior*. (Hillbrand, M. and Spitz, R. T., eds.), American Psychological Association, Washington, DC, pp. 139–165.

Kashani, J., Dahlmeier, J., Burduin, C., Soltrys, M., and Reid J. (1995) Characteristics of anger expression in depressed children. *J. Acad. Child Adolesc. Psychiatry* **34**, 322–326.

Kashani, J., Suarez, L., Allen, W., and Reid, J. (1997) Hopelessness in inpatient youths: a closer look at behavior, emotional expression, and social support. *J. Acad Child Adolesc. Psychiatry* **36**, 1625–1631.

Katz, M. M., Koslow, S., and Maas, J. (1991) Identifying the specific clinical actions of amitriptyline: interrelationships of behavior, affect, and plasma levels in depression. *Psychol. Med.* **21**, 599–611.

Kavoussi, R., Liu, J., and Coccaro, E. (1994) An open trial of sertraline in borderline personality disordered patients with impulse aggression. *J. Clin. Psychiatry* **55**, 137–141.

Kazdin, A. E., Esweldt-Dawson, K., Unis, A., and Rancurello, M. D. (1983) Child and parent evaluations of depression and aggression in psychiatric inpatient children. *J. Abnorm. Child Psychol.* **11,** 4011–413.

Kessler, R. C., Sonnega, A., Bromet, E., Hughes, M., and Nelson, C. B. (1995) Posttraumatic stress disorder in the National Comorbidity Survey. *Arch. Gen. Psychiatry* **52,** 1048–1060.

Kho, K., Sensky, T., Mortimer, A., and Corcos, C. (1998) Prospective study into factors associated with aggressive incidents in psychiatric acute admission wards. *Br. J. Psychiatry* **172,** 38–43.

Klassen, D. and O'Connor, W. (1990) Assessing the risk of violence in released mental patients: a cross validation study. Psychological assessment. *J. Consult. Clin. Psychol.* **1,** 75–81.

Knox, M., King, C., Hanna, G. L., Logan D., and Ghaziuddin, N. (2000) Aggressive behavior in clinically depressed adolescents. *J. Am. Adolesc. Psychiatry* **39,** 611–618.

Kovaks, M. and Pollack, M. (1995) Bipolar disorder and comorbid conduct disorder in childhood and adolescence. *J. Am. Acad. Children Adolesc. Psychiatry* **34,** 715–723.

Kruesi, M. J., Rapoport, J. L., Hamburger, S., et al. (1990) Cerebrospinal fluid monoamine metabolites, aggression, and inoulsivity in disruptive behavior disorders of children and adolescents. *Arch. Gen. Psychiatry* **47,** 419–426.

Kruesi, M. J., Hibbs, E. D., Zahn, T. P., et al. (1992) A 2-year prospective follow-up study of children and adolescents with disruptive behavior disorders. Prediction by cerebrospinal fluid 5-hydroxyindoleacetic acid, homovanillic acid, and autonomic measures? *Arch. Gen. Psychiatry* **49,** 429–435.

Kulka, R. A., Schlenger, W. E., Fairbank, J. A., et al. (1990) *Trauma and the Vietnam War Generation.* Brunner/Maze, New York.

Lachman, H. M., Nolan, K. A., Mohr, P., Saito, T., and Volavka, J. (1998. Association between catechol O-methyltransferase genotype and violence in schizophrenia and schizoaffective disorder. *Am. J. Psychiatry* **155,** 835–837.

Lewinsohn, P. M., Klein, D. N., and Seeley, J. R. (1995) Bipolar disorders in a community sample of older adolescents: prevalence, phenomenology, comorbidity and course. *J. Am. Acad. Children Adolesc. Psychiatry* **34,** 454–463.

Lindqvist, P. and Allebeck, P. (1990) Schizophrenia and crime. *Br. J. Psychiatry* **162,** 87–92.

Link, B. G., Andrews, H., and Cullen, F. T. (1992) The violent and illegal behavior of mental patients reconsidered. *Am. Sociol. Rev.* **57,** 275–292.

Loeber, R. (1982) The stability of antisocial and delinquent child behavior: a review. *Child Dev.* **53,** 1431–1446.

Loeber, R. and Dishion, T. (1983) Early predictors of male delinquency: a review. *Psychol. Bull.* **94,** 68–98.

Lyketsos, C. G., Baker, L., Warren, A., et al. (1997) Major and minor depression in Alzheimer's disease: prevalence and impact. *J. Neuropsychiatry Clin. Neurosci.* **9,** 556–561.

Lyketsos, C. G., Steele, C., Galik, E., et al. (1999) Physical aggression in dementia patients and its relationship to depression. *Am. J. Psychiatry* **156,** 66–71.

Marazziti, D., Conti, L., Presta, S., et al. (2001) No correlation between aggression and platelet ^{3}H-paroxetine binding in obsessive-compulsive disorder patients. *Neuropsychobiology* **43,** 117–122.

Marquis, P. (1992) Family dysfunction as a risk factor in the development of antisocial behaviour. *Psychol. Rep.* **71,** 468–470.

McFall, M., Fontana, A., Raskind, M., and Rosenheck, R. (1999) Analysis of violent behavior in Vietnam combat veteran psychiatric inpatients with posttraumatic stress disorder. *Int. Soc. Traumatic Stress Studies* 501–517.

Meltzer, H. Y. and Lowy, M. T. (1987) The serotonin hypothesis of depression, in *Psychopharmacology: Third Generation of Progress.* (Meltzer, H. Y., ed.), Raven Press, New York, pp. 513–526.

Menon, A. S., Gruber-Baldini, A. L., Hebel, R., et al. (2001) Relationship between aggressive behaviors and depression among nursing home residents with dementia. *Int. J. Geriatr. Psychiatry* **16,** 139–146.

Murdoch, D., Pihl, R. O., and Ross, D. (1990) Alcohol and crimes of violence: present issues. *Int. J. Addic.* **25,** 1065–1081.

Nestor, P. G., Haycock, J., Doiron, S., Kelly, J., and Kelly, D. (1995) Lethal violence and psychosis: a clinical profile. *Bull. Am. Acad. Psychiatry Law* **23,** 331–334.

New, A. S., Sevin, E. M., Mitropoulou, V., et al. (1999) Serum cholesterol and impulsivity in personality disorders. *Psychiatry Res.* **85,** 145–150.

Nolan, K. A., Volavka, J., Mohr, P., and Czobor, P. (1999) Psychopathy and violent behavior among patients with schizophrenia or schizoaffective disorder. *Psychiatr. Serv.* **50,** 787–792.

Nolan, K. A., Volavka, J., Lachman, H. M., and Saito, T. (2000) An association between a polymorphism of the tryptophan hydroxylase gene and aggression in schizophrenia and schizoaffective disororder. *Psychiatr. Genet.* **10,** 109–115.

Palmstierna, T., Lassenius, R., and Wistedt, B. (1989) Evaluation of the brief psychopathological rating scale in relation to aggressive behavior by acute involuntarily admitted patients. *Acta Psychiatr. Scand.* **79,** 313–316.

Procter, A. W., Francis, P. T., Startmann, G. C., and Bowen, D. M. (1992) Serotonergic pathology not widespread in Alzheimer's patients without prominent aggressive symptoms. *Neurochem. Res.* **17,** 917–922.

Rabkin, J. G. (1979) Criminal behavior of discharged mental patients: a critical appraisal of the literature. *Psychol. Bull.* **86,** 1–27.

Rappoprt, M. J., Reekum, R. V., Freedman, M., et al. (2001) Relationship of psychosis to aggression, apathy and function in dementia. *Int. J. Geriatr. Psychiatry* **16,** 123–130.

Rasanen, P., Tiihonen, J., Isohanni, M., Rantakallio, P., Lehtonen, J., and Moring, J. (1998) Schizophrenia, alcohol abuse, and violence behavior: a 26-year follow-up study of an unselected birth cohort. *Schizophr. Bull.* **24,** 437–441.

Reid, J. B. and Patterson, G. R. (1989) The development of antisocial behaviour patterns in childhood and adolescence. *Eur. J. Personality* **3,** 107–119.

Richelson, E. (1991) Biological basis of depression and therapeutic relevance. *J. Clin. Psychiatry* **52(Suppl.),** 4–10.

Robbins, L. (1966) *Deviant Children Grown Up.* Williams and Wilkens, Baltimore.

Roy, S., Herrera, J., Parent, M., and Costs, J. (1987) Violent and nonviolent schizophrenic patients. Clinical and developmental characteristics. *Psychol. Rep.* **61,** 855–861.

Ryden, M. B. (1988) Aggressive behavior in persons with dementia who live in the community. *Alzheimer Dis. Ass. Disord.* **2,** 342–355.

Scholte, E. M. (1992) Identification of children at risk at the police station and the prevention of delinquency. *Psychiatry* **55,** 354–369.

Schubiner, H., Scott, R., and Tzelpis, A. (1993) Exposure to violence among inner city youth. *J. Adolesc. Health* **14,** 214–219.

Schwartz, S. M. and Ketterer, M. W. (1997) Cholesterol lowering and emotional distress: current status and future directions, in *Lipids Health and Behavior.* (Hillbrand, M. and Spitz, R. T., eds.), American Psychological Association, Washington, DC, pp. 113–123.

Sheridan, M., Henrion, R. E., Robinson, L., and Baxter, V. (1990) Precipitants of violence in a psychiatric inpatient setting. *Hosp. Community Psychiatry* **41,** 776–780.

Silva, J. A., Derecho, D. V., Leong, G. B., Weinstock, R., and Ferrari, M. M. (2001) A classification of psychological factors leading to violent behavior in posttraumatic stress disorder. *J. Forensic Sci.* **46,** 309–316.

Soloff, P. H., Lynch, K. G., Kelly, T. M., Malone, K. M., and Mann, J. J. (2000) Characteristics of suicide attempts of patients with major depressive episode and borderline personality disorder: a comparative study. *Am. J. Psychiatry* **157,** 601–608.

Spencer, T. J., Biederman, J., Wozniak, J., Faraone, S. V., Wilens, T., and Mick, E. (2001) Parsing pediatric bipolar disorder from its associated comorbidity with the disruptive behavior disorder. *Soc. Biol. Psychiatry* **49,** 1062–1070.

Steadman, H. J., Monahan, J., Appelbaum, P. S., et al. (1994) Designing a new generation of risk assessment research, in *Violence and Mental Disorder: Developments in Risk Assessment.* (Monahan, J. and Steadman, H. J., eds.), University of Chicago Press, Chicago, pp. 297–318.

Steinert, T., Wiebe, C., and Gebhardt, R. P. (1999) Aggressive behavior against self and others among first admission patients with schizophrenia. *Psychiatr. Serv.* **50,** 85–90.

Steinert, T., Woelfle, M., and Gebhardt, R. P. (1999) No correlation of serum cholesterol levels with measures of violence in patients with schizophrenia and non-psychotic disorders. *Eur. Psychiatry* **14,** 346–348.

Stone, M. H. (1989) Long-term follow up of narcissistic/borderline patients. *Psychiatr. Clin. N. Am.* **12,** 621–641.

Strous, R.D., Bark, N., Woerner, M., and Lachman, H. M. (1997) Analyses of functional catechol O-methyltransferase gene polymorphism in schizophrenia: evidence for association with aggressive and antisocial behavior. *Psychiatry Res.* **69,** 71–77.

Swann, A. C. (1999) Treatment of aggression in patients with bipolar disorder. *J. Clin. Psychiatry* **60(Suppl. 15),** 25–28.

Swann, A. C., Secunda, S. K., Katz, M. M., et al. (1993) Specificity of mixed affective states: clinical comparison of dysphoric mania and agitated depression. *J. Affect. Disord.* **28,** 81–89.

Swann, A. C., Stokes, P. E., Secunda, S. K., et al. (1994) Depressive mania versus agitated depression: biogenic amine and hypothalamic-pituitary-adrenocorticol function. *Biol. Psychiatry* **35,** 803–813.

Swanson, J. W., Holzer, C. E., Ganju, V. K., and Jono, R. T. (1990) Violence and psychiatric disorder in the community: evidence from the Epidemiologic Catchment Area Surveys. *Hosp. Community Psychiatry* **41,** 761–770.

Swartz, M. S., Swanson, J. W., Hiday, V. A., Borum, R., Wagner, H. R., and Burns, B. J. (1998) Violence and severe mental illness: the effects of substance abuse and nonadherence to medication. *Am. J. Psychiatry* **155,** 226–231.

Swett, C. and Mills, T. (1997) Use of the NOSIE to predict assaults among acute psychiatric patients. *Psychiatr. Serv.* **48,** 1177–1180.

Tam, E., Engelsmann, F., and Fugere, R. (1996) Patterns of violent incidents by patients in a general hospital psychiatric facility. *Psychiatr. Serv.* **47,** 86–88.

Tardiff, K. and Sweillam, A. (1980) Assault, suicide and mental illness. *Arch. Gen. Psychiatry* **37,** 164–169.

Tardiff, K. and Sweillam, A. (1982) Assaultive behavior among chronic psychiatric inpatients. *Am. J. Psychiatry* **139,** 212–215.

Tardiff, K., Marzuk, P. M., Leon, A. C., Portera, L., and Weiner, C. (1997) Violence in patients admitted to a private psychiatric hospital. *Am. J. Psychiatry* **154,** 88–93.

Torrey, E. F. (1994) Violent behavior by individuals with serious mental illness. *Hosp. Community Psychiatry* **45,** 653–661.

Unis, A. S., Cook, E. H., Vincent, J. G., et al. (1997) Platelet serotonin measures in adolescents with conduct disorder. *Biol. Psychiatry* **42,** 553–559.

van Praag, H. M. (1998) Anxiety and increased aggression as pacemaker of depressions. *Acta Psychiatr. Scand. Suppl.* **393,** 81–88.

Wessely, S. C., Castle, D., Douglas, A. J., and Taylor, P. J. (1994) The criminal cases of schizophrenia. *Psychol. Med.* **24,** 483–502.

Wicki, W. and Angst, J. (1991) The Zurich study: X. Hypomania in a 28-30 year old cohort. *Eur. Arch. Psychiatry Clin. Neurosci.* **240,** 339–348.

Yesavage, J. A. (1983) Inpatient violence and the schizophrenic patient. *Acta Psychiatr. Scand.* **67,** 353–357.

Zohar, J. and Insel, T. R. (1987) Obsessive-compulsive disorder: psychological approaches to diagnosis, treatment and pathophysiology. *Biol. Psychiatry* **22,** 667–687.

9
Aggression in Brain Injury, Aging, and Neurodegenerative Disorders

Mark P. Mattson

INTRODUCTION

Among the most prominent and devastating neurological disorders are those that involve degeneration of neurons as the result of acute injuries (traumatic brain injury and stroke) or age-related diseases including Alzheimer's disease (AD), Parkinson's disease (PD), and Huntington's disease (HD). Not only are these disorders major causes of death, but each typically involves long-term morbidity that exacts great tolls on relatives and healthcare systems. These disorders manifest degeneration of nerve cells in particular brain regions, resulting in deficits in the functions controlled by those brain regions: learning and memory in AD, body movements in PD and HD, and speech and language in many cases of stroke and head trauma (Table 1). The purpose of this chapter is to provide a brief overview of the pathogenesis of acute and chronic neurodegenerative conditions that manifest aberrant aggressive behaviors and to consider the neurobiological substrates of aggression in each disorder.

Two general causes of aggression in patients with neurodegenerative conditions are the frustration and anger associated with the patient's awareness of their predicament, and the structural damage and neurochemical alterations caused by damage and death of neurons. This chapter focuses on the latter cause, namely, how degeneration of certain populations of neurons can result in aggressive behaviors. Studies of patients with traumatic brain injuries, and of animals in which specific brain structures are lesioned or stimulated, have revealed neuronal circuits that play particularly important roles in the control of aggression. In addition, functional brain imaging studies in humans and pharmacological and genetic manipulations of laboratory animals have identified neuronal circuits and neurotransmitter and neuropeptide systems that are central to the regulation of aggressive behaviors. For example, single positron emission computed tomography (SPECT) analyses of people who exhibited aggressive behaviors and non-aggressive controls revealed the following differences in the aggressive individuals: (*i*) decreased activity in the prefrontal cortex; (*ii*) increased activity in the anteromedial portions of the frontal lobes; (*iii*) increased activity in the left basal ganglia and/or limbic system; and (*iv*) focal abnormalities in the left temporal lobe (Amen et al., 1996). An example from studies of genetically manipulated mice showed that mice deficient in monoamine oxidase-A exhibit increased levels of serotonin, norepinephrine, and

From: *Neurobiology of Aggression: Understanding and Preventing Violence*
Edited by: M. Mattson © Humana Press Inc., Totowa, NJ

dopamine and increased aggressiveness (Shih and Chen, 1999). Because details of these kinds of studies that have revealed the cellular and molecular substrates of aggression can be found elsewhere in this book, they will not be reviewed in this chapter.

TRAUMATIC BRAIN INJURY

Traumatic brain injury (TBI) is a major cause of death and disability, particularly in young active individuals. Motor vehicle and bicycle accidents are the most common causes of traumatic brain injury in the United States (McNair, 1999). Dysfunction and death of neurons following TBI appears to occur in two different time windows; some neurons may die as a direct result of the injury during the first hours to days following the injury, while other neurons die in the ensuing days, weeks, and months. The neurons may die as the result of physical trauma resulting in damage to membranes and a disruption of cellular ion homeostasis and/or because of impaired energy metabolism and oxyradical production (Fig. 1). Injury-induced cytokine production and overactivation of excitatory amino acid receptors may contribute to the death of neurons in head injury patients (Lynch and Dawson, 1994; Allan and Rothwell, 2001). At the same time the neurodegenerative cascades are activated, adaptive neuroprotective responses occur including the expression of neurotrophic factors (Mattson and Scheff, 1994), protein chaperones (Sharp et al., 1999), and transcription factors that activate genes encoding anti-apoptotic proteins, such as Bcl-2 and manganese superoxide dismutase (Mattson and Camandola, 2001). Treatments that have proven effective in protecting neurons against dysfunction and degeneration in animal models of TBI include vitamin E (Hall et al., 1992), creatine (Sullivan et al., 2000), and cyclosporin A (Albensi et al., 2000).

Patients that have suffered a TBI often exhibit changes in aggression. Most commonly there is increased impulsive aggression, which is correlated with increased irritability and antisocial behavior (Greve et al., 2001). Although it is difficult to identify the specific structural and functional alterations that result in aggression following TBI, the available data are consistent with damage to neuronal circuits of the frontal cortex and limbic system. For example, a patient with trauma to the right frontal region, including the orbitofrontal cortex, exhibited high levels of aggression and sociopathy (Blair and Cipolotti, 2000). More specifically, he had severe difficulty in emotional expression recognition, autonomic responding, and social cognition. He also showed impairment in the recognition of, and autonomic responding to, angry and disgusted expressions; attributing the emotions of fear, anger, and embarrassment to story protagonists; and the identification of violations of social behavior. Important clinical data concerning TBI and aggression have come from studies of patients who suffered their injuries in combat. For example, the Vietnam Head Injury Study revealed that frontal lobe lesions resulting from perforating head injury most commonly resulted in increased aggressive behavior when the lesion affected the frontal ventromedial cortex (Grafman et al., 1996).

The central roles of the amygdala and frontal lobes in controlling anger and aggression in humans was established in studies of patients who suffered a TBI that involved a selective lesion of the amygdala and/or frontal cortex, and resulted in loss of affect including the inability to express anger or aggression (Eichelman, 1983). Experimental lesions of the amygdala in rodents results in a similar nonaggressive phenotype (Chozick, 1986). Subsequent clinical studies have shown that head trauma and epilepsy patients with severe rage and aggression can be effectively treated by microsurgical amygdalo-hippocampectomy (Sachdev et al., 1992). In a clever study of the effects of brain damage on social aggression, Desjardins et al. (2001) housed normal male rats with rats that had seizure-induced brain damage in groups of six, with varying ratios of normal to injured rats. They found that aggressive behaviors increased with increasing numbers of brain-injured rats in a cage, and that the numbers of the aggressive behaviors was correlated with amount of neuronal loss in the amygdala and hippocampus. The neuropeptide corticotrophin-releasing hormone (CRH) is a key player in the neuronal circuits that mediate stress responses, including responses involving aggression (Van Praag, 2001). It was reported that administration of prednisolone, which suppresses CRH production, can ameliorate social aggression in brain-injured rats (Bedard and Persinger, 1995).

The kinds of treatments that have proven effective in reducing aggressive behaviors in brain-injured patients are those that are effective in other aggressive patients (Table 2) and include cognitive–behavioral therapies and the use of benzodiazepines, antidepressants, and antipsychotics (Teichner et al., 1999; Wroblewski et al., 1997). In addition, lithium treatment was reported to be effective in reducing aggressive behaviors in head trauma patients (Glenn et al., 1989).

STROKE

A stroke occurs when a cerebral blood vessel becomes occluded or ruptures; the most common antecedent vascular alteration leading to a stroke is atherosclerosis (Lusis, 2000). Cells in the core of the infracted brain tissue die rapidly by necrosis, whereas many neurons in the surrounding penumbra undergo a delayed apoptotic cell death that occurs over a period of days to weeks (Graham and Chen, 2001). A reduced availability of oxygen (hypoxia) and glucose (hypoglycemia) initiates neurodegenerative cascades in stroke, with impaired ion homeostasis and oxidative stress being important events in the cell death process. Many neurons in the ischemic penumbra may die by apoptosis. The apoptotic cascade likely involves increased expression and activation of prostate apoptosis response-4 (Par-4) (Culmsee et al., 2001), Bax and related pro-apoptotic Bcl-2 family members (Krajewski et al., 1995), mitochondrial membrane permeability changes and cytochrome c release (Sims and Anderson, 2002), and activation of caspases (Ni et al., 1998). Studies of rodent models of stroke have shown that antioxidants (Clark et al., 2001; Keller et al., 1998; Yu et al., 1998), agents that suppress calcium influx (Horn and Limburg, 2001), the overexpression of anti-apoptotic Bcl-2 family members (Martinou et al., 1994), drugs that stabilize mitochondrial membranes (Li et al., 2000; Liu et al., 2002), and caspase inhibitors (Schulz et al., 1999) can reduce neuronal death and may improve behavioral outcome.

An inability to control anger and aggression has been reported to occur in up to one-third of stroke patients and is associated with motor dysfunction, emotional incontinence, and lesions affecting the frontal-lenticulocapsular-pontine base areas (Kim et al., 2002). In a study of aggressive behaviors in stroke patients, based on self-reporting of episodes of anger and aggression, it was concluded that increased aggression is associated with anterior hemisphere lesions and greater cognitive impairment (Paradiso et al., 1996). A less common alteration in anger and aggression is the so-called catastrophic reaction in which patients exhibit dramatic outbursts of frustration and anger when confronted with a task; this syndrome is associated with nonfluent aphasias and lesions of the left opercular region (Carota et al., 2001). On the other hand, many stroke patients become withdrawn and less aggressive, which may be related to impaired cognitive function (Wang and Smyers, 1977).

IMPACT OF AGING ON AGGRESSION

During normal aging of rodents and humans, there is typically a decrease in aggressive behaviors. Interestingly, during normal aging there is also a decrease in the prevalence of anxiety and depression (Jorm, 2000). However, anxiety disorders remain very common in the elderly, even in the absence of chronic disease, and result in a considerable burden on family members and healthcare providers (Carmin et al., 2000). The decrease in levels of aggression in the elderly may result from an inability or unwillingness to act upon aggressive impulses rather than a reduction in levels of anger and frustration. There are many changes in brain neurochemistry and neuroendocrine function that occur during aging that may contribute to qualitative and quantitative changes in aggressive behaviors. For example, changes in monoaminergic circuits that control aggressive behaviors have been documented during aging in humans and rodents, including decreases in levels of type 2 dopamine receptors (Roth and Joseph, 1994). Mice lacking the type 2 dopamine receptors exhibit lower levels of aggression than wild-type mice (Vukhac et al., 2001), suggesting a contribution of reduced dopaminergic signaling to the age-related decrease in aggressive behaviors. A study of the relationships of dominance (aggression) and anger to hormones in aging men revealed increased aggression in men with a hormonal pattern expected to increase the bioavailability of androgens (Gray et al., 1991). However, it should be noted that not all studies have revealed changes in aggression during aging. For example, in a study of two different strains of rats that differed in their levels of aggressiveness it was concluded that male aggression did not change during aging up to 600 d of age (Blanchard et al., 1984).

Aggressive behaviors in animals and humans often occur during stressful situations, be it psychosocial or physical stress. A number of alterations in the neuronal circuits and neuroendocrine systems that mediate stress responses have been documented in aging and age-related neurodegenerative disorders (*see* Pedersen et al., 2001, for review). Brain regions involved in learning and memory processes and emotion, such as the hippocampus, amygdala, and associated limbic and cortical structures, play major roles in stress responses. Studies of baboons have shown that social subordinance is associated with elevated levels of cortisol that may promote cognitive decline during aging (Sapolsky et al., 1997). Such subordinate individuals withdraw from social interactions during aging, which is associated with decreased amounts of aggression (Veenema et al., 1997).

Studies of rodents have attempted to elucidate the neurochemical basis for age-related changes in emotional behaviors. For example, levels of CRH following chronic restraint stress were significantly decreased in 12- and 24-mo-old rats, compared to 4-mo-old rats, and this was correlated with decreased anxiety-like behaviors in the older rats (Pisarska et al., 2000). Immunohistochemical analyses of serotonergic fibers in rats of different ages revealed that there is a marked increase in the number of immunoreactive axons that exhibit varicosities and a tortuous course, suggestive of degeneration (Nishimura et al., 1998). These abnormalities were prominent in the frontoparietal cortex, striatum, cingulate cortex, dentate gyrus of the hippocampus, amygdala, and hypothalamus. Electrophysiological studies of brain slices from rats of different ages revealed an age-related impairment of synaptic interactions between the amygdala and hippocampus that are required for long-term potentiation of synaptic transmission in the hippocampus, which is a cellular correlate of learning and memory (Almaguer et al., 2002). Because aggression involves an interplay between brain structures involved in cognitive and instinctive–emotional processes, it is likely that age-related impairments in cognitive function may also modify aggressive behaviors.

ALZHEIMER'S DISEASE

AD currently afflicts more than 4 million Americans. These patients exhibit a progressive impairment in short-term memory (DeKosky and Orgogozo, 2001). Degeneration of neurons in limbic structures involved in aggression, including the amygdala and associated prefrontal cortical regions, have been documented in AD patients (Braak et al., 1996; Kromer Vogt et al., 1990). Magnetic resonance imaging analyses have revealed significant atrophy of brain structures that control aggressive behaviors including the amygdala, hippocampus, and septal area (Callen et al., 2001). Two major histopathological alterations in AD are the accumulation of insoluble aggregates of the amyloid β-protein (Aβ) in the form of plaques, and the degeneration and death of neurons, which typically manifest "tangles" consisting of fibrillar aggregates of the microtuble-associated protein tau (Yankner, 1996). Risk factors for AD have begun to be identified and may include a high calorie intake (Mattson et al., 2001), elevated homocysteine levels and folate deficiency (Kruman et al., 2002; Seshadri et al., 2002), and reduced levels of intellectual and physical activities in midlife (Friedland et al., 2001).

Although most cases of AD are sporadic with no clear genetic basis, some cases are caused by mutations in the amyloid precursor protein (APP), presenilin-1, or presenilin-2 (Hardy, 1997). The latter genetic mutations and increased oxidative stress and metabolic impairment resulting from the aging process, result in an alteration in the proteolytic processing of APP such that increased amounts of a long (42 amino acid) form of Aβ are produced (Fig. 1). Studies of cultured neurons and transgenic mice expressing AD-linked APP and presenilin-1 mutations suggest that Aβ plays a central role in the neuronal dysfunction and death in AD by inducing oxidative stress and disrupting cellular calcium homeostasis in neurons (Mattson, 1997). Synapses appear to be particularly sensitive to the adverse effects of Aβ (Keller et al., 1997; Mark et al., 1995), and it has been shown that Aβ can activate apoptotic biochemical cascades in synapses that may ultimately kill the neurons (Guo et al., 1998; Mattson et al., 1998).

Aggression and other behavioral abnormalities are very frequent in patients with AD (Aarsland et al., 1996; Devanand et al., 1997). Indeed, behavioral disorders (aggression, agitation, and paranoid delusions) are major reasons why patients with dementia are placed in a nursing home. A variety of drugs are commonly used to alleviate such symptoms, including antipsychotics, sedatives, and antidepressants (Stoppe et al., 1999). Both major and minor depression are very common in patients with AD and represent a major factor affecting the daily lives of the patients and their caregivers (Lyketsos et al., 1997). A study of over 5000 elderly residents of Cache County, UT revealed that aggression–agitation was four times more common in residents with dementia, regardless of the severity of the dementia, suggesting that this behavioral disturbance may occur very early in the course of the disease process (Lyketsos et al., 2000).

Because dysfunction of serotonergic neuronal circuits has been linked to aggressive behavior in patients with AD, recent molecular genetic studies have aimed to identify genetic factors related to serotonin neurotransmission that may contribute to such aggressive behavior. In a study of a geriatric psychiatry inpatient clinic and AD research center, it was shown that AD patients homozygous for the long variant of a biallelic polymorphism of the serotonin transporter promoter region exhibited increased risk of aggressive behavior (Sukonick et al., 2001). Another study was designed to determine whether polymorphisms in dopamine receptor genes DRD1–DRD4 are associated with aggressive behavior or psychosis in patients with AD (Sweet et al., 1998). Aggression and psychosis were both significantly more frequent in patients homozygous for the DRD1 B2/B2 polymorphism. Studies of a community-based cohort of patients with late-onset AD revealed an association between aggressive behavior and polymorphisms in dopamine receptor genes, such that carriers of a polymorphism in the type 1 dopamine receptor gene were more likely to be aggressive (Holmes et al., 2001).

Studies of transgenic mouse models of AD and related disorders have revealed behavioral alterations consistent with increased aggression. Transgenic mice expressing AD-linked mutations in both APP and presenilin-1 exhibit deficits in spatial learning, but also exhibit an increase in open field activity and motor dysfunction (Arendash et al., 2001). Presenilin-1 mutant transgenic mice exhibit enhanced hippocampal synaptic potentiation, which can be normalized by treatment with a γ-amino butyric acid (GABA) receptor agonist (Zaman et al., 2000). Mice overexpressing mutant forms of APP exhibited pronounced behavioral abnormalities without extensive deposition of amyloid in the brain (Kumar-Singh et al., 2000; Moechars et al., 1998). In one study, the increased aggressiveness of APP mutant transgenic mice was alleviated with 8-hydroxy-2(di-n-propyl-amino) tetraline (8-OH-DPAT) and buspirone, which are two serotonergic agonists (Moechars et al., 1998). Mutations in the microtubule-associated protein tau are responsible for some cases of frontotemporal dementia and Parkinsonism linked to chromosome 17 (FTDP-17). Transgenic mice expressing an FTDP-17 tau mutation exhibit hyperactivity in a novel environment and an impairment in contextual fear conditioning (Ikegami et al., 2000).

Prolonged stress, particularly uncontrollable stress, can promote neuronal degeneration and may contribute to neuronal degeneration in AD (Stein-Behrens et al., 1994). Conversely, the degeneration of neurons in AD can result in the dysregulation of stress responses and abnormal emotional responses to stress. Recent studies have documented abnormal stress responses in transgenic mouse models of familial AD. For example, APP mutant mice exhibit an abnormal response to stress characterized by an enhanced glucocorticoid response and hypoglycemia following restraint stress (Pedersen et al., 1999).

As is the case in other patient populations, aggressive behaviors, anxiety, and depression in Alzheimer's patients can be successfully treated with antidepressants, benzodiazepines, and antiepileptic drugs, such as valproate (Rojas-Fernandez et al., 2001). For example, the benzisoxazole derivative risperidone, which has antagonistic activity at serotonin 5-HT2A and dopamine D2 receptors, exhibited efficacy in reducing aggressive behavior in patients with dementia (Bhana and Spencer, 2000). Enhancement of cholinergic neurotransmission with drugs such as cholinersterase inhibitors can improve cognitive function in some patients with AD (Bullock, 2002). It was reported that metrifonate, a drug that enhances cholinergic transmission can also reduce aggressive behaviors in Alzheimer's patients (Cummings et al., 2001).

PARKINSON'S DISEASE

PD is the second most common age-related neurodegenerative disorder. The most prominent clinical signs of PD are akinesia and tremor. In addition to degeneration of nigrostriatal dopaminergic neurons, neurons in limbic structures involved in aggression, including the amygdala and associated prefrontal cortical regions, have been documented in PD patients (Mattila et al., 1999). The vast majority of PD cases occur late in life and in a sporadic manner, suggesting that the aging process and yet-to-be-identified environmental factors are involved. Serendipitous discoveries have shown that specific environmental toxins can induce PD-like neuropathology and motor dysfunction in humans, monkeys, and rodents. One prominent example is 1-methyl-4-phenyl-1, 2, 3, 6-tetrahydropyridine (MPTP), which was identified as the pathogenic agent contaminating a drug supply of several drug users in California who developed PD-like symptoms (Langston, 1996). Another example is the pesticide rotenone, which causes PD-like pathology and motor dysfunction in rodents (Betarbet et al., 2002). Studies of rodents and monkeys exposed to these toxins have provided major advances in our understanding of the cellular and molecular alterations responsible for the dysfunction and degeneration of dopaminergic neurons in PD. Although most cases of PD are sporadic, families have been identified in which PD is inherited, and disease-causing mutations have been localized to the genes encoding α-synuclein and Parkin (Gasser, 2001). Recent studies of mice expressing mutant forms of the latter two proteins are revealing novel aspects of PD pathophysiology.

Data suggest that oxidative stress and DNA damage, metabolic impairment, aberrant protein degradation and aggregation, and activation of apoptotic cascades are central to the neurodegenerative process in PD (Duan et al., 1999). Accordingly, animal studies have shown that antioxidants such as vitamin E (Halliwell, 2001), agents such as creatine that enhance cellular energy efficiency (Persky and Brazeau, 2001) and agents that block apoptotic cascades such as pifithrin-α (PFTalpha) (Duan et al., 2002a) can reduce neuronal damage and improve motor function in models of PD. Interestingly, recent findings suggest that dietary factors may influence one's vulnerability to PD For example, studies using the MPTP model of PD in mice have shown that dietary restriction (periodic fasting) can decrease the vulnerability of dopaminergic neurons and can improve behavioral outcome (Duan and Mattson, 1999). Dietary folate can also protect dopaminergic neurons against environmental toxins relevant to PD (Duan et al., 2002b).

Increased levels of anger and hostility occur in some PD patients. Increased physical and mental fatigue may contribute to such emotional disturbances (Lou et al., 2001). However, there is also reason to believe that alterations in aggressive behaviors in PD patients can result from dysfunction and degeneration of the nigrostriatal dopaminergic pathway and perhaps other neuronal circuits as well. Many patients with PD exhibit a predisposition to depressive illness and suppressed aggressiveness prior to becoming symptomatic (Todes and Lees, 1985). Interestingly, side effects of chronic use of neuroleptic drugs include movement disorders with symptoms of PD and tardive dyskinesia

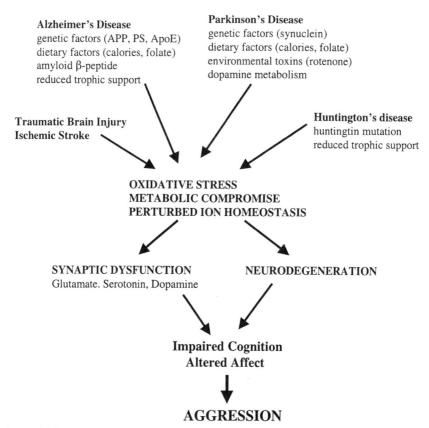

Fig. 1. Model for the mechanisms underlying abnormal aggressive behaviors is neurodegenerative conditions. Both acute insults such as traumatic brain injury and stroke, and chronic neurodegenerative disorders such as AD, PD, and HD result in oxidative stress and impaired energy metabolism and ion homeostasis in neurons. The latter alterations result in synaptic dysfunction and neuronal degeneration. Damage to brain regions involved in cognition and emotional behaviors, such as the hippocampus, amygdala, and connected structures results in increased aggression.

(Richardson et al., 1991), further strengthening structural and functional relationships between neuronal circuits that control mood and those that control motor function.

Neuropathological alterations in PD occur not only in the substantia nigra and striatum, but also in brain structures known to be involved in regulating emotions and cognition, including the amygdala, hippocampus, and basal forebrain (Braak and Braak, 2000; Wszolek et al., 2002).

HUNTINGTON'S DISEASE

HD is an inherited disorder caused by trinucleotide (CAG) expansions in the *huntingtin* gene resulting in a corresponding polyglutamine expansion in the huntingtin protein (Ho et al., 2001). The most profoundly affected brain region in HD is the striatum,

Table 1
Brain Structures and Neurotransmitter Systems Affected in Neurodegenerative Disorders

Disorder	Brain region	Neurotransmitter systems
Alzheimer's	Hippocampus (CA1), basal forebrain amygdala, entorhinal cortex	Glutamate, acetylcholine, serotonin
Parkinson's	Substantia nigra	Dopamine, serotonin
Huntington's	Caudate and putamen	Glutamate, GABA, acetylcholine

Table 2
Examples of Drugs Used to Treat Aggressive Symptoms in Patients with Neurodenerative Conditions

Drug	Mechanism of action	References
Haloperidol	D2 receptor antagonist	Sultzer et al., 2001
Risperidone	D2 receptor antagonist	Rainer et al., 2001
Olanzapine	D2 and 5-HT2A antagonist	Street et al., 2001
Loxapine	D2 antagonist; h5-HT2C-INI inverse agonist	Carlyle et al., 1993
Metrifonate	Acetylcholinesterase inhibitor	Cummings et al., 2001
Carbamazepine	Sodium channel blocker	Tariot et al., 1998
Valproate	Sodium channel blocker	Wroblewski et al., 1997
Lithium	Multiple actions have been proposed	Glenn et al., 1989

wherein medium spiny neurons degenerate. Degeneration of these neurons accounts for the inability to control body movements that characterizes HD (Sieradzan and Mann, 2001). It has not yet been established how huntingtin mutations damage and kill neurons. The mutant protein may alter synaptic function, induce oxidative stress, and trigger apoptosis (Mattson, 2002).

While the caudate nucleus, putamen, and globus pallidus are the brain regions in which the most severe degeneration of neurons occurs in HD, many other brain regions are affected. Atrophy of frontal, temporal, and parietal lobes has been described, and degeneration in the amygdala and thalamus also occur in most patients with HD (Mann et al., 1993). An analysis of different neuropeptides in brain tissues from HD patients revealed significantly elevated concentrations of neurotensin, somatostatin, and thyrotropin-releasing hormone in the caudate nucleus (Nemeroff et al., 1983). In addition, levels of somatostatin were increased in the nucleus accumbens and thyrotropin-releasing hormone levels were increased in the amygdala.

HD patients typically manifest abnormalities of mood and behavior, including depression and aggression (Rosenblatt and Leroi, 2000). Patients with HD can become extremely aggressive, and it has been suggested that this may result from a premorbid trait of "bad temper" in individuals with HD (Burns et al., 1990).

Analyses of behaviors in transgenic mice expressing mutant huntingtin revealed hypoactivity and impaired motor function (Schilling et al., 2001). In contrast, transgenic mice expressing mutant atrophin-1 (a causative gene for dentatorubral and pallidoluysian atrophy in humans) exhibit hyperactivity and tremors (Schilling et al., 2001). In another line of mice expressing a HD mutation, the mice exhibited abnormal social behaviors, including irritability and aggression, suggesting that brain circuits that regulate aggression are affected in the mice in a manner similar to patients with HD (Shelbourne et al., 1999). On the other hand, the testing of huntingtin mutant mice in the elevated plus maze test suggested a marked reduction in anxiety (File et al., 1998).

CONCLUSIONS

After suffering a traumatic injury to the brain or a stroke, many patients exhibit aggression, as do patients with chronic neurodegenerative disorders, including those with AD, PD, and HD. The aggression results from damage to neuronal circuits involved in cognition and emotional behaviors, including the hippocampus, amygdala, and related brain regions. Neurotransmitter systems affected include those that employ the neurotransmitters glutamate, serotonin, dopamine, GABA, and acetylcholine. Treatments that have proven effective in treating aggression in patients with neurodegenerative conditions include dopamine (D2) receptor antagonists, serotonin reuptake inhibitors, and anticonvulsants (De Marchi et al., 2001).

REFERENCES

Aarsland, D., Cummings, J. L., Yenner, G., and Miller, B. (1996) Relationship of aggressive behavior to other neuropsychiatric symptoms in patients with Alzheimer's disease. *Am. J. Psychiatry* **153,** 243–247.

Albensi, B. C., Sullivan, P. G., Thompson, M. B., Scheff, S. W., and Mattson, M. P. (2000) Cyclosporin ameliorates traumatic brain-injury-induced alterations of hippocampal synaptic plasticity. *Exp. Neurol.* **162,** 385–389.

Allan, S. M. and Rothwell, N. J. (2001) Cytokines and acute neurodegeneration. *Nat. Rev. Neurosci.* **2,** 734–744.

Almaguer, W., Estupinan, B., Uwe Frey, J., and Bergado, J. A. (2002) Aging impairs amygdala-hippocampus interactions involved in hippocampal LTP. *Neurobiol. Aging* **23,** 319–324.

Amen, D. G., Stubblefield, M., Carmicheal, B., and Thisted, R. (1996) Brain SPECT findings and aggressiveness. *Ann. Clin. Psychiatry* **8,** 129–137.

Bedard, A. W. and Persinger, M. A. (1995) Prednisolone blocks extreme intermale social aggression in seizure-induced, brain-damaged rats: implications for the amygdaloid central nucleus, corticotrophin-releasing factor, and electrical seizures. *Psychol. Rep.* **77,** 3–9.

Betarbet, R., Sherer, T. B., and Greenamyre, J. T. (2002) Animal models of Parkinson's disease. *Bioessays* **24,** 308–318.

Bhana, N. and Spencer, C. M. (2000) Risperidone: a review of its use in the management of the behavioural and psychological symptoms of dementia. *Drugs Aging* **16,** 451–471.

Blair, R. J. and Cipolotti, L. (2000) Impaired social response reversal. A case of 'acquired sociopathy'. *Brain* **123,** 1122–1141.

Blanchard, R. J., Flannelly, K. J., Layng, M., and Blanchard, D. C. (1984) The effects of age and strain on aggression in male rats. *Physiol. Behav.* **33,** 857–861.

Braak, H., Braak, E., Yilmazer, D., de Vos, R. A., Jansen, E. N., and Bohl, J. (1996) Pattern of brain destruction in Parkinson's and Alzheimer's diseases. *J. Neural. Transm.* **103,** 455–490.

Braak, H. and Braak, E. (2000) Pathoanatomy of Parkinson's disease. *J. Neurol.* **247,** II3–II10.

Bullock, R. (2002) New drugs for Alzheimer's disease and other dementias. *Br. J. Psychiatry* **180,** 135–139.

Burns, A., Folstein, S., Brandt, J., and Folstein, M. (1990) Clinical assessment of irritability, aggression, and apathy in Huntington and Alzheimer disease. *J. Nerv. Ment. Dis.* **178,** 20–26.

Callen, D. J., Black, S. E., Gao, F., Caldwell, C. B., and Szalai, J. P. (2001) Beyond the hippocampus: MRI volumetry confirms widespread limbic atrophy in AD. *Neurology* **57,** 1669–1674.

Carlyle, W., Ancill, R. J., and Sheldon, L. (1993) Aggression in the demented patient: a double-blind study of loxapine versus haloperidol. *Int. Clin. Psychopharmacol.* **8,** 103–108.

Carmin, C. N., Wiegartz, P. S., and Scher, C. (2000) Anxiety disorders in the elderly. *Curr. Psychiatry Rep.* **2,** 13–19.

Carota, A., Rossetti, A. O., Karapanayiotides, T., and Bogousslavsky, J. (2001) Catastrophic reaction in acute stroke: a reflex behavior in aphasic patients. *Neurology* **57,** 1902–1905.

Chozick, B. S. (1986) The behavioral effects of lesions of the amygdala: a review. *Int. J. Neurosci.* **29,** 205–221.

Clark, W. M., Rinker, L. G., Lessov, N. S., Lowery, S. L., and Cipolla, M. J. (2001) Efficacy of antioxidant therapies in transient focal ischemia in mice. *Stroke* **32,** 1000–1004.

Culmsee, C., Zhu, Y., Krieglstein, J., and Mattson, M. P. (2001) Evidence for the involvement of Par-4 in ischemic neuron cell death. *J. Cereb. Blood Flow Metab.* **21,** 334–433.

Cummings, J. L., Nadel, A., Masterman, D., and Cyrus, P. A. (2001) Efficacy of metrifonate in improving the psychiatric and behavioral disturbances of patients with Alzheimer's disease. *J. Geriatr. Psychiatry Neurol.* **14,** 101–108.

DeKosky, S. T. and Orgogozo, J. M. (2001) Alzheimer disease: diagnosis, costs, and dimensions of treatment. *Alzheimer Dis. Assoc. Disord.* **15,** S3–S7.

De Marchi, N., Daniele, F., and Ragone, M. A. (2001) Fluoxetine in the treatment of Huntington's disease. *Psychopharmacology (Berl).* **153,** 264–266.

Desjardins, D., Parker, G., Cook, L. L., and Persinger, M. A. (2001) Agonistic behavior in groups of limbic epileptic male rats: pattern of brain damage and moderating effects from normal rats. *Brain Res.* **905,** 26–33.

Devanand, D. P., Jacobs, D. M., Tang, M. X., et al. (1997) The course of psychopathologic features in mild to moderate Alzheimer disease. *Arch. Gen. Psychiatry* **54,** 257–263.

Duan, W. and Mattson, M. P. (1999) Dietary restriction and 2-deoxyglucose administration improve behavioral outcome and reduce degeneration of dopaminergic neurons in models of Parkinson's disease. *J. Neurosci. Res.* **57,** 195–206.

Duan, W., Zhu, Z., Ladenheim, B., et al. (2002a) Synthetic p53 inhibitors preserve dopaminergic neurons and motor function in experimental Parkinsonism. *Ann. Neurol.* **52,** 597–606.

Duan, W., Ladenheim, B., Cutler, R. G., Kruman, I. I., Cadet, J. L., and Mattson, M. P. (2002b) Dietary folate deficiency and elevated homocysteine levels endanger dopaminergic neurons in models of Parkinson's disease. *J. Neurochem.* **80,** 101–110.

Eichelman, B. (1983) The limbic system and aggression in humans. *Neurosci. Biobehav. Rev.* **7,** 391–394.

File, S. E., Mahal, A., Mangiarini, L., and Bates, G. P. (1998) Striking changes in anxiety in Huntington's disease transgenic mice. *Brain Res.* **805,** 234–240.

Friedland, R. P., Fritsch, T., Smyth, K. A., et al. (2001) Patients with Alzheimer's disease have reduced activities in midlife compared with healthy control-group members. *Proc. Natl. Acad. Sci. USA* **98,** 3440–3445.

Gasser, T. (2001) Molecular genetics of Parkinson's disease. *Adv. Neurol.* **86,** 23–32.

Glenn, M. B., Wroblewski, B., Parziale, J., Levine, L., Whyte, J., and Rosenthal, M. (1989) Lithium carbonate for aggressive behavior or affective instability in ten brain-injured patients. *Am. J. Phys. Med. Rehabil.* **68,** 221–226.

Grafman, J., Schwab, K., Warden, D., Pridgen, A., Brown, H. R., and Salazar, A. M. (1996) Frontal lobe injuries, violence, and aggression: a report of the Vietnam Head Injury Study. *Neurology* **46,** 1231–1238.

Graham, S. H. and Chen, J. (2001) Programmed cell death in cerebral ischemia. *J. Cereb. Blood Flow Metab.* **21,** 99–109.

Gray, A., Jackson, D. N., and McKinlay, J. B. (1991) The relation between dominance, anger, and hormones in normally aging men: results from the Massachusetts Male Aging Study. *Psychosom. Med.* **53,** 375–385.

Greve, K. W., Sherwin, E., Stanford, M. S., Mathias, C., Love, J., and Ramzinski, P. (2001) Personality and neurocognitive correlates of impulsive aggression in long-term survivors of severe traumatic brain injury. *Brain Inj.* **15,** 255–262.

Guo, Q., Fu, W., Xie, J., et al. (1998) Par-4 is a mediator of neuronal degeneration associated with the pathogenesis of Alzheimer disease. *Nat. Med.* **4,** 957–962.

Hall, E. D., Braughler, J. M., and McCall, J. M. (1992) Antioxidant effects in brain and spinal cord injury. *J. Neurotrauma* **9,** S165–S172.

Halliwell, B. (2001) Role of free radicals in the neurodegenerative diseases: therapeutic implications for antioxidant treatment. *Drugs Aging* **18,** 685–716.

Hardy, J. (1997) Amyloid, the presenilins and Alzheimer's disease. *Trends Neurosci.* **20,** 154–159.

Ho, L. W., Carmichael, J., Swartz, J., Wyttenbach, A., Rankin, J., and Rubinsztein, D. C. (2001) The molecular biology of Huntington's disease. *Psychol. Med.* **31,** 3–14.

Holmes, C., Smith, H., Ganderton, R., et al. (2001) Psychosis and aggression in Alzheimer's disease: the effect of dopamine receptor gene variation. *J. Neurol. Neurosurg. Psychiatry* **71,** 777–779.

Horn, J. and Limburg, M. (2001) Calcium antagonists for ischemic stroke: a systematic review. *Stroke* **32,** 570–576.

Ikegami, S., Harada, A., and Hirokawa, N. (2000) Muscle weakness, hyperactivity, and impairment in fear conditioning in tau-deficient mice. *Neurosci. Lett.* **279,** 129–132.

Jorm, A. F. (2000) Does old age reduce the risk of anxiety and depression? A review of epidemiological studies across the adult life span. *Psychol. Med.* **30,** 11–22.

Kim, J. S., Choi, S., Kwon, S. U., and Seo, Y. S. (2002) Inability to control anger or aggression after stroke. *Neurology* **58,** 1106–1108.

Keller, J. N., Pang, Z., Geddes, J. W., et al. (1997) Impairment of glucose and glutamate transport and induction of mitochondrial oxidative stress and dysfunction in synaptosomes by amyloid beta-peptide: role of the lipid peroxidation product 4-hydroxynonenal. *J. Neurochem.* **69,** 273–284.

Keller, J. N., Kindy, M. S., Holtsberg, F. W., et al. (1998) Mitochondrial manganese superoxide dismutase prevents neural apoptosis and reduces ischemic brain injury: suppression of peroxynitrite production, lipid peroxidation, and mitochondrial dysfunction. *J. Neurosci.* **18,** 687–697.

Krajewski, S., Mai, J. K., Krajewska, M., Sikorska, M., Mossakowski, M. J., and Reed, J. C. (1995) Upregulation of bax protein levels in neurons following cerebral ischemia. *J. Neurosci.* **15,** 6364–6376.

Kromer Vogt, L. J., Hyman, B. T., Van Hoesen, G. W., and Damasio, A. R. (1990) Pathological alterations in the amygdala in Alzheimer's disease. *Neuroscience* **37,** 377–385.

Kruman, I. I., Kumaravel, T. S., Lohani, A., et al. (2002) Folic acid deficiency and homocysteine impair DNA repair in hippocampal neurons and sensitize them to amyloid toxicity in experimental models of Alzheimer's disease. *J. Neurosci.* **22,** 1752–1762.

Kumar-Singh, S., Dewachter, I., Moechars, D., et al. (2000) Behavioral disturbances without amyloid deposits in mice overexpressing human amyloid precursor protein with Flemish (A692G) or Dutch (E693Q) mutation. *Neurobiol. Dis.* **7,** 9–22.

Langston, J. W. (1996) The etiology of Parkinson's disease with emphasis on the MPTP story. *Neurology* **47,** S153–S160.

Li, P. A., Kristian, T., He, Q. P., and Siesjo, B. K. (2000) Cyclosporin A enhances survival, ameliorates brain damage, and prevents secondary mitochondrial dysfunction after a 30-minute period of transient cerebral ischemia. *Exp. Neurol.* **165,** 153–163.

Liu, D., Lu, C., Wan, R., Auyeung, W. W., and Mattson, M. P. (2002) Activation of mitochondrial ATP dependent potassium channels protects neurons against ischemia-induced death by a mechanism involving suppression of Bax translocation and cytochrome c release. *J. Cereb. Blood Flow Metab.* **22,** 431–443.

Lou, J. S., Kearns, G., Oken, B., Sexton, G., and Nutt, J. (2001) Exacerbated physical fatigue and mental fatigue in Parkinson's disease. *Mov. Disord.* **16,** 190–196.

Lusis, A. J. (2000) Atherosclerosis. *Nature* **407,** 233–241.

Lyketsos, C. G., Steele, C., Baker, L., et al. (1997) Major and minor depression in Alzheimer's disease: prevalence and impact. *J. Neuropsychiatry Clin. Neurosci.* **9,** 556–561.

Lyketsos, C. G., Steinberg, M., Tschanz, J. T., Norton, M. C., Steffens, D. C., and Breitner, J. C. (2000) Mental and behavioral disturbances in dementia: findings from the Cache County Study on Memory in Aging. *Am. J. Psychiatry* **157,** 708–714.

Lynch, D. R. and Dawson, T. M. (1994) Secondary mechanisms in neuronal trauma. *Curr. Opin. Neurol.* **7,** 510–516.

Mann, D. M., Oliver, R., and Snowden, J. S. (1993) The topographic distribution of brain atrophy in Huntington's disease and progressive supranuclear palsy. *Acta Neuropathol.* **85,** 553–559.

Mark, R. J., Hensley, K., Butterfield, D. A., and Mattson, M. P. (1995) Amyloid beta-peptide impairs ion-motive ATPase activities: evidence for a role in loss of neuronal Ca^{2+} homeostasis and cell death. *J. Neurosci.* **15,** 6239–6249.

Martinou, J. C., Dubois-Dauphin, M., Staple, J. K., et al. (1994) Overexpression of BCL-2 in transgenic mice protects neurons from naturally occurring cell death and experimental ischemia. *Neuron* **13,** 1017–1030.

Mattila, P. M., Rinne, J. O., Helenius, H., and Roytta, M. (1999) Neuritic degeneration in the hippocampus and amygdala in Parkinson's disease in relation to Alzheimer pathology. *Acta Neuropathol.* **98,** 157–164.

Mattson, M. P. and Scheff, S. W. (1994) Endogenous neuroprotection factors and traumatic brain injury: mechanisms of action and implications for therapy. *J. Neurotrauma* **11,** 3–33.

Mattson, M. P. (1997) Cellular actions of beta-amyloid precursor protein and its soluble and fibrillogenic derivatives. *Physiol. Rev.* **77,** 1081–1132.

Mattson, M. P., Partin, J., and Begley, J. G. (1998) Amyloid beta-peptide induces apoptosis related events in synapses and dendrites. *Brain Res.* **807,** 167–176.

Mattson, M. P. and Camandola, S. (2001) NF-kappaB in neuronal plasticity and neurodegenerative disorders. *J. Clin. Invest.* **107,** 247–254.

Mattson, M. P., Duan, W., Lee, J., and Guo, Z. (2001) Suppression of brain aging and neurodegenerative disorders by dietary restriction and environmental enrichment: molecular mechanisms. *Mech. Ageing Dev.* **122,** 757–778.

Mattson, M. P. (2002) Accomplices to neuronal death. *Nature* **415,** 377–379.

McNair, N. D. (1999) Traumatic brain injury. *Nurs. Clin. North Am.* **34,** 637–659.

Moechars, D., Gilis, M., Kuiperi, C., Laenen, I., and Van Leuven, F. (1998) Aggressive behaviour in transgenic mice expressing APP is alleviated by serotonergic drugs. *Neuroreport* **9,** 3561–3564.

Nemeroff, C. B., Youngblood, W. W., Manberg, P. J., Prange, A. J., Jr., and Kizer, J. S. (1983) Regional brain concentrations of neuropeptides in Huntington's chorea and schizophrenia. *Science* **221,** 972–975.

Ni, B., Wu, X., Su, Y., et al. (1998) Transient global forebrain ischemia induces a prolonged expression of the caspase-3 mRNA in rat hippocampal CA1 pyramidal neurons. *J. Cereb. Blood Flow Metab.* **18,** 248–256.

Nishimura, A., Ueda, S., Takeuchi, Y., Matsushita, H., Sawada, T., and Kawata, M. (1998) Vulnerability to aging in the rat serotonergic system. *Acta Neuropathol.* **96,** 581–595.

Paradiso, S., Robinson, R. G., and Arndt, S. (1996) Self-reported aggressive behavior in patients with stroke. *J. Nerv. Ment. Dis.* **184,** 746–753.

Pedersen, W. A., Wan, R., and Mattson, M. (2001) Impact of aging on stress-responsive neuroendocrine systems. *Mech. Ageing Dev.* **122,** 963–983.

Pedersen, W. A., Culmsee, C., Ziegler, D., Herman, J. P., and Mattson, M.P. (1999) Aberrant stress response associated with severe hypoglycemia in a transgenic mouse model of Alzheimer's disease. *J. Mol. Neurosci.* **13**, 159–165.

Persky, A. M. and Brazeau, G. A. (2001) Clinical pharmacology of the dietary supplement creatine monohydrate. *Pharmacol. Rev.* **53**, 161–176.

Pisarska, M., Mulchahey, J. J., Welge, J. A., Geracioti, T. D., Jr., and Kasckow, J. W. (2000) Age-related alterations in emotional behaviors and amygdalar corticotropin-releasing factor (CRF) and CRF-binding protein expression in aged Fischer 344 rats. *Brain Res.* **877**, 184–190.

Rainer, M. K., Masching, A. J., Ertl, M. G., Kraxberger, E., and Haushofer, M. (2001) Effect of risperidone on behavioral and psychological symptoms and cognitive function in dementia. *J. Clin. Psychiatry* **62**, 894–900.

Richardson, M. A., Haugland, G., and Craig, T. J. (1991) Neuroleptic use, parkinsonian symptoms, tardive dyskinesia, and associated factors in child and adolescent psychiatric patients. *Am. J. Psychiatry* **148**, 1322–1328.

Rojas-Fernandez, C. H., Lanctot, K. L., Allen, D. D., and MacKnight, C. (2001) Pharmacotherapy of behavioral and psychological symptoms of dementia: time for a different paradigm? *Pharmacotherapy* **21**, 74–102.

Rosenblatt, A. and Leroi, I. (2000) Neuropsychiatry of Huntington's disease and other basal ganglia disorders. *Psychosomatics* **41**, 24–30.

Roth, G. S. and Joseph, J. A. (1994) Cellular and molecular mechanisms of impaired dopaminergic function during aging. *Ann. NY Acad. Sci.* **719**, 129–135.

Sachdev, P., Smith, J. S., Matheson, J., Last, P., and Blumbergs, P. (1992) Amygdalo-hippocampectomy for pathological aggression. *Aust. NZ J. Psychiatry* **26**, 671–676.

Sapolsky, R. M., Alberts, S. C., and Altmann, J. (1997) Hypercortisolism associated with social subordinance or social isolation among wild baboons. *Arch. Gen. Psychiatry* **54**, 1137–1143.

Seshadri, S., Beiser, A., Selhub, J., et al. (2002) Plasma homocysteine as a risk factor for dementia and Alzheimer's disease. *N. Engl. J. Med.* **346**, 476–483.

Schilling, G., Jinnah, H. A., Gonzales, V., et al. (2001) Distinct behavioral and neuropathological abnormalities in transgenic mouse models of HD and DRPLA. *Neurobiol. Dis.* **8**, 405–418.

Schulz, J. B., Weller, M., and Moskowitz, M. A. (1999) Caspases as treatment targets in stroke and neurodegenerative diseases. *Ann. Neurol.* **45**, 421–429.

Sharp, F. R., Massa, S. M., and Swanson, R A. (1999) Heat-shock protein protection. *Trends Neurosci.* **22**, 97–99.

Shelbourne, P. F., Killeen, N., Hevner, R. F., et al. (1999) A Huntington's disease CAG expansion at the murine Hdh locus is unstable and associated with behavioural abnormalities in mice. *Hum. Mol. Genet.* **8**, 763–774.

Shih, J. C. and Chen, K. (1999) MAO-A and -B gene knock-out mice exhibit distinctly different behavior. *Neurobiology* **7**, 235–246.

Sieradzan, K. A. and Mann, D. M. (2001) The selective vulnerability of nerve cells in Huntington's disease. *Neuropathol. Appl. Neurobiol.* **27**, 1–21.

Sims, N. R. and Anderson, M. F. (2002) Mitochondrial contributions to tissue damage in stroke. *Neurochem. Int.* **40**, 511–526.

Stein-Behrens, B., Mattson, M. P., Chang, I., Yeh, M., and Sapolsky, R. (1994) Stress exacerbates neuron loss and cytoskeletal pathology in the hippocampus. *J. Neurosci.* **14**, 5373–5380.

Stoppe, G., Brandt, C. A., and Staedt, J. H. (1999) Behavioural problems associated with dementia: the role of newer antipsychotics. *Drugs Aging* **14**, 41–54.

Street, J. S., Clark, W. S., Kadam, D. L., et al. (2001) Long-term efficacy of olanzapine in the control of psychotic and behavioral symptoms in nursing home patients with Alzheimer's dementia. *Int. J. Geriatr. Psychiatry* **16**, S62–S70.

Sukonick, D. L., Pollock, B. G., Sweet, R. A., et al. (2001) The 5-HTTPR*S/*L polymorphism and aggressive behavior in Alzheimer disease. *Arch. Neurol.* **58**, 1425–1428.

Sullivan, P. G., Geiger, J. D., Mattson, M. P., and Scheff, S. W. (2000) Dietary supplement creatine protects against traumatic brain injury. *Ann. Neurol.* **48,** 723–729.

Sultzer, D. L., Gray, K. F., Gunay, I., Wheatley, M. V., and Mahler, M. E. (2001) Does behavioral improvement with haloperidol or trazodone treatment depend on psychosis or mood symptoms in patients with dementia? *J. Am. Geriatr. Soc.* **49,** 1294–1300.

Sweet, R. A., Nimgaonkar, V. L., Kamboh, M. I., Lopez, O. L., Zhang, F., and DeKosky, S. T. (1998) Dopamine receptor genetic variation, psychosis, and aggression in Alzheimer disease. *Arch. Neurol.* **55,** 1335–1340.

Tariot, P. N., Erb, R., Podgorski, C. A., et al. (1998) Efficacy and tolerability of carbamazepine for agitation and aggression in dementia. *Am. J. Psychiatry* **155,** 54–61.

Teichner, G., Golden, C. J., and Giannaris, W. J. (1999) A multimodal approach to treatment of aggression in a severely brain-injured adolescent. *Rehabil. Nurs.* **24,** 207–211.

Todes, C. J. and Lees, A. J. (1985) The pre-morbid personality of patients with Parkinson's disease. *J. Neurol. Neurosurg. Psychiatry* **48,** 97–100.

Van Praag, H. M. (2001) Anxiety/aggression—driven depression. A paradigm of functionalization and verticalization of psychiatric diagnosis. *Prog. Neuropsychopharmacol. Biol. Psychiatry* **25,** 893–924.

Veenema, H. C., Spruijt, B. M., Gispen, W. H., and van Hooff, J. A. (1997) Aging, dominance history, and social behavior in Java-monkeys (*Macaca fascicularis*). *Neurobiol. Aging* **18,** 509–515.

Vukhac, K. L., Sankoorikal, E. B., and Wang, Y. (2001) Dopamine D2L receptor- and age-related reduction in offensive aggression. *Neuroreport* **12,** 1035–1038.

Wang, P. L. and Smyers, P. L. (1977) Psychological status after stroke as measured by the Hand Test. *J. Clin. Psychol.* **33,** 879–882.

Wroblewski, B. A., Joseph, A. B., Kupfer, J., and Kalliel, K. (1997) Effectiveness of valproic acid on destructive and aggressive behaviours in patients with acquired brain injury. *Brain Inj.* **11,** 37–47.

Wszolek, Z. K., Gwinn-Hardy, K., Wszolek, E. K., et al. (2002) Neuropathology of two members of a German-American kindred (Family C) with late onset parkinsonism. *Acta Neuropathol.* **103,** 344–350.

Yankner, B. A. (1996) Mechanisms of neuronal degeneration in Alzheimer's disease. *Neuron* **16,** 921–932.

Yu, Z. F., Bruce-Keller, A. J., Goodman, Y., and Mattson, M. P. (1998) Uric acid protects neurons against excitotoxic and metabolic insults in cell culture, and against focal ischemic brain injury in vivo. *J. Neurosci. Res.* **53,** 613–625.

Zaman, S. H., Parent, A., Laskey, A., et al. (2000) Enhanced synaptic potentiation in transgenic mice expressing presenilin 1 familial Alzheimer's disease mutation is normalized with a benzodiazepine. *Neurobiol. Dis.* **7,** 54–63.

10
Environmental Factors and Aggression in Nonhuman Primates

Michael Lawrence Wilson

INTRODUCTION

Our primate relatives provide excellent models for understanding human aggression. Like us, most other primates are long-lived relatively brainy animals that live in complex societies in which aggression plays important roles. Unlike the favorite laboratory animals, rats and mice, which live in a nocturnal, underground world, primates are generally diurnal and readily observable in the wild. A few hours spent watching a troop of baboons or rhesus macaques should reward even the casual observer with examples of various aggressive interactions, from mild threats to squabbles and chases and perhaps even sustained attacks. Severe aggression, though rarely observed, marks many individuals with scars, disfiguring gashes, and open wounds. Our closest living relatives, chimpanzees, resemble humans in that they defend group territories and sometimes kill members of neighboring communities.

Many early studies of primates focused on aggression (e.g. Hall, 1964; Holloway, 1974; Zuckerman, 1932), and aggression remains a central topic of primate studies today. A search of Zoological Abstracts (1990–1998) found 1666 articles on nonhuman primate behavior, of which more than a quarter (446) concerned aggression (Howell, 1999). I will not attempt to review this entire literature here. Instead, I will focus on a few of the most important adaptive goals of primate aggression, reviewing some of the extensive literature on how primates use aggression strategically in response to environmental factors.

Definitions of Aggression

Aggression has been defined in numerous ways with some definitions including seemingly mild forms such as verbal aggression (*see* reviews in Eibl-Eibesfelt, 1979; Volavaka, 1995). Here, I will attempt to focus on the least ambiguous cases. I will pay particular attention to the small fraction of aggressive interactions that result in killing, which is an unambiguous outcome that provides a particularly clear assay on factors responsible for aggression (Daly and Wilson, 1996). As an additional consideration, killing is the outcome of aggression that most concerns researchers interested in understanding and preventing human violence such as child abuse, homicide, deadly ethnic riots, genocide, and warfare.

Kinds of Explanation

Many studies, including many chapters in this volume, focus on physiological mechanisms underlying aggression, such as brain structures and endocrine systems. Understanding physiological mechanisms is crucial, of course, but we now know that many of these mechanisms vary with environmental factors. Testosterone levels in male chimpanzees, for example, respond to social factors such as the number of fertile females available (Muller, 2002). A full understanding of any biological trait, including aggression, requires finding answers to four distinct kinds of explanation: ultimate (function), proximate (mechanism), development (ontogeny), and evolutionary history (phylogeny) (Daly and Wilson, 1983; Tinbergen, 1963).

Consider, for example, a male baboon about to start a fight. The male is a young adult who recently joined the troop, and his opponent is the troop's aging alpha male, recently weakened by disease. A functional explanation would focus on how the fight will affect the newcomer's reproductive success. By attacking the alpha male, the newcomer may gain alpha rank himself, enabling him to increase his share of matings with the troop's females. Proximate explanations might focus on the young male's neuroendocrine system, such as changes in serotonin and testosterone levels. Proximate explanations could also focus on the newcomer's assessment of the costs and benefits of fighting: in this case, the costs are unusually low, because the alpha male is weakened, and the potential benefits (gaining alpha rank) are enormous. Development also plays a role, especially life stage: the newcomer is now in his physical prime. A year or two ago he would have been too weak to challenge the alpha male, and in another few years his own physical powers will start to decline. The newcomer's current physiology may depend on many other factors in his development. Was his growth interrupted by harsh seasons of drought? Was his mother high ranking enough to compete successfully for food for herself and her offspring? Considering evolutionary history, male baboon psychology has been shaped by the costs and benefits of aggression over many generations. Over the years, it has paid for males to emigrate from the troop of their birth, join a new troop of strangers, and fight to attain high rank while still in their physical prime. Had he been born a male chimpanzee instead, he would have stayed in his natal group instead of seeking his fortunes elsewhere.

Keeping in mind that a complete understanding of aggression requires all four kinds of explanation, here I focus on the functional goals and economics of aggression. I briefly examine five potential goals of aggression: mates, food, land, killing of adults, and killing of infants. Aggression may serve various additional functions, such as defense of self and offspring from predators, but limited space prohibits an exhaustive survey. For each goal, I discuss potential benefits and costs involved in attaining that goal. I then describe how those costs and benefits might vary depending on environmental factors.

Costs and Benefits of Aggression

In a recent review of primates as models of human aggression, Kalin (1999) observed that "all forms of aggression in rhesus monkeys appear to be modulated by environmental factors." This observation applies equally to other primates and indeed most animals. Like any other product of natural selection, aggressive behavior should be designed to benefit the aggressor's "inclusive fitness," which is evolutionary jargon for the reproductive success of an individual (and the individual's kin, insofar as they are affected)

(Daly and Wilson, 1983). Because the costs and benefits of aggression vary according to circumstances, individuals should not perform aggressive acts blindly or automatically, but should, instead, modulate their aggression according to relevant environmental factors.

Modern behavioral biology thus views aggression as a strategic option to be used when assessment of the odds indicates that the fitness benefits will outweigh the costs (for excellent reviews, see Archer, 1988; Huntingford and Turner, 1987). Important benefits of aggression include access to resources essential for reproductive success, especially mates, food, and territory. Costs of aggression include energy expenditure, the risk of injury or death, and opportunity costs, in that time spent fighting could be spent looking for food or mates instead. The relative balance of costs to benefits varies with environmental factors. For example, theoretical considerations suggest that animals should avoid fighting opponents that are larger, better armed, or more numerous than themselves (Parker, 1974). Animals also should be more likely to fight over valuable resources, such as fertile females or rich food sources (Parker, 1974). Such assessment of environmental factors has been demonstrated experimentally in diverse species including spiders (Leimar et al., 1991), toads (Davies and Halliday, 1978), and lions (McComb et al., 1994).

In practice, the actual costs and benefits involved are usually difficult to measure, especially in the wild, where many factors vary simultaneously and interact in complicated ways. Fortunately, our understanding of aggression in wild primates now benefits both from technological advances and from the maturation of long-term observational studies. Molecular technology enables field workers to collect genetic and endocrine data from hair, feces, and urine without harming or even directly contacting the study subjects. Using these methods, we can now test whether types of aggression, such as infanticide, yield reproductive benefits (Borries et al., 1999) and how environmental factors influence hormone levels (Muller, 2002). Studies of intergroup aggression now rely on playback experiments, using portable speakers and high-quality recordings to simulate intruders (Kitchen, 2000; Mitani, 1990; Wilson et al., 2001). Maturing long-term studies of many species and multiple populations of the same species enable rich comparative tests of how and why aggression varies over space and time. Computerizing long-term data on ranging and other behavior enables quantitative tests, such as the specific benefits obtained by territorial expansion (Williams, 2000; Williams and Pusey, submitted).

Aggression Controversies: Function Vs Dysfunction

Explanations of aggression in terms of inclusive fitness calculations continue to provoke controversy, especially when applied to lethal aggression, and even more so when applied to humans (Lewontin, 1999; Sommer, 2000; Sussman, 1999; Wrangham, 1999). Much of this controversy stems from enduring hostility toward any biological explanation of human behavior (reviewed in Niehoff, 1999; Sommer, 2000; Tooby and Cosmides, 1992). The controversy also has roots in the attempt by early ethologists, such as Konrad Lorenz, to counter the Victorian view that nature is "red in tooth and claw." Lorenz regarded aggression as an adaptation, but he asserted that animals do not intentionally kill one another in the wild owing to a supposed inhibition against killing members of their own species (1966).

Early field studies supported Lorenz's view. Most animals are peaceful most of the time; long studies are needed to observe severe fights among long-lived slow-reproducing animals in the wild. Schaller, for example, emphasized that during the year he spent observing gorillas in the wild, they appeared to be gentle giants (Schaller, 1964). Longer field studies, however, found that male gorillas can fight fiercely and inflict severe wounds (Fossey, 1983). Male gorilla skulls frequently show evidence of healed bite wounds (Jurmain, 1997), including canines embedded in the skull (Fossey, 1983). Moreover, both male and female gorillas sometimes kill the infants of other gorillas, and such infanticidal attacks account for up to 37% of infant mortality (Fossey, 1984; Watts, 1989).

Despite the continued popularity of the view that animals are "naturally pacifists" (Lorenz, 1966; Montagu, 1976; Power, 1991), accumulating evidence permits us to discard this hypothesis. The consensus in animal behavior studies is that aggression, including deadly aggression, is simply part of ongoing competition in a world of limited resources (Archer, 1988; Huntingford and Turner, 1987).

Despite the growing consensus in behavioral biology, studies of human aggression rarely examine fitness costs and benefits (for notable exceptions, *see* Daly and Wilson, 1988; Wilson and Daly, 1998). Instead, many discussions of environmental influences view aggression as a dysfunctional response to factors such as overcrowding (Ostfeld and D'Atri, 1975), poverty (Pagani et al., 1999), too much television (Johnson et al., 2002), or abusive parents (Barnow et al., 2001). Although such factors clearly have important effects on aggression in humans, we should be cautious of the implicit assumption that aggression is bad, and that a bad environment leads to bad behavior. Focusing on variables perceived as "bad" can distract attention from factors with a more direct causal link to aggression.

For example, a widespread view is that high levels of aggression result from high population densities. Although, in general, individually measured rates of aggression do increase with population density, many studies have found only a weak effect or an inverse relation (Judge and de Waal, 1997; Moore, 1999). From an evolutionary perspective, there is no reason to suppose that animals should require an ideal population density to behave adaptively. Indeed, population density varies greatly among populations in the wild and, by definition, should be highest in habitats most suitable for the species in question. Focusing on population density obscures the influence of variables more likely to have a direct influence on aggression, such as intruder pressure, population age structure, and the abundance and distribution of food (Janson and van Schaik, 2000; Moore, 1999). In general, examining the functional goals and related costs and benefits of aggression should result in sharpened evolutionarily relevant hypotheses.

MATES

Benefits

In any sexual species, reproductive success obviously requires mating success. The benefits of competing for mates, however, differ between the sexes. Mating among mammals involves a highly asymmetric bargain (Trivers, 1972). Females provide an egg and a commitment to gestate, lactate, and raise the young. Males provide sperm and usually little else. Because females invest so much in each pregnancy, they tend to be choosy about their mates. Because males stand to gain so much from each mating, males tend to mate with as many females as possible, and compete intensely with other males for mating opportunities. Numerous exceptions exist to the general trend of coy choosy females and promiscuous caddish males. Female chimpanzees, for example, are far from coy; instead, they often attempt to mate with every male in their community, probably as an anti-infanticide strategy (Wrangham, 2002). Male tamarin monkeys help feed and carry their young (Goldizen, 1988), and males in many species defend their young from predators (van Schaik and Hörstermann, 1994) and potentially infanticidal rival males (van Schaik, 2000b). Nevertheless, even in the exceptional cases, females invest far more in their offspring than males. Because of this asymmetry, males are much more likely to fight for a mate than females, and fights between males are much more likely to lead to injury or death.

Costs

Aggressors face high potential costs when competing for mates, including energy expenditure and risk of injury or death from fighting. Competing for mates also can incur more subtle costs. For example, male baboons often "mate guard" by traveling closely with a fertile female, mating with her, and repelling rival males. While mate guarding, males reduce their daily travel distance to match that of the females they guard, reducing their food intake (Alberts et al., 1996). In species with intense contest competition for mates, males face the additional cost of growing the large body size, muscle mass, and canine weaponry needed to compete successfully. Male gorillas, for example, grow to twice the size of female gorillas. In most mammals, the only environmental limitation on growth to full adult size is sufficient nutrition. Among orangutans, however, male growth also depends on the social environment (*see* Operational Sex Ratio).

Environmental Factors

When competing for mates, males fight most intensely when the net benefits appear to be highest. The primary environmental factor favoring aggressive competition for mates is, of course, the presence of fertile females. The intensity of male competition also depends on two additional factors: the operational sex ratio and the ratio of young males to older males in the population.

Fertile Females

The frequency of male aggression varies closely with the number of mating opportunities. In seasonally breeding species, such as rhesus macaques, males injure one another most often during the mating season. For example, in the colony of free-ranging rhesus macaques on Cayo Santiago, Puerto Rico, 87% of male deaths occurred during the mating season (Wilson and Boelkins, 1970). Male rhesus macaques exhibit seasonal changes in sexual behavior, testes mass, and scrotal color. Experiments with ovariectomized females artificially brought into estrus (with estradiol benzoate) revealed that the presence of an estrus female is sufficient to induce all of these physiological changes in sexually quiescent males (Vandenbergh, 1969)

In species that breed year round, aggression varies depending on the cycles of potentially fertile females. Among chimpanzees, males modulate their fighting effort according to the likelihood of conception. Cues that indicate likelihood of conception include sexual swellings and individual reproductive history. Female chimpanzees advertise their fertility with a large, bright pink anogenital swelling, and may provide other (perhaps olfactory) cues to the timing of ovulation (Wrangham, 2002). Females who have had at least one offspring (parous) are more likely to conceive than females who have never given birth (nulliparous). Males therefore compete intensely for access to parous females with sexual swellings, but do not fight over (and may even ignore) fully swollen nulliparous females. Male aggression also varies within each female's cycle. Early in a female's cycle, before she is likely to conceive, she often mates with all available males. As probability of conception increases, however, tensions rise. Muller (2002) found that when parous females were maximally swollen, males had higher testosterone levels, and aggression was more frequent and more severe. High-ranking males may guard the female, threatening or fighting any male who approaches her. Males may even team up with one or more partners in coalitionary mate guarding (Watts, 1998).

Operational Sex Ratio

Male mating opportunities depend not only on the number of females in the population, but how often those females are fertile, which in turn depends on interbirth intervals. A statistic designed to capture this information is the operational sex ratio, which is defined as the number of breeding females that will be available to each male each year (Emlen and Oring, 1977; Mitani, 1990). If the operational sex ratio is particularly low, intense fighting can result. A horrific unintended experiment in extremely low operational sex ratios occurred in the 1920s, when zookeepers at the London Zoo formed a group of 100 hamadryas baboons that contained only six females (de Waal, 1989; Zuckerman, 1932). In the wild, hamadryas baboons live in multilevel societies based on "one male units," which are groups of females fiercely defended by a single male (Kummer, 1968). At the London Zoo, the males fought brutally with one another and herded the females mercilessly. Fighting continued even after 30 additional females were added. "Six and a half years later, the few surviving females were removed. Sixty-two males and thirty-two females, over two-thirds of the original population, had died of stress and injuries" (de Waal, 1989). Although such a bloodbath has never been recorded from the wild, skewed sex ratios can lead to fatal fighting. In Budongo Forest, Uganda, male chimpanzees ganged up on and killed a male of their own community during a time of particularly intense mating competition (Fawcett and Muhumuza, 2000).

Over evolutionary time, low operational sex ratio favors the evolution of large male body size (Mitani et al., 1996). In rare cases, development of large male body size depends on the social environment. Orangutan males, for example, occur in two morphs: big males and small males (Maggioncalda et al., 1999; Rodman and Mitani, 1987). Big males are nearly twice the weight of females, with conspicuous fibrous pads (flanges) on the sides of their face. Small males are about the same size as adult females and lack flanges and other secondary sexual traits. Small size also may serve as a sort of sexual mimicry, in that big males tolerate small males, but challenge other big males who enter their range (Mitani, 1985). Body size also relates to strategies for aggression: whereas big males fight other big males, small males adopt a "sneaking" strategy, mating with females (often coercively), but avoiding fights with other males.

These two morphs appear to be the result of developmental flexibility rather than fixed genetic differences. Small males can stay small for many years, but then grow big when circumstances are more favorable. Male orangutans thus delay the costs of growing to large body size until social circumstances suggest the investment in growth will pay off.

Relative Number of Young Males

While most primate studies focus on the number of adult males in a population, the age structure of the male population may also affect rates of aggression. This factor appears relatively neglected in primate studies, but has been addressed in several human studies. Human males, like other primates, are most aggressive as young adults (e.g., 15–30 yr old). In the United States, regions that had unusually large numbers of young males, such as the western frontier, experienced high crime rates (Courtwright, 1996). Mesquida and Wiener (1996) argue that, in humans, male coalitionary aggression is best understood as a "reproductive-fitness-enhancing social behavior" and that such aggression should be most frequent when a society contains a relatively large number of young males. Analysis of collective aggression in a wide variety of modern states found that societies with relatively many young males had more conflict-related deaths (Mesquida and Wiener, 1996). A large ratio of young to old males is a feature of rapidly growing populations. This finding suggests that intense male–male aggression should be a widespread trait of rapidly growing populations.

FOOD

Benefits

Primates in the wild spend much of their time either feeding or looking for food. Food is, of course, essential for both males and females, but is particularly important for females who gestate, nurse, and carry their offspring. Therefore, food is a crucial limiting factor for female reproductive success (Trivers, 1972). Food quality and distribution can affect grouping patterns and intergroup relations and have figured prominently in discussions of primate social evolution (Isbell, 1991; van Schaik, 1989; Wrangham, 1980).

Costs

Fights over food are generally less severe than fights over mates. Although food is essential for survival and reproduction, a given piece of fruit or leaf is usually low in value compared to the risk of injury from fighting. Overt aggression, therefore, should be limited to cases in which the contested item is particularly rare and valuable (e.g., Stevens and Stephens, 2002).

Environmental Factors

In general, animals are more likely to fight over food that is high quality and defendable. Chimpanzees, for example, mainly eat fruit and other plant parts, but sometimes hunt monkeys and other vertebrates. Meat obtained in such hunts elicits considerable excitement, feeding competition, and begging from chimpanzees (Goodall, 1986). Clumps of abundant food may influence aggression in at least three ways: (*i*) they serve as a focus for competition; (*ii*) they provide energy for fighting; and (*iii*) in species with fission–fusion societies, such as chimpanzees, abundant food brings together large parties, which may then be more likely to visit borders with neighboring groups.

Focus for Competition

Although natural foods, such as meat, can elicit aggression, an especially pronounced effect occurs with the introduction of human foods. In many early studies, researchers provisioned primates with cultivated foods to facilitate observations. Such provisioning provided exactly the sort of food distribution most likely to lead to increased aggression. For example, at Gombe, rates of aggression, including wounding, were higher during the early 1960s, when provisioning was intense, than in the late 1960s and 1970s, when provisioning was reduced (Wrangham, 1974). Other human sources of food, such as garbage pits, may produce similar effects on wild primate populations (Altmann and Muruthi, 1988). A study of rhesus monkeys found that rates of aggression were highest in captivity and in provisioned sites, such as farms and temples, with aggression occurring much less frequently in forest sites (Southwick, 1969). Although Southwick attributed the variation in rates of aggression to population density, food distribution may have been a more salient factor, given that the high-density populations all received food from humans.

Energy for Fighting

In general, aggression intensifies when resources are scarce or limited. Among wedge-capped capuchins, for example, fighting in large groups increases when food is scarce (Miller, 1996). Nonetheless, extremely limited food supplies can actually reduce aggression, as individuals lack the energy to waste on fighting. Southwick (1969) found that imposing food shortages on captive rhesus monkeys led to reduced rates of agonistic behavior. Abundant food can also free up energy for high-risk high-gain activities, such as hunting. For example, in contrast to the expectation that chimpanzees hunt animals when plant foods are scarce, chimpanzees in Kibale National Park, Uganda, hunt more often when fruit is more abundant (Watts and Mitani, 2002).

Food for Coalitionary Power

In chimpanzees, success in intergroup encounters depends on relative party size. Abundant food sources could increase the likelihood of border patrols, both by providing energy for long-distance travel and by bringing together the many males needed for safe patrolling. Testing this hypothesis is one goal of the current analyses of long-term data records of the chimpanzees of Gombe National Park, Tanzania.

STATUS

Benefits

In addition to fighting over specific resources, such as food and mates, primates frequently compete over status. In most primate species, individuals of one or both sexes can be ordered in a linear dominance hierarchy. Individuals generally appear keenly aware of their own place in the hierarchy. Chimpanzees, for example, give a specific call, the "pant–grunt," when approaching higher ranking individuals. High rank generally provides priority of access to food, mates, and other resources. Recent analyses of DNA from wild populations supported the prediction of observational studies that high-ranking males father more offspring in baboons (Altmann et al., 1996), chimpanzees (Constable et al., 2001), and pygmy chimpanzees or bonobos (Gerloff et al., 1999).

High rank may also provide some protection from stress, depending on the stability of the hierarchy (Virgin and Sapolsky, 1997).

Even when rank is difficult for human observers to assess, it can have important fitness consequences. For example, dominance interactions between female chimpanzees are subtle, and for many years, researchers assumed social status did not have great importance for them. In contrast to this assumption, analysis of long-term records from Gombe National Park, Tanzania, found important effects (Pusey et al., 1997). Higher-ranking females had "higher infant survival, faster maturing daughters, and more rapid production of young." High-ranking females appeared able to acquire better home ranges within the community's territory and, thus, had access to a better food supply.

Costs

Status provides a cue to the cost of attacking a given individual. High-status individuals are dangerous to attack, either because they are big and strong or because they have many powerful allies. Loss of status can be especially costly. Falling from top rank can lead to a rapid and irreversible decline. With such high stakes, fights for alpha status can be especially fierce and are sometimes fatal (Kitopeni et al., 1995; Nishida, 1996). During times of intense competition for ranks, high-ranking individuals may suffer from increased stress (Alberts et al., 1992).

Environmental Factors

Status depends to some extent on individual traits, such as age, fighting ability, and political skills. Status also depends on factors in the individual's environment, particularly the number of kin and the number and quality of rivals.

Kin

Many species of *Cercopithecines*, the group that includes baboons, rhesus macaques, and vervet monkeys, live in groups in which most females are related to one another, and most males are unrelated immigrants (reviewed in Dunbar, 1988). Female status depends largely on the number and rank of her kin. Large matrilines generally outrank small matrilines. Younger daughters rank higher than their older sisters, because the younger ones are born with more kin. Individuals with more kin win more fights and may be more likely to initiate fights with members of rival matrilines.

Among chimpanzees and bonobos, males stay in their natal group and females usually emigrate at adolescence (Pusey, 1979; Gerloff et al., 1999). Because males stay in their natal group, average relatedness among males may be on the order of half-siblings (Morin et al., 1994), facilitating within-group cooperation. Male chimpanzees may sometimes form coalitions with brothers, but because of long birth intervals, chimpanzees rarely have brothers close enough in age to form useful coalition partners. Mitochondrial DNA studies show that in chimpanzees many coalition partners are not maternal siblings (Goldberg and Wrangham, 1997; Mitani et al., 2000). Among bonobos, male rank and reproductive success may depend on support from their mothers (Gerloff et al., 1999).

Number and Quality of Rivals

Dominance hierarchies are most stable when the top-ranking individual is clearly more powerful than any rivals. In chimpanzees, a powerful alpha male can maintain his rank for many years. In Mahale National Park, Tanzania, the male Ntologi maintained his alpha status for 16 yr before being deposed and killed by his rivals (Kitopeni et al., 1995; Nishida, 1996). Status striving is much more intense when many equally matched contenders are jostling for rank (Maynard Smith, 1982).

LAND

Benefits

Many primate species defend territories, warning outsiders with loud calls and attacking intruders. Either or both sexes may participate in territory defense. For females, the primary benefits of a territory are food for self and offspring. Males may benefit both by defending a feeding territory for self, females, and offspring, and by defending females (Fashing, 2001).

Costs

Territory defense is energetically expensive and can result in costly fights with rival groups. The loud calls primates commonly produce to advertise territory ownership can also attract the attention of unintended audiences such as predators (Wilson et al., in review; Zuberbuhler et al., 1997).

Environmental Factors

Primates defend territories when it is economically feasible to do so (Mitani and Rodman, 1979; Lowen and Dunbar, 1994). Mitani and Rodman (1979) proposed an "index of defendability (D), which is the ratio of observed daily path length (d) to an area equal to the diameter (d') of a circle with area equal to home range area of the animal." Species with an index of $D = 1.0$ or greater, thus had day ranges that were long compared to the size of their territory. Mitani and Rodman found that all territorial species and few nonterritorial species had an index of $D = 1.0$ or greater. Lowen and Dunbar (1994) developed a refined version of this model that generally supported Mitani and Rodman's earlier findings. Although these models focused on comparisons among species, territory defense varies within species as well. For example, chimpanzees defend territories in forest sites, but the enormous home ranges occupied in drier savanna sites are probably not economical to defend.

Territorial behavior may vary among members of a group, depending on an individual's stakes in defending the territory. For example, if high-ranking males gain a disproportionate share of matings, low-ranking males may have little incentive to join in intergroup fights (Nunn, 2000).

In forest habitats, male chimpanzees patrol the boundaries of their territories and show hostility toward any members of other communities except for estrous females without offspring (Goodall, 1986; Williams and Pusey, submitted). Chimpanzee intergroup relations include fatal attacks, which will be discussed more in the following section.

KILLING ADULTS

Perhaps the most widespread concern underlying studies of aggression is the question, "Why do people kill one another?" Enormous attention has focused on questions such as whether such killing is uniquely human, and if other animals do kill conspecifics, why they do so. Despite early assertions to the contrary, intraspecific killing is not uniquely human. Animals sometimes kill both infants and adults of their own species. Because the risks involved and underlying evolutionary logic differ depending on the age of the victim, I discuss infanticide in the next section and focus on killing of adults here.

Lethal fighting is readily observed in small animals such as arthropods, including many species of ants (Holldobler and Wilson, 1990), fig wasps (West et al., 2001), and spiders (Leimar et al., 1991). Among larger animals, intraspecific killing occurs infrequently, but can account for a large fraction of adult mortality (reviewed in Gat, 1999; Wrangham, 1999). For example, despite Montagu's assertion that wolves "do not attack other wolves" (Montagu, 1979), field observations reveal that intraspecific fighting accounts for 39–65% of adult wolf mortality (Mech et al., 1998).

Intraspecific killing has been reported for at least eight primate species. Most attention has focused on intergroup "wars" in our closest relatives, chimpanzees (Manson and Wrangham, 1991). Killing also occurs occasionally in other apes, including orangutans (Knott, 1998) and gibbons (Palombit, 1993). In addition to the apes, fights sometimes result in fatal injuries in several monkey species, including baboons (Starin, 1994) and rhesus monkeys (Lindburg, 1971; Westergarrd et al., 1999). In red colobus monkeys, coalitions of male and female residents have killed potentially infanticidal male intruders (Starin, 1994). Recent observations have revealed fatal fighting in two species of capuchin monkeys (Miller, 1998; Perry, Manson, and Gros-Louis, unpublished data). Reports of intraspecific killing in other species will likely emerge as additional long-term studies mature.

Benefits

Killing is widely regarded as the result of accidental or unusual circumstances, rather than the goal of aggression. Lorenz, for example, described numerous cases of killing among fishes in captivity, but attributed these deaths to the inability of the victims to escape in confined quarters (Lorenz, 1966). In support of this view, many deaths in the wild do result from wounds that become infected rather than from a concerted effort by the attacker to ensure his rival's death (e.g., Drews, 1996).

Nevertheless, attackers may obtain at least four benefits from killing their rivals. First, attackers may gain immediate nutritional benefits by eating their rival. Such predatory attacks are widespread among fish and arthropods, but are rare among mammals (Polis, 1981). Chimpanzees, for example, often eat victims of infanticide, but rarely if ever eat adult victims (Watts and Mitani, 2000). Second, killing rivals directly eliminates competitors for status or for resources, such as food and mates. Third, killing unrelated rivals reduces the genetic contribution of rivals to the population. A fourth benefit applies to species that defend group territories. In general, larger groups tend to defeat smaller groups in territorial fights. By killing rivals, attackers reduce the numerical strength of rival coalitions, thereby increasing their chances of success in future territorial contests (Wrangham, 1999).

In some cases, the unintended death of the victim may even cost the attacker. For example, Sapolsky describes a case in which a female baboon died following infection of a bite wound inflicted by a male of her troop (Sapolsky, 2001). It seems unlikely that this male benefited from killing a potential mate.

Costs

The apparent inhibition against killing conspecifics results not from an urge to preserve one's species, but from the fact that killing rivals is usually difficult and dangerous. The costs of killing can be reduced dramatically, however, in species with coalitionary bonds (Manson and Wrangham, 1991). In a discussion of the costs of killing, Wrangham (1999) distinguished three categories of species. In the first category, which includes most species, killing occurs infrequently if at all. The second category includes a smaller number of species in which fights are dyadic, but frequently result in death. For example, 12% of male pronghorn antelope and 5–10% of male musk oxen may die from fighting during the annual mating season (Byers, 1997; Wilkinson and Shank, 1976). For such species, individuals compete over such high stakes (such as access to many fertile females) that even costly fights may pay off (Enquist and Leimar, 1990). The third category consists of species in which killing is frequent, but involves coalitions or "gang attacks." If members of one coalition greatly outnumber their rival, they can kill the rival without much risk of injury for themselves. Wrangham (1999) included two primate species in this category: humans and chimpanzees (with western red colobus a potential candidate).

Manson and Wrangham (1991) argue that gang attacks should occur most frequently in species with fission–fusion social organization, in which group members travel in subgroups (parties) of variable size and composition. Many primates, such as baboons and rhesus monkeys, live in stable troops in which opportunities for gang attacks rarely occur. Coalitionary killing has been reported, however, for some populations of baboons (Popp, 1978), rhesus macaques (Lindburg, 1971), and white-faced capuchins (Perry, Manson, and Gros-Louis, unpublished data).

Environmental Factors

Dyads

In species with only dyadic fights, killing should be rare except when the potential benefits are unusually high. Among primates, such fights appear most common during competition for mates and when males attempt to join a new group (Wilson and Boelkins, 1970; Brain, 1992).

Coalitions

Among species with coalitionary fighting, factors likely to influence the frequency of gang attacks include the following: (*i*) numerical imbalances; (*ii*) distribution and availability of food; and (*iii*) estrous females.

NUMERICAL IMBALANCES

Observations and field experiments show that chimpanzees assess numerical imbalances before approaching intruders (Watts and Mitani, 2001; Wilson et al., 2001). Watts and Mitani found that chimpanzees were more likely to approach the calls of neighboring groups when in parties with many males. In these observations, however, the party composition of neighboring groups was generally unknown. Wilson and colleagues conducted a series of playback experiments for a more controlled test (Wilson et al., 2001). Playback experiments found that chimpanzees were more willing to call to and approach a simulated intruder the more they outnumbered the intruder (Wilson et al., 2001). Parties with one or two males remained quiet, approached in only half the cases, and approached more slowly when they did approach. In contrast, parties with three or more males gave a loud vocal response and rapidly approached the speaker. Chimpanzees are more likely to visit borders when in larger parties (Bauer, 1980; Wilson, 2001) and boundary patrols tend to contain large numbers of males (Watts and Mitani, 2001).

Playback experiments with howler monkeys also show evidence of numerical assessment (Kitchen, 2000). Kitchen played recordings of the howls of one or three male intruders to defenders in groups with one, two, or three males. Alpha males gave a weaker howling and approach response when played the howls of larger groups. Low-ranking males only howled and approached if their group outnumbered the simulated intruders.

Gang attacks should be more likely in cases in which a large group neighbors a small group. Although party size should vary for both groups, the larger group has a greater maximum party size and, therefore, has better odds of meeting the neighbors with an overwhelming numerical advantage. At Gombe, the main study community (Kasakela) now greatly outnumbers its two neighboring communities. In recent years, the Kasakela males have expanded their range considerably and have brutally attacked members of both neighboring communities (Wilson et al., submitted).

Even in cases where neighboring groups are of similar size, males in large groups may visit borders more often than males in small groups. The largest chimpanzee community studied to date, the Ngogo community of Kibale National Park, Uganda, contains at least 144 chimpanzees (Watts and Mitani, 2001). Ngogo chimpanzees patrol their boundaries at an unusually high rate. Only one of the communities neighboring Ngogo, the Kanyanchu community, has been habituated. Kanyanchu appears to be a very large community and may be similar in size to Ngogo.

In captive situations, management practices can create opportunities for gang attacks. For example, rhesus monkeys live in troops composed of one or more female kin groups (matrilines). Related females frequently join in fights against members of rival matrilines. Such fights rarely lead to severe injuries under normal circumstances. In captive groups, however, members of matrilines are frequently separated for management purposes. Under such circumstances, members of rival matrilines may attack and kill a female separated from her kin (Westergarrd et al., 1999).

FOOD AVAILABILITY AND DISTRIBUTION

In fission–fusion societies, party size depends largely on food availability. When more food is available, larger parties form. Chimpanzees depend on ripe fruit, and the availability of ripe fruit can vary greatly over short distances. The possibility exists that one community could enjoy a bumper crop of fruit while their neighbors suffer through a lean season. The community with more fruit would travel in larger parties and would have an increased chance of meeting their neighbors with overwhelming numerical superiority.

Food availability also affects opportunities for killing, if it enables group members to travel in more stable parties. Among bonobos and some populations of chimpanzees, the costs of grouping appear to be reduced by the availability of high-quality herbaceous plants that enable individuals to continue feeding when traveling between fruit trees (Wrangham et al., 1996). Increased stability of parties may be responsible for the lack of observed coalitionary killing in both bonobos and chimpanzees of Taï National

Park, Côte d'Ivoire, despite hostile intergroup relations in both populations (Wrangham, 1999).

ESTROUS FEMALES

In chimpanzees, party size also varies with the number of estrous females. Estrous females attract many males, even when fruit supplies are poor. One intergroup killing at Kibale occurred when many males traveled with an estrous female into border regions (Kibale Chimpanzee Project, unpublished data).

KILLING INFANTS

In many primate species, infants face a serious risk of being killed by conspecific adults. Sociobiological explanations of infanticide have generated a large and contentious literature, with controversy continuing today (Bartlett et al., 1993; Curtin and Dolhinow, 1978; Hausfater and Hrdy, 1984; Hrdy, 1974; Schubert, 1982; Sussman, 1997; Sommer, 2000). Infanticide has been directly observed in the wild in 17 primate species and is known or strongly suspected to occur in a total of 39 primate species (van Schaik, 2000a). Infanticide also occurs in many other mammals, mainly carnivores and rodents, but also dolphins, horses, and hippos (van Schaik, 2000a). Infanticide rates vary among species and populations, but can account for a large portion of infant mortality. At least 30–40% of infant mortality results from infanticide in mountain gorillas (Watts, 1989), chacma baboons (Palombit et al., 2000), and howler monkeys (Clarke and Glander, 1984). In most cases, the infant killers are male, though in chimpanzees, gorillas, and others, including some rodents, females may commit infanticide (Goodall, 1977; Digby, 2000).

Benefits

Both nonadaptive and adaptive hypotheses have been proposed for infanticide. The two major nonadaptive hypothesis are that infanticide is (*i*) a social pathology caused by some disturbance, such as unusually high population density (Curtin and Dolhinow, 1978) or (*ii*) an accidental byproduct of generalized aggression (Bartlett et al., 1993; Sussman et al., 1995). Neither of the nonadaptive hypotheses explains the species distribution of infanticide or its occurrence in undisturbed environments (e.g., Enstam et al., 2002). While disturbance or pathology may account for some cases, current evidence supports the view that, in general, infanticide is a behavioral strategy that provides fitness benefits to infant killers (Palombit, 1999; van Schaik, 2000b). The particular benefits obtained may differ between the sexes and among species with different social organization.

The most prominent adaptive hypotheses for infanticide is the sexual selection hypothesis (Hrdy, 1974). This hypothesis proposes that males kill the infants sired by rival males to bring the infant's mother into reproductive status faster. Such killings usually take place after the infanticidal male has taken over a troop (e.g., hanuman langurs) (Hrdy, 1977) or entered a new troop (e.g., baboons) (Palombit, 1999). The sexual selection hypothesis now enjoys strong support (van Schaik, 2000c). Genetic testing has shown that male hanuman langurs selectively killed unrelated infants and were the most likely father of the mother's next infant (Borries et al., 1999). A similar hypothesized benefit is that infant killing may induce females to join the attacker's group, as the attacker has demonstrated the inability of the female's current male to defend her (e.g., gorillas) (Fossey, 1983; Watts, 1989; Wrangham and Peterson, 1996).

The sexual selection hypothesis does not explain all cases of infanticide. Among chimpanzees, for example, many infanticides occur during intercommunity conflict, with little evidence that infant-deprived mothers later mate with the infant's killers. In addition to increased opportunities for mating, killing infants may provide at least four additional benefits to attackers, comparable to the benefits of killing adults (Hrdy, 1977; van Schaik, 2000c). First, attackers may gain immediate nutritional benefits by eating their rival. Among chimpanzees, for example, attackers often do eat infant victims (Watts and Mitani, 2000), but such cannibalism is rare among other primates and, even in chimpanzees, does not appear to be the primary motive for infanticide. Second, killing unrelated infants eliminates future competitors for resources such as food and mates. Third, killing unrelated infants reduces the genetic contribution of rivals to the population. Fourth, in species that defend group territories, killing infants of rival groups can reduce the rival group's coalitionary strength. Among chimpanzees, infanticide may depend on some combination of these factors (Arcadi and Wrangham, 1999; Watts and Mitani, 2000).

Costs

The primary cost to infanticide is that mothers, potential fathers, and others may defend infants. Among baboons, for example, female baboons form "friendships" with males who defend them against attack by potentially infanticidal male immigrants (Palombit, 1999). Another potential cost of infant killing is that in cases of uncertain paternity, males could kill their own offspring. Male chimpanzees of M-group in Mahale National Park, Tanzania, killed the infant of a female who had been absent from the group for several months prior to birth (Takahata, 1985). Takahata believed the infant was sired by M-group males, raising the possibility that a father mistakenly killed his infant. Though such mistakes may well happen, the possibility exists that, in this case, the chimpanzees evaluated paternity more accurately than the researchers. In the great majority of cases in which observers can estimate paternity with confidence, the killers are in fact unrelated to their infant victims (van Schaik, 2000c).

Environmental Factors

Environmental factors proposed to affect infanticide rates include population density, intruder pressure, the number of males in groups, and female dispersal patterns.

Population Density

A popular explanation holds that infanticide (along with other forms of aggression) is a pathological response to high population density (Bartlett et al., 1993; Curtin and Dolhinow, 1978; Judge and de Waal, 1997; Moore, 1999). Population density, however, is a poor predictor of infanticide frequency (Moore, 1999; van Schaik, 2000c). Infanticide occurs in populations with low density, such as patas monkeys (Enstam et al., 2002) and some langur populations (Newton, 1986). Among blue monkeys, comparison of high- and low-density populations found that infanticide rates were actually higher at low population density (Butynski, 1990).

Intruder Pressure

In most cases, male intruders represent the greatest threat of infanticide. Rates of infanticide should therefore vary with the rate at which lone males attempt to join or take over troops (Moore, 1999; Janson and van Schaik, 2000). In a comparative study of 16 primate species, Janson and van Schaik (2000) found that one measure of intruder pressure, the "relative rate of male replacement," was the most important factor affecting infanticide rate.

Number of Males in Groups

In general, infanticide rate declines with increasing number of males per group (Janson and van Schaik, 2000). This result depends on two factors. First, males in multi-male groups may cooperate to defend against intruder males. Second, multi-male groups present a less attractive target to lone intruders, because, if they joined such a group, they would have to share matings with the group's current males. In contrast, intruders joining one-male groups often expel the former resident male (Janson and van Schaik, 2000).

Female Dispersal Patterns

At Gombe, the mother–daughter pair of Passion and Pom were observed to kill three infants and were suspected of killing up to six more during a 4-yr period (Goodall, 1990). Goodall initially interpreted these killings as aberrant behavior resulting from Passion's strange personality and poor mothering skills. In recent years, however, observers have witnessed additional attempts by females to kill newborns (Pusey et al., 1997). These observations suggest that Passion and Pom's behavior was not the result of social pathology, but, instead, simply be an extreme example of ordinary female–female competition. All observed female infanticide attempts have involved pairs of females (usually mother–daughter) attacking lone mothers. This suggests that female infanticide at Gombe depends not on individual pathology, but on the numerical advantage gained by females with grown daughters or other close allies.

At most study sites, female chimpanzees leave their natal group at adolescence, thereby avoiding the risk of inbreeding (Pusey, 1979). At Gombe, roughly half of females stay in their natal community or return there to have infants. This unusually high percentage likely results from the few choices available to females. The park contains only two groups besides the main study group, and destruction of forest outside the park has severed connections to any other chimpanzee communities. A number of Gombe females benefit from the unusual advantage of having an adult daughter in her group, which is a natural ally during fights with other females.

CONCLUSIONS

Studies of primates have explored many ways in which aggression varies with environmental factors. Though our understanding of primate aggression is far from complete, current studies support the view that primates, like other animals, use aggression strategically. Primates assess various environmental factors, such as the relative fighting ability of their opponents and the value of the resource in question, to determine whether aggression is likely to pay.

Early observers of primate aggression, especially infanticide, cannibalism, and intergroup killing, regarded these behaviors as pathological or dysfunctional behaviors. Though some critics persist in this view, current evidence indicates that in most cases aggression follows evolutionary logic. Animals attack other animals when the costs of attacking are low or when the benefits are likely to be high. In a complicated world, animals may not always correctly calculate the odds of success. Moreover, pathologies do occur, and some behaviors that proved adaptive in ancestral environments may prove unsuitable to novel environments. Nevertheless, the evidence from field and laboratory studies supports the view that aggression occurs when it is likely to benefit the reproductive success of the aggressor and/or the aggressor's kin.

The focus on pathology and dysfunction includes a long history of searching for some single biological trait intrinsic to criminals and other aggressors, such as head shape, body type, or XYY genotype (reviewed in Niehoff, 1999). Recent studies have focused on physiological traits such as serotinergic function (Wallman, 1999). Evolutionary considerations suggest that any such single trait is likely to provide, at best, a partial explanation.

In nature, individuals must contend with frequent changes in their environment and in themselves. Evolutionary considerations suggest that individuals should be designed to use aggression strategically. Rather than predicting individual fate, physiological mechanisms may provide a means for tracking the environment and motivating the individual to use aggression appropriately under particular circumstances. The studies reviewed above provide some examples of how physiology depends on environmental factors, such as the intensity of competition for mates. Our understanding of the relation between physiological mechanisms and functional goals should continue to improve with advances in technology for physiological assays of free-ranging animals.

Current evidence indicates that humans use aggression much as other primates do, to obtain fitness-enhancing resources such as mates, food, status, and territory (Chagnon, 1988; Daly and Wilson, 1988). For example, just as male rhesus monkeys, baboons, and chimpanzees are more likely than females to kill conspecifics, men commit murder far more often than women. Male sexual jealousy may account for 30–50% of all homicides (Daly and Wilson, 1988). Just as infant gorillas and langur monkeys risk being killed by new males, human children are up to 100 times more likely to be killed if living with a step-parent (Daly and Wilson, 1988). Intergroup conflict, from street gangs to world wars, involves territorial disputes. Members of street gangs mark territorial boundaries with graffiti (Alonso, 1999), and gangland homicides are frequently described as battles over "turf" (Decker and van Winkle, 1996). Just as howler monkeys and chimpanzees assess the odds of winning before joining an intergroup fight, people around the world participate in deadly ethnic riots when they perceive that the costs of attacking their enemies will be low (Horowitz, 2001).

Human violence does not result from the release of a built-up aggression drive, but neither does violence among other primates. Like other primates, we are sensitive to environmental factors that affect the likely payoffs of aggression. Moreover, unlike any other primate, we have some hope of learning lessons from history and science. Human societies show enormous variation in rates of aggression across space and time. Regions, such as Western Europe and Japan, once plagued with war, are now among the most peaceful in the world. Rates of violent crime and warfare respond quickly to changes in demography and other environmental factors. A better understanding of how human violence responds to changes in costs and benefits should provide us with better tools to make our own societies safer from violence. Any hope of such solutions will require answers to all four of Tinbergen's questions of function, mechanism, ontogeny, and phylogeny.

ACKNOWLEDGMENTS

I thank Mark Mattson for the invitation to contribute this chapter. For helpful critiques, I am grateful to Seth Blackshaw, Bob Montgomery, Anne Pusey, Jeff Stevens, and Becky Sun. Dawn Kitchen and Julie Gros-Louis generously provided access to unpublished material.

REFERENCES

Alberts, S. C., Altmann, J., and Wilson, M. L. (1996) Mate guarding constrains foraging activity of male baboons. *Anim. Behav.* **51,** 1269–1277.

Alberts, S. C., Sapolsky, R. M., and Altmann, J. (1992) Behavioral, endocrine, and immunological correlates of immigration by an aggressive male into a natural primate group. *Horm. Behav.* **26,** 167–178.

Alonso, A. A. (1999) Territoriality among African-American street gangs in Los Angeles. Unpublished MA Thesis, University of Southern California, Los Angeles.

Altmann, J., Alberts, S. C., Haines, S., et al. (1996) Behavior predicts genetic structure in a wild primate group. *Proc. Natl. Acad. Sci. USA* **93,** 5797–5801.

Altmann, J. and Muruthi, P. (1988) Differences in daily life between semiprovisioned and wild-feeding baboons. *Am. J. Primatol.* **15,** 213–221.

Arcadi, A. C. and Wrangham, R. W. (1999) Infanticide in chimpanzees: review of cases and a new within-group observation from the Kanyawara study group in Kibale National Park. *Primates* **40,** 337–351.

Archer, J. (1988) *The Behavioural Biology of Aggression.* Cambridge University Press, Cambridge.

Barnow, S., Lucht, M., and Freyberger, H. J. (2001) Influence of punishment, emotional rejection, child abuse, and broken home on aggression in adolescence: an examination of aggressive adolescents in Germany. *Psychopathology* **34,** 167–173.

Bartlett, T. Q., Sussman, R. W., and Cheverud, J. M. (1993) Infant killing in primates: a review of observed cases with specific reference to the sexual selection hypothesis. *Am. Anthropologist* **95,** 958–990.

Bauer, H. R. (1980) Chimpanzee society and social dominance in evolutionary perspective, in *Dominance Relations: Ethological Perspectives of Human Conflict.* (Omark, D. R., Strayer, F. F., and Freedman, D., eds.), Garland, New York, pp. 97–119.

Borries, C., Launhardt, K., Epplen, C., Epplen, J. T., and Winkler, P. (1999) DNA analyses support the hypothesis that infanticide is adaptive in langur monkeys. *Proc. R. Soc. Lond. B Biol. Sci.* **266,** 901–904.

Brain, C. (1992) Deaths in a desert baboon troop. *Int. J. Primatol.* **13,** 593–599.

Butynski, T. M. (1990) Comparative ecology of blue monkeys (*Cercopithecus mitis*) in high and low density subpopulations. *Ecological Monographs* **60,** 1–26.

Byers, J. (1997) *American Pronghorn: Social Adaptations and the Ghosts of Predators Past.* University of Chicago Press, Chicago.

Chagnon, N. A. (1988) Life histories, blood revenge, and warfare in a tribal population. *Science* **239,** 985–992.

Clarke, M. R. and Glander, K. E. (1984) Female reproductive success in a group of free-ranging howling monkeys (*Alouatta palliata*) in Costa Rica, in *Female Primates: Studies by Women Primatologists.* (Small, M. F., eds.), Alan R. Liss, New York, pp. 111–126.

Constable, J. L., Ashley, M. V., and Goodall, J. (2001) Noninvasive paternity assignment in Gombe chimpanzees. *Mol. Ecol.* **10,** 1279–1300.

Courtwright, D. T. (1996) *Violent Land: Single Men and Social Disorder from the Frontier to the Inner City.* Harvard University Press, Cambridge, MA.

Curtin, R. and Dolhinow, P. (1978) Primate social behavior in a changing world. *Am. Sci.* **66,** 468–475.

Daly, M. and Wilson, M. (1983) *Sex, Evolution, and Behavior.* Wadsworth Publishing Company, Belmont, CA.

Daly, M. and Wilson, M. (1988) *Homicide.* Aldine, Hawthorne.

Daly, M. and Wilson, M. (1996) The evolutionary psychology of homicide. *Demos* 39–45.

Davies, N. B. and Halliday, T. R. (1978) Deep croaks and fighting assessment in toads (*Bufo bufo*). *Nature* **274,** 683–685.

de Waal, F. B. M. (1989) *Peacemaking among Primates.* Harvard University Press, Cambridge, MA.

Decker, S. H. and van Winkle, B. (1996) *Life in the Gang: Family, Friends, and Violence.* Cambridge University Press, Cambridge, UK.

Digby, L. (2000) Infanticide by female mammals: implications for the evolution of social systems, in *Infanticide by Males and Its Implications.* (van Schaik, C. P. and Janson, C. H., eds.), Cambridge University Press, Cambridge, UK, pp. 423–446.

Drews, C. (1996) Contexts and patterns of injuries in free-ranging male baboons (*Papio cynocephalus*). *Behaviour* **133,** 443–474.

Dunbar, R. (1988) *Primate Social Systems.* Cornell University Press, Ithaca, NY.

Eibl-Eibesfelt, I. (1979) *The Biology of War and Peace.* Viking, New York.

Emlen, S. T. and Oring, L. (1977) Ecology, sexual selection and the evolution of mating systems. *Science* **197,** 215–223.

Enquist, M. and Leimar, O. (1990) The evolution of fatal fighting. *Anim. Behav.* **39,** 1–9.

Enstam, K. L., Isbell, L. A., and De Maar, T. W. (2002) Male demography, female mating behavior, and infanticide in wild patas monkeys (*Erythrocebus patas*). *Int. J. Primatol.* **23,** 85–104.

Fashing, P. (2001) Male and female strategies during intergroup encounters in guerezas (*Colobus guereza*): evidence for resource defense mediated through males and a comparison with other primates. *Behav. Ecol. Sociobiol.* **50,** 219–230.

Fawcett, K. and Muhumuza, G. (2000) Death of a wild chimpanzee community member: possible outcome of intense sexual competition. *Am. J. Primatol.* **51,** 243–247.

Fossey, D. (1983) *Gorillas in the Mist.* Houghton Mifflin Company, Boston.

Fossey, D. (1984) Infanticide in mountain gorillas (*Gorilla gorilla beringei*) with comparative notes on chimpanzees, in *Infanticide: Comparative and Evolutionary Perspectives.* (Hausfater, G. and Hrdy, S. B., eds.), Aldine Publishing, New York, pp. 217–235.

Gat, A. (1999) The pattern of fighting in simple, small-scale, prestate societies. *J. Anthropol. Res.* **55,** 563–583.

Gerloff, U., Hartung, B., Fruth, B., Hohman, G., and Tautz, D. (1999) Intracommunity relationships, dispersal pattern and paternity success in a wild living community of bonobos (*Pan paniscus*) determined from DNA analysis of faecal samples. *Proc. R. Soc. Lond. B* **266,** 1189–1195.

Goldberg, T. L. and Wrangham, R. W. (1997) Genetic correlates of social behaviour in wild chimpanzees: evidence from mitrochondrial DNA. *Anim. Behav.* **54,** 559–570.

Goldizen, A. W. (1988) Tamarin and marmoset mating systems: unusual flexibility. *Trends Ecol. Evol.* **3,** 36–40.

Goodall, J. (1977) Infant killing and cannibalism in free-living chimpanzees. *Folia Primatol.* **22,** 259–282.

Goodall, J. (1986) *The Chimpanzees of Gombe: Patterns of Behavior.* Belknap Press, Cambridge, MA.

Goodall, J. (1990) *Through a Window: My Thirty Years with the Chimpanzees of Gombe.* Houghton Mifflin Company, Boston.

Hall, K. R. L. (1964) Aggression in monkey and ape societies, in *The Natural History of Aggression*. (McCarthy, J. D. and Ebling, F. J., eds.), Academic Press, London, pp. 51–64.

Hausfater, G. and Hrdy, S. B. (1984) *Infanticide: comparative and evolutionary perspectives*, in *Biological Foundations of Human Behavior*. Aldine Publishing, New York, p. 598.

Holldobler, B. and Wilson, E. O. (1990) *The Ants*. Belknap Press of Harvard University Press, Cambridge, MA.

Holloway, R. L. (1974) *Primate Aggression, Territoriality, and Xenophobia: A Comparative Perspective*. Academic Press, New York.

Horowitz, D. L. (2001) *The Deadly Ethnic Riot*. University of California Press, Berkeley.

Howell, S. (1999) An assessment of primatology in the 1990s. (http://www.primate.wisc.edu/pin/careers/howell.html).

Hrdy, S. B. (1974) Male-male competition and infanticide among the langurs (*Presbytis entellus*) of Abu, Rajasthan. *Folia Primatol.* **22,** 19–58.

Hrdy, S. B. (1977) *The Langurs of Abu*. Harvard University Press, Cambridge, MA.

Huntingford, F. and Turner, A. (1987) *Animal Conflict*. Chapman and Hall, Ltd, New York.

Isbell, L. A. (1991) Contest and scramble competition: patterns of female aggression and ranging behavior among primates. *Behav. Ecol.* **2,** 143–155.

Janson, C. H. and van Schaik, C. P. (2000) The behavioral ecology of infanticide by males, in *Infanticide by Males and its Implications*. (van Schaik, C. P. and Janson, C. H., eds.), Cambridge University Press, Cambridge, UK, pp. 469–494.

Johnson, J. G., Cohen, P., Smailes, E. M., Kasen, S., and Brook, J. S. (2002) Television viewing and aggressive behavior during adolescence and adulthood. *Science* **295,** 2468–2471.

Judge, P. G. and de Waal, F. B. M. (1997) Rhesus monkey behaviour under diverse population densities: coping with long-term crowding. *Anim. Behav.* **54,** 643–662.

Jurmain, R. (1997) Skeletal evidence of trauma in African apes, with special reference to the Gombe chimpanzees. *Primates* **38,** 1–14.

Kalin, N. H. (1999) Primate models to understand human aggression. *J. Clin. Psychiatry* **60,** 29–32.

Kitchen, D. (2000) Aggression and assessment among social groups of Belizean black howler monkeys (*Alouatta pigra*). Unpublished Ph.D. Thesis, University of Minnesota, St. Paul, MN.

Kitopeni, R., Kasagula, M., and Turner, L. (1995) Ntologi falls??! *Pan Africa News* **2(2),** 9–11.

Knott, C. D. (1998) Orangutans in the wild. *Natl. Geographic,* August, 30–57.

Kummer, H. (1968) *Social Organization of Hamadryas Baboons*. University of Chicago Press, Chicago.

Leimar, O., Austad, S., and Enquist, M. (1991) A test of the sequential assessment game: fighting in the bowl and doily spider *Frontinella pyramitela*. *Evolution* **45,** 862–874.

Lewontin, R. C. (1999) The problem with an evolutionary answer. *Nature* **400,** 728–729.

Lindburg, D. G. (1971) The rhesus monkey in North India: an ecological and behavioral study, in *Primate Behavior*. (Rosenblum, L. A., ed.), Academic Press, New York, pp. 2–106.

Lorenz, K. (1966) *On Aggression*. Harcourt Brace, New York.

Lowen, C. and Dunbar, R. I. M. (1994) Territory size and defendability in primates. *Behav. Ecol. Sociobiol.* **35,** 347–354.

Maggioncalda, A. N., Sapolsky, R. M., and Czekala, N. M. (1999) Reproductive hormone profiles in captive male orangutans: implications for understanding developmental arrest. *Am. J. Phys. Anthropol.* **109,** 19–32.

Manson, J. H. and Wrangham, R. W. (1991) Intergroup aggression in chimpanzees and humans. *Curr. Anthropol.* **32,** 369–390.

Maynard Smith, J. (1982) *Evolution and the Theory of Games*. Cambridge University Press, Cambridge, UK.

McComb, K., Packer, C., and Pusey, A. (1994) Roaring and numerical assessment in contests between groups of female lions, *Panthera leo*. *Anim. Behav.* **47,** 379–387.

Mech, L. D., Adams, L. G., Meier, T. J., Burch, J. W., and Dale, B. W. (1998) *The Wolves of Denali*. University of Minnesota Press, Minneapolis.

Mesquida, C. G. and Wiener, N. I. (1996) Human collective aggression: a behavioral ecology perspective. *Ethol. Sociobiol.* **17,** 247–262.

Miller, L. E. (1996) Behavioral ecology of wedge-capped capuchin monkeys (*Cebus olivaceus*), in *Adaptive Radiations in Neotropical Primates*. (Garber, P., Norconk, M., and Rosenberger, A., eds.), Plenum Press, New York, pp. 271–288.

Miller, L. E. (1998) Fatal attack among wedge-capped capuchins. *Folia Primatol. (Basel)* **69,** 89–92.

Mitani, J. C. (1985) Mating behavior of male orangutans in the Kutai Game Reserve, Indonesia. *Anim. Behav.* **33,** 392–402.

Mitani, J. C. (1990) Experimental field studies of Asian ape social systems. *Int. J. Primatol.* **11,** 103–126.

Mitani, J. C., Gros-Louis, J., and Richards, A. F. (1996) Sexual dimorphism, the operational sex ratio, and the intensity of male competition in polygynous primates. *Am. Naturalist* **147,** 966–980.

Mitani, J. C., Merriwether, A., and Zhang, C. (2000) Male affiliation, cooperation, and kinship in wild chimpanzees. *Anim. Behav.* **59,** 885–893.

Mitani, J. C. and Rodman, P. S. (1979) Territoriality: the relation of ranging pattern and home range size to defendability, with an analysis of territoriality among primate species. *Behav. Ecol. Sociobiol.* **5,** 241–251.

Montagu, A. (1976) *The Nature of Human Aggression*. Oxford University Press, New York.

Moore, J. (1999) Population density, social pathology, and behavioral ecology. *Primates* **40,** 1–22.

Morin, P. A., Moore, J. J., Chakraborty, R., Jin, L., Goodall, J., and Woodruff, D. S. (1994) Kin selection, social structure, gene flow, and the evolution of chimpanzees. *Science* **265,** 1193–1201.

Muller, M. (2002) Testosterone and reproductive aggression in wild chimpanzees. *Am. J. Phys. Anthropol.* **34,** 116–117.

Newton, P. N. (1986) Infanticide in an undisturbed forest population of hanuman langurs, *Presbytis entellus*. *Anim. Behav.* **34,** 785–789.

Niehoff, D. (1999) *The Biology of Violence: How Understanding the Brain, Behavior, and Environment Can Break the Vicious Circle of Aggression*. Free Press, New York.

Nishida, T. (1996) The death of Ntologi, the unparalleled leader of M Group. *Pan Africa News* **3(1)**.

Nunn, C. L. (2000) Collective action, free-riders, and male extragroup conflict, in *Primate Males*. (Kappeler, P. M., eds.), Cambridge University Press, Cambridge, pp. 192–204.

Ostfeld, A. M. and D'Atri, D. A. (1975) Psychophysiological responses to the urban environment. *Int. J. Psychiatry Med.* **6,** 15–28.

Pagani, L., Boulerice, B., Vitaro, F., and Tremblay, R. E. (1999) Effects of poverty on academic failure and delinquency in boys: a change and process model approach. *J. Child Psychol. Psychiatry* **40,** 1209–1219.

Palombit, R. A. (1993) Lethal territorial aggression in a white-handed gibbon. *Am. J. Primatol.* **31,** 311–318.

Palombit, R. A. (1999) Infanticide and the evolution of pair bonds in nonhuman primates. *Evol. Anthropol.* **7,** 117–129.

Palombit, R. A., Cheney, D. L., Fischer, J., et al. (2000) Male infanticide and defense of infants in chacma baboons, in *Infanticide by Males and its Implications*. (van Schaik, C. P. and Janson, C. H., eds.), Cambridge University Press, Cambridge, UK, pp. 123–151.

Parker, G. A. (1974) Assessment strategy and the evolution of fighting behavior. *J. Theoret. Biol.* **47,** 223–243.

Polis, G. A. (1981) The evolution and dynamics of intraspecific predation. *Ann. Rev. Ecol. Syst.* **12,** 255–251.

Popp, J. L. (1978) Male baboons and evolutionary principles. Unpublished PhD thesis, Harvard University, Cambridge, MA.

Power, M. (1991) *The Egalitarians—Human and Chimpanzee: An Anthropological View of Social Organization.* Cambridge University Press, Cambridge.

Pusey, A., Williams, J. M., and Goodall, J. (1997) The influence of dominance rank on the reproductive success of female chimpanzees. *Science* **277**, 828–831.

Pusey, A. E. (1979) Intercommunity transfer of chimpanzees in Gombe National Park, in *The Great Apes.* (Hamburg, D. A. and McCown, E. R., eds.), Benjamin/Cummings, Menlo Park, CA, pp. 464–479.

Rodman, P. S. and Mitani, J. C. (1987) Orangutans: sexual dimorphism in a solitary species, in *Primate Societies.* (Smuts, B. B., Cheney, D. L., Seyfarth, R. M., Wrangham, R. W., and Struhsaker, T. T., eds.), University of Chicago Press, Chicago, pp. 146–154.

Sapolsky, R. M. (2001) *A Primate's Memoir: A Neuroscientist's Unconventional Life among the Baboons.* Scribner, New York.

Schaller, G. B. (1964) *The Year of the Gorilla.* The University of Chicago Press, Chicago.

Schubert, G. (1982) Infanticide by usurper Hanuman langur males: a sociobiological myth. *Social Science Information* **21**, 199–244.

Sommer, V. (2000) The holy wars about infanticide. Which side are you on? and why? in *Infanticide by Males and its Implications.* (van Schaik, C. P. and Janson, C. H., eds.), Cambridge University Press, Cambridge, UK, pp. 9–26.

Southwick, C. H. (1969) Aggressive behaviour of rhesus monkeys in natural and captive groups, in *Aggressive Behaviour.* (Garattini, S. and Sigg, E. B., eds.), John Wiley & Sons, New York, pp. 32–43.

Starin, E. D. (1994) Philopatry and affiliation among red colobus. *Behaviour* **130**, 253–269.

Stevens, J. R. and Stephens, D. W. (2002) Food sharing: a model of manipulation by harassment. *Behav. Ecol.* **13**, 393–400.

Sussman, R. W. (1997) *The Biological Basis of Human Behavior: A Critical Review.* Prentice Hall, New York, p. 382.

Sussman, R. W. (1999) The myth of man the hunter, man the killer and the evolution of human morality (evolutionary and religious perspectives on morality). *Zygon* **34**, 453–472.

Sussman, R. W., Cheverud, J. M., and Bartlett, T. Q. (1995) Infant killing as an evolutionary strategy: reality or myth? *Evol. Anthropol.* **3**, 149–151.

Takahata, Y. (1985) Adult male chimpanzees kill and eat a newborn infant: newly observed intragroup infanticide and cannibalism in Mahale National Park, Tanzania. *Folia Primatol.* **44**, 161–170.

Tinbergen, N. (1963) On aims and methods of Ethology. *Zeitschrift fur Tierpyschologie* **20**, 410–433.

Tooby, J. and Cosmides, L. (1992) The psychological foundations of culture, in *The Adapted Mind: Evolutionary Psychology and the Generation of Culture.* (Barkow, J. H., Cosmides, L., and Tooby, J., eds.), Oxford University Press, Oxford, pp. 19–136.

Trivers, R. L. (1972) Parental investment and sexual selection, in *Sexual Selection and the Descent of Man 1871–1971.* (Campbell, B., ed.), Heinemann, London, pp. 136–179.

van Schaik, C. P. (1989) The ecology of social relationships amongst female primates, in *Comparative Socioecology: The Behavioral Ecology of Humans and Other Mammals.* (Standen, V. and Foley, R. A., eds.), Blackwell, Oxford, pp. 195–218.

van Schaik, C. P. (2000a) Infanticide by male primates: the sexual selection hypothesis revisited, in *Infanticide by Males and its Implications.* (van Schaik, C. P. and Janson, C. H., eds.), Cambridge University Press, Cambridge, UK, pp. 27–60.

van Schaik, C. P. (2000b) Social counterstrategies against infanticide by males in primates and other mammals, in *Primate Males: Causes and Consequences of Variation in Group Composition.* (Kappeler, P., eds.), Cambridge University Press, Cambridge, pp. 34–54.

van Schaik, C. P. (2000c) Vulnerability to infanticide by males: patterns among mammals, in *Infanticide by Males and Its Implications.* (van Schaik, C. P. and Janson, C. H., eds.), Cambridge University Press, Cambridge, UK, pp. 61–71.

van Schaik, C. P. and Hörstermann, M. (1994) Predation risk and the number of adult males in a primate group: a comparative test. *Behav. Ecol. Sociobiol.* **35,** 261–272.

Vandenbergh, J. G. (1969) Endocrine coordination in monkeys: male sexual response to the female. *Physiol. Behav.* **4,** 261–264.

Virgin, C. E. and Sapolsky, R. M. (1997) Styles of male social behavior and their endocrine correlates among low-ranking baboons. *Am. J. Primatol.* **42,** 25–39.

Volavaka, J. (1995) *Neurobiology of Violence.* American Psychiatric Press, Washington, DC.

Wallman, J. (1999) Serotonin and impulse aggression: not so fast. *HFG Rev.* **3**.

Watts, D. P. (1989) Infanticide in mountain gorillas: new cases and a reconsideration of the evidence. *Ethology* **81,** 1–18.

Watts, D. P. (1998) Coalitionary mate guarding by male chimpanzees at Ngogo, Kibale National Park, Uganda. *Behav. Ecol. Sociobiol.* **44,** 43–55.

Watts, D. P. and Mitani, J. C. (2000) Infanticide and cannibalism by male chimpanzees at Ngogo, Kibale National Park, Uganda. *Primates* **41,** 357–365.

Watts, D. P. and Mitani, J. C. (2001) Boundary patrols and intergroup encounters in wild chimpanzees. *Behaviour* **138,** 299–327.

Watts, D. P. and Mitani, J. C. (2002) Hunting behavior of chimpanzees at Ngogo, Kibale National Park, Uganda. *Int. J. Primatol.* **23,** 1–28.

West, S. A., Murray, M. G., Machado, C. A., Griffin, A. S., and Herre, E. A. (2001) Testing Hamilton's rule with competition between relatives. *Nature* **409,** 510–513.

Westergarrd, G. C., Izard, M. K., Drake, J. H., Suomi, S. J., and Higley, J. D. (1999) Rhesus macaque (*Macaca mulatta*) group formation and housing: wounding and reproduction in a specific pathogen free (SPF) colony. *Am. J. Primatol.* **49,** 339–347.

Wilkinson, P. F. and Shank, C. C. (1976) Rutting fight mortality among musk oxen on Banks Island, Northwest Territories, Canada. *Anim. Behav.* **24,** 756–758.

Williams, J. M. (2000) Female Strategies and the Reasons for Territoriality in Chimpanzees: Lessons from Three Decades of Research at Gombe. Unpublished Ph. D. Thesis, University of Minnesota, St. Paul, MN.

Williams, J. M. and Pusey, A. E. Why do male chimpanzees defend a group range? Reassessing male territoriality. *Anim. Behav.*, submitted.

Wilson, A. P. and Boelkins, R. C. (1970) Evidence for seasonal variation in aggressive behaviour by Macaca mulatta. *Anim. Behav.* **18,** 719–724.

Wilson, M. and Daly, M. (1998) Sexual rivalry and sexual conflict: recurring themes in fatal conflicts. *Theor. Criminol.* **2,** 291–310.

Wilson, M. L. (2001) Imbalances of power: how chimpanzees respond to the threat of intergroup aggression. Unpublished Ph. D. Thesis, Harvard University, Cambridge, MA.

Wilson, M. L., Hauser, M. D., and Wrangham, R. W. (2001) Does participation in intergroup conflict depend on numerical assessment, range location, or rank for wild chimpanzees? *Anim. Behav.* **61,** 1203–1216.

Wilson, M. L., Hauser, M. D., and Wrangham, R. W. Vocal suppression in wild chimpanzees, submitted.

Wilson, M. L., Wallauer, W., and Pusey, A. E. Intergroup aggression and territory expansion in the chimpanzees of Gombe National Park, Tanzania, submitted.

Wrangham, R. (1974) Artificial feeding of chimpanzees and baboons in their natural habitat. *Anim. Behav.* **22,** 83–93.

Wrangham, R. W. (1980) An ecological model of female-bonded primate groups. *Behaviour* **75,** 262–300.

Wrangham, R. W. (1999) The evolution of coalitionary killing. *Yearbook Phys. Anthropol.* **42,** 1–30.

Wrangham, R. W. (2002) The cost of sexual attraction: is there a trade-off in female *Pan* between sex appeal and received coercion? in *Behavioural Diversity in Chimpanzees and Bonobos.* (Boesch, C., Hohmann, G., and Marchant, L., eds.), Cambridge University Press, Cambridge.

Wrangham, R. W., Chapman, C. A., Clark-Arcadi, A. P., and Isabirye-Basuta, G. (1996) Socioecology of Kanyawara chimpanzees: implications for understanding the costs of great ape groups, in *Great Ape Societies.* (McGrew, W. C., Marchant, L. F., and Nishida, T., eds.), Cambridge University Press, Cambridge, pp. 45–57.

Wrangham, R. W. and Peterson, D. (1996) *Demonic Males: Apes and the Origins of Human Violence.* Houghton Mifflin, Boston.

Zuberbuhler, K., Noe, R., and Seyfarth, R. M. (1997) Diana monkey long-distance calls: messages for conspecifics and predators. *Anim. Behav.* **53,** 589–604.

Zuckerman, S. (1932) *The Social Life of Monkeys and Apes.* K. Paul Trench Trubner & Co. Ltd., London.

11
Aggression, Biology, and Context
Dejá-Vù All Over Again?

Rebecca M. Young and Evan Balaban

"We have probably already found the genotype that, in a statistical sense, predicts violent crime better than any gene to be discovered in the future. This is simply an XY genotype. Women are, of course, perpetrators of some violence, but from both self-report and official records, men tend to commit many more acts of assault, robbery, and homicide than women. At this point the skeptical reader may react with considerable incredulity. How is this genetic? Surely, one might protest, the socialization process that gives us male linebackers but female cheerleaders cannot be overlooked in its contribution to violent crime.

Our whole point is that processes like socialization cannot be overlooked in biology just as gonadotropins cannot be overlooked by the social scientist. Consider XX, XY, and violence more deeply. No one would seriously contend that the physical differences between an X and a Y chromosome are social and cultural. These differences are genetic. Similarly, no one can seriously doubt that ever since statistics have been kept in Western society, males commit more homicides than females. Hence, there are individual differences that are genetic in origin and statistically predict violence." (p. 89)

Carey, G. and Gottesman, I. (1996) Genetics and antisocial behavior: substance versus sound bites. *Politics and the Life Sciences* 15, 88–90.

INTRODUCTION

Attempts to gain deeper insights into the biology underlying aggression have been characterized by many false starts, particularly with inappropriate attributions of highly specific causative roles to particular neurotransmitters and hormones (Balaban et al., 1996; Lee and Coccaro, 2001). To highlight persistent biases and shortcomings that may be actively hindering progress in understanding biological correlates of aggression, we address a ubiquitous piece of folk-wisdom that keeps reappearing in the biology-of-aggression literature: male–female differences and the assumption that differences in aggression are somehow biologically driven by the Y chromosome.

Here, we develop the idea that aggressive phenotypes implicitly include contextual variables and discuss the complications this introduces for biological research. We consider the genetic evidence linking aggression to the Y chromosome, including observational and experimental data from both animals and humans. We examine influential

Table 1
Various Interpretations of Contextual Cues

Situation	Interpretation	Link to action	Performance
1	same	same	same
2	different	same	different
3	same	different	different
4	different	different	different

evolutionary theories that serve to "sharpen" relatively scattered and sometimes contradictory data on sex differences in aggression into a familiar and sensible-looking scientific story. In all of these domains, there has been a tendency to minimize the role that context plays in the evolution, development, and expression of aggression. We argue that little progress in genuine empirical understanding of biological correlates of aggressive behavior will be made until this situation is remedied.

AGGRESSION AND THE PROBLEM OF CONTEXT

A previous review (Balaban et al., 1996) discussed many of the problems encountered when trying to transform aggressive phenotypes into suitable objects for biological research. One of these problems, the fact that the context in which a behavior occurs is part of what defines it as "aggressive," was not discussed in detail; we consider it more fully here.

Behavioral context is problematic in two different ways. First, defining a context requires a subjective point of view, which can differ among the participants in a situation, as well as to an external observer (and among multiple external observers). We will call this aspect of the problem "context as interpretation." Second, conducting research on aggressive behavior generally requires manipulation of the context in a repeatable manner. In most experimental aggression studies, this is achieved by deliberately placing subjects in highly abnormal circumstances. In human correlative studies, populations of subjects whose lives have been conducted in highly abnormal circumstances are frequently the objects of scientific scrutiny. We will call this second type of contextual problem "representiveness."

CONTEXT AS INTERPRETATION

In a social situation that potentially leads to aggressive behavior, subjects can differ in their interpretations of contextual cues, and/or can differ in their linkage of one particular interpretation of these cues to aggressive behavior. This dichotomy creates an ambiguity for biological linkages between aggressive behavior and genetic or physiological variables.

As reflected in Table 1, individuals could differ in their aggressive behavior in a particular context for three reasons. They could have a similar scheme in their brains relating particular aggressive or nonaggressive behaviors to particular contexts, but differ in their interpretation of which context is present (situation 2). They could agree in their interpretation of which context is present, but have different schemes in their brains relating particular behaviors to particular contexts (situation 3). Finally, they could differ in both interpretations and schemes for relating contexts to behavior (situation 4).

Situation 1 is perhaps the clearest situation for a biological research program. If linkage between one particular interpretation of a set of social cues and a particular type of aggressive behavior is the same across many different individuals, then behavioral, genetic, pharmacological, developmental, and neurobiological manipulations could be used to meaningfully unravel biological correlates of this linkage. It would also be reasonable to discuss how the evolutionary process might have favored the association between context and behavior. We would argue that this situation exemplifies what biology is after when it studies complex entities like aggression.

By the same token, Table 1 tells us that we should be less sure what we are studying when we use behavioral differences to tell us about the biology of aggressive behavior. For example, situation 2 would predict different neurobiological correlates of observed differences in aggressive behavior from situation 3. In situation 2, individuals differ in their assessment of a context, but not in how their brains link contexts to aggressive behavior (done in the same way). Behavioral differences would be caused by differences in perception or evaluation of social cues. In situation 3, individuals agree on their assessment of context, but differ in how their brains link context and aggression. The brain correlates of these behavioral differences would presumably have a lot to do with the generation (and/or selection) of aggression.

This example makes the important point that observed differences in aggressive behavior may have little or much to do with the brain systems conventionally pictured as directly controlling aggression. While we agree that the processes shaping an individual's interpretation of social cues are a part of what leads to aggressive behavior, we see this as being once-removed in the pathway of causation and as being much less realistic to study using present experimental techniques. The distinction between making an interpretation of a social situation and selecting a response given one particular interpretation (akin to the standard psychological distinction between perception and performance) has been and continues to be widely ignored in aggression work on both humans and animals. Phenotypic assays that confound these two alternatives will almost certainly produce conflicting and seemingly contradictory results.

A recent study of human subjects with an altered form of an allele for the enzyme monoamine oxidase A (MAOA), which is involved in the metabolic processing of a number of neurotransmitters in the brain, provides a concrete illustration of this kind of contextual pitfall. Caspi et al. (2002) studied a large sample of male children to see if there was any correlation between the biological activity of different forms of the enzyme, "violence," and a contextual variable, the amount of maltreatment experienced while growing up. The study suggested that "low-enzyme-activity" MAOA individuals, who had experienced maltreatment, were more likely to have conduct disorders, be convicted of violent offenses, to have a higher disposition toward violence, and to have antisocial personality disorder symptoms than "high-enzyme-activity" MAOA individuals. Unfortunately, the study did not provide any measures of nonviolence-related mental health problems to allow readers to judge the provenance of this suggested linkage. If it turns out that low MAOA enzyme activity puts an individual at risk for a variety of mental illnesses (and socially inappropriate behaviors) that are not necessarily related to violence, this interpretation may be inappropriate. A combination of abnormal brain chemistry and childhood maltreatment is as likely to seriously warp one's perception and interpretation of social signals, as it is to alter the specific physiology of aggressive behavior.

CONTEXT AND REPRESENTATIVENESS

Research on the physiological and genetic correlates of aggression has predominantly been carried out in rodents and nonhuman primates. Mice are rapidly becoming the organism of choice for research in this area (Brodkin et al., 2002; Miczek et al., 2001; Stowers et al., 2002). As a recent review of aggressive behavior in house mice (*Mus musculus*) emphasizes (Miczek et al., 2001), normal laboratory housing conditions prevent these subjects from establishing any semblance of the social organization they exhibit in the wild. Additionally, the testing procedures used to elicit aggressive behavior (isolating animals, applying painful or noxious stimuli, engineered social provocation, drug or alcohol administration) often involve extreme experiences that wild mice are, historically, unlikely to have regularly encountered.

Is this abnormal research context a trivial or an important concern? When looking for biological correlates, it may be reasonable to first examine them in "exaggerated" contexts, in which relationships between basic dependent variables can be more easily established. We think that there is one good biological reason why this may not be the case for aggressive behavior. To paraphrase Dostoyevsky, "typical" social situations (like Dostoyevsky's happy families) have a sameness about them and can, therefore, be systematically studied in an interpretable way. "Atypical" social situations (like unhappy families) are harder to interpret, because each one may be atypical in a different way. Why is this the case?

Biologists think that evolutionary processes can exert direct influences on behavioral situations in particular species, if the behavioral situations (*i*) occur reliably in each generation and (*ii*) have a measurable impact on an individual's ability to survive and reproduce. The result of such evolutionary influence is called an adaptation. If there are certain regularly occurring situations of impending harm to an individual's body or resources that physical force can effectively counter, it makes sense that, over evolutionary time, neurobiological mechanisms that promote the association between particular contexts and particular "aggressive" acts will arise. Both the ability to produce effective responses and to match these responses to the proper situation may require considerable learning, and the evocation of both evolutionary history and individual learning during development in the biology underlying complex behavioral choices, such as aggression, is surely not controversial.

In typical situations, we would expect evolutionary and developmental factors to be working synergistically in ways that selection has presumably acted upon within the development of each individual subject. Thus, we would expect subjects to be using similar physiological mechanisms in similar ways to develop and express aggressive phenotypes. Groups of subjects with a particular gene difference, or who have been reared in different environments, could be experimentally exploited to provide details about the common mechanisms through which a particular gene product or a particular environmental difference gets translated into brain differences that result in behavioral differences.

One consistent feature of populations reared in historically atypical environments is their greater anatomical and physiological variability (Pigliucci, 2001; Schlichting and Pigliucci, 1998). Developmental sources of variation (genomic, chance, and external) that are "factored out" or "developmentally buffered" in typical environments (resulting in no consistent individual phenotypic differences) may now yield substantial differences between individuals. Groups of subjects may, therefore, no longer be uniform from the point of view of the biological processes that are the object of "biological correlates" research. The methods of experimental biology deal very well with groups of subjects that share particular mechanisms or comparisons of a group of subjects that deviate from everyone else in the same way. These same methods deal poorly with subjects that are idiosyncratically different. Research designs that predominantly use atypical conditions without assessing the developmental implications of these conditions potentially shoot themselves in the foot.

Miczek et al. (2001) try to mitigate the idea that the popular isolation paradigms for inducing aggression in male laboratory mice are pathological or abnormal by claiming that "wild, dominant male mice that exclude other males from their territory are *behaviorally* isolated and their aggression may be an adaptive behavior that increases their relative fitness" (p. 168, emphasis theirs). This conveniently ignores the fact that the wild dominant mice are still presumably encountering females who have home ranges within their territories on a regular basis. If their relative fitness is to increase, they have to interact with the females (who, not being caged, can run away) at least enough to mate with them (and react to predators enough to avoid getting eaten). We would contend that mice in the wild emphatically do not have the kind of behavioral isolation analogous to being kept in solitary confinement in a cage.

SEX DIFFERENCES IN AGGRESSION?

The complexities introduced by context being a part of the biological object of study are of special importance for understanding the development of sex differences in aggression. Given that males and females live very different social lives in most human and animal societies, the biology resulting from the combination of contextual differences and noncontextual factors during individual development would appear to be seamlessly intertwined.

Using the distinction between evaluating situations and linking particular evaluations to aggressive behavior, we could simplistically imagine two basic ways in which males and females differ: (A) males and females could evaluate particular situations differently; or (B) males and females could differ in the way that particular evaluations are linked to aggressive behaviors. Because this discussion assumes some difference in the way males and females do things, the two possibilities lead to three potential avenues of explanation: A not B, B not A, A and B.

All three avenues of explanation could result from similar sets of developmental mechanisms. For instance, all three could predominantly reflect the effects of social experience during development or predominantly reflect the effects of differences in gene expression between males and females that are not experience-dependent or, more realistically, reflect everything in between these two extremes.

Unlike Carey and Gottesman (1996), we believe that the point of correlating behavioral differences with biological ones is to understand differences in the developmental processes through which the behaviors become manifest. If we are to take their statement about X and Y chromosomes as something other than a sound bite, we have to translate it into a claim about a pathway of causation. Do physical differences in X and Y chromosomes lead men and women to have similarly "aggressive" brains but different social cues and contexts, differently "aggressive" brains given similar social cues and contexts, or a mixture of both, and how could we know?

AGGRESSION AS A "Y CHROMOSOMAL" TRAIT: EVIDENCE FROM ANIMAL STUDIES

One way of knowing is to use the tools of quantitative genetics and molecular biology to examine the association between aggression and male- or femaleness at a more detailed level.

The most basic way of doing this is to see if a population of animals has any genetic variation that influences the degree to which they show aggressive behavior. This can be ascertained by taking the most or least aggressive animals in the population, mating them to each other, and continuing to do so for as many generations as possible—an "artificial selection" experiment. If this results in a population of animals that is on the average more or less aggressive than the original population, it implies that there is genetic variation in the original population that may be related to individual differences in aggression. We use the word "implies" rather than "demonstrates" because there are nongenetic factors (maternally transmitted molecules and/or behaviors that influence fetal and neonatal development) that could also produce these results, and sufficient control experiments for both uterine and postnatal effects must be carried out.

Selection experiments have been carried out on both male and (more rarely) female mice (reviewed in Miczek et al., 2001). While the contextual conditions eliciting fighting vary differently between the sexes (and vary among laboratory strains, and between wild-caught mice and laboratory strains), both male and female selected populations show similar responses to selection. In males of three selected lines (Turku Aggressive and Nonaggressive, North Carolina 900 and 100, Netherlands Short- and Long-Attack Latency), cross-fostering experiments suggested that the selected differences do not result from the postnatal maternal environment, and control experiments for the effect of the uterine environment in the Netherlands Attack-Latency strain were also negative (Miczek et al., 2001; Sluyter et al., 1996). Similar controls were not conducted for female selection experiments (Ebert, 1983; Hood, 1992), although Hood (1992) found effects of rearing context on subsequent behavior. Interestingly, selection for female (Ebert, 1983) or male (Van Oortmerssen and Bakker, 1981) aggression did not produce any consistent change in aggressive behavior in the unselected sexes of the same strains, suggesting that any genetic variation (or any uncontrolled nongenetic factor) mediating the effects of selection was acting in a sex- and/or situation-specific manner.

A "genetically" mediated sex/situational specificity could manifest itself in several ways. Males and females may express different forms of many genes, or some genes may be silent in one sex and activated in the other. These genes would not necessarily be located on the sex chromosomes, but genes for molecules that regulate their expression would be. Or, less subtly, according to Carey and Gottesman (1996), genes on sex chromosomes could themselves directly affect aggressive behavior. Sex chromosomes consist of "pseudoautosomal" parts that can engage in recombination with the X or the nonsex chromosomes, and "nonpseudoautosomal" parts that cannot. To obtain sex-limited behavioral effects in humans, one would need to posit the presence of genes affecting aggression on the Y chromosome or on a nonpseudoautosomal part of the X chromosome with a mechanism for Y chromosome sequences to shut off the effects of the X chromosome genes in males (otherwise, selection for aggression in females would be expected to produce a correlated response in males, if X chromosome sequences were involved in female aggression). Alternatively, the genes could exist anywhere in the genome, with sequences on the X and/or Y chromosome controlling sexually dimorphic transcription or processing of the gene products. Simplistic hormonal theories hardwiring male aggression to the production of gonadal steroids like testosterone, or less simplistic theories that propose a modulating role of gonadal steroids on aggression, would also predict that the Y chromosome genes that influence male gonadal characteristics should strongly influence male aggression.

Here, experimental results are contradictory. Two research groups have claimed to find Y chromosomal locations that affect male aggressive behavior, one in the nonpseudoautosomal region (Maxson, 1992, 1996, 2000) and one in the pseudoautosomal region (Roubertoux et al., 1994). The group that found the pseudoautosomal region effect initially excluded the nonpseudoautosmal region from having any effect (Roubertoux et al., 1994), but later claimed to also find a context-dependent nonpseudoautosomal effect when they changed the rearing and testing conditions of their mice (Guillot et al., 1995). This context-dependence has not been generally commented on. A third group looked for a Y chromosomal effect in mice selected for male aggression (Netherlands Attack-Latency strains) and mice selected for differences in thermoregulatory nest building. Y chromosome involvement in aggression differences was suggested for the aggression-selected strains, but no effect on aggression was found for the nest-building-selected strains (Sluyter et al., 1997). This raises the possibility that Y chromosomal covariation with aggressive behavior may be a by-product of laboratory selection for certain (unrepresentative?) forms of male aggressive behavior. Finally, Brodkin et al. (2002) conducted the most complete genome-wide screen to date looking for chromosomal regions that co-vary with variation in male aggressive behavior. This study found no evidence for any genes on the Y chromosome affecting male aggression, instead finding one locus on an autosome (chromosome 10) and another locus on the X chromosome (not in the pseudoautosomal region, so it could not recombine with the Y chromosome). If the X chromosome locus is generally linked with aggressive behavior in mice, it is puzzling why selection experiments failed to find correlated responses to selection between male and female aggression.

Thus, both males and females appear to show similar responses to selection on aggressive phenotypes, but the responses are not correlated between sexes. There is preliminary evidence for alleles of genes that may correlate (either directly or very indirectly—the tests do not allow this determination) with variation in male aggressive behavior on both sex chromosomes. We characterize this evidence as preliminary, because studies suggesting chromosomal locations on the Y chromosome were not replicated in a genome screen, and the two chromosomal locations suggested by the genome screen have not been independently replicated.

A recent genetic study (Stowers et al., 2002) has highlighted the importance of contextual cues on the types of male aggressive behavior utilized in mouse work. Mice were engineered to contain a mutation in an ion channel necessary for sensory transduction in the vomeronasal organ, a special chemosensory structure in the ventral nasal septum. The ion channel is only expressed in this sensory structure and not in the rest of the olfactory system. Homozygous mutant male mice showed no aggressive responses using a common behavioral paradigm that robustly elicited aggressive behavior in heterozygous nestmates. According to the distinction between perception of context and linkage of aggression to a particular context that we have defined above, this would constitute a difference in the detection or interpretation of cues, not a difference in the triggering of aggressive responses based on a similar perception or interpretation. Interestingly, the researchers who believe they have identified Y chromosomal regions correlated to aggression posit that one of the ways the nonpseudoautosomal region mediates its effects is to change the odor characteristics emitted by the animals, changing their capacity to elicit attack (Guillot et al., 1995). It would be ironic if work using aggressive behavioral paradigms in rodents end up telling us more about olfactory signaling than about the behaviors "targeted" by the assays.

Above all else, the extant genetic work suggests that little is to be gained from Carey and Gottesman's statistical thought experiment, and reemphasizes the overwhelming importance of the context in which subjects develop and the situations in which they are tested, a theme we will return to.

SEX DIFFERENCES IN HUMAN AGGRESSION

Sex differences in what? As basic as it may seem, human aggression is an extremely slippery concept. Harre and Lamb (1983) note two common elements found in more than 200 different definitions of aggressive behavior in the psychological literature: aggressive behavior is intended to harm and is perceived as hurtful by the target (quoted in Underwood et al., 2001). Yet, Underwood and colleagues identify crucial difficulties with these criteria:

> [A]s clear as these criteria may appear, they are difficult to define operationally, because *neither intentions nor perceptions of harm can be directly observed*. What counts as aggressive behavior depends heavily on social judgments, which are sensitive to the social context and heavily influenced by the values of both the aggressor and the perceiver (Underwood et al., 2001, p. 249; emphasis added).

Hewing to the broad definition encompassing both intention and perception of harm, human aggression takes many forms beyond physical attacks, and researchers are currently engaged in a thoughtful process of devising conceptual and operational means for distinguishing among types of aggression. Social aggression, for example, has been described as "directed toward damaging another's self-esteem, social status, or both, and may take such direct forms as verbal rejection, negative facial expressions or body movements, or more indirect forms such as slanderous rumors or social exclusion" (Galen and Underwood, 1997). Yet others suggest that broad definitions of aggression lump too many disparate constructs under one term. For example, Robarchek, whose anthropological studies of the relatively violent Waorani and the relatively nonviolent Semai provide cross-cultural comparisons of "male" and "female" violence, prefers the term "physical violence," because the term "[aggression] typically encompasses things that I think are very different, e.g., the killing of animals, physical attacks on people, verbal attack, passive–aggressive behavior, gossip, backbiting, and so on" (Robarchek, personal communication). Certainly, adopting a definition for aggression that goes beyond the physical domain, especially one which relies on intentions and perceptions, creates special difficulties for identifying nonhuman homologues for studying biological correlates of aggression. We return to this point below.

Careful attention to definitions and measures is especially critical for understanding male–female differences in aggression (Underwood et al., 2001). Even within single populations, sex differences in aggression are not generalizable across all types of aggression, but show different patterns for specific subtypes of aggression[1] (Österman et al., 1998; Underwood et al., 2001). While most studies have found that nonphysical kinds of aggression are more common among girls than physical aggression, nonphysical forms of aggression are also more common than physical aggression for boys and men in most cultures, especially as age increases (Österman et al., 1998; Ramirez et al., 2001). Underwood and colleagues (2001) caution that "different forms of aggression may have different correlates in different cultures," and present preliminary evidence indicating that gender is among the correlates that display cultural variation. It is therefore important to carefully specify a number of covariates when considering sex differences in aggression (e.g., cultural group, age, and the relationship between aggressor and target) as well as specifying the particular form of aggression of interest.

We believe that this multidimensionality of human aggression poses particular problems for a program of biological research. Much of the current research on biological correlates of human aggression does not adequately specify what forms of aggression have been assessed, and great caution must therefore be exercised in interpreting patterns of correlations across studies, because they may not be measuring comparable entities. Second, and more fundamentally, we are unaware of any biological research on aggression using homologues for nonphysical aggression (the most common human form) in other animals.

A program of biological research on aggression by necessity relies on nonhuman data for all experimental evidence related to the development, physiology, and neurobiology of aggression, because human studies in behavioral biology are quasi-experiments rather than experiments (Young, 2000). Ethics as well as pragmatic issues (such as extremely long developmental periods) make it impossible to experimentally manipulate biological or environmental variables that are hypothesized to affect the development of human behaviors or traits. Instead, we must be content with creatively piecing together observational data, experiments that test short-term effects of physiological or environmental interventions (such as "hormone challenge" experiments or studies of physiological sequelae of human encounters), or correlational studies that look for associations between physiological characteristics (e.g., genotypes, neurotransmitter characteristics) and behaviors. The critical issue is that quasi-experiments are much more partial and tentative than actual experiments, and the interpretation of any individual study is particularly dependent on other related research (including relevant data

---

[1] Indirect, social, verbal and relational are the main subtypes of nonphysical aggression discussed in the literature, though there is disagreement regarding the extent to which these are redundant and/or overlapping concepts (see Underwood et al., 2001; Archer, 2001; Bjorkqvist, 2001, for a clear review of the issues).

on nonhuman animals). Thus, in all areas of behavioral biology, there is (or should be) great concern to find adequate means for investigating the same construct in different species.

Imposing a strict definition of physical aggression for the dependent variable in the biological paradigm is probably an unavoidable decision to safeguard validity. Yet restricting the focus to physical aggression may be deceptive in its appearance of resolving contextual problems. While physical aggression certainly exhibits the most convincing homology between humans and other animals, the functions (both in terms of aims and effects) of aggressive behaviors might not neatly and universally divide along physical and nonphysical lines. As an example, consider the diversity of motivations (fear, anger, irritability, territoriality) for an aggressive attack, as well as the diversity of salient contextual factors (especially the relationships between attacker and target, and the respective sexes of attacker and target). Add to this the diversity of stimuli that produce the motivations for aggression in the first place, and the question of appropriate homology quickly becomes quite murky. Should the model focus on the motivation (fear-based attacks) vs form of response (physical vs nonphysical)? Ideally, of course, it would focus on both, at least in order to establish valid boundaries for the construct that the overall research program is trying to explain. The state of research on aggression at this point falls far short of reasonable conditions for adequate conceptualization and measurement.

With these cautions in mind, we turn to two sorts of evidence for sex differences in aggression among humans. First, there are data regarding sex differences in "normal" populations, including cross-cultural data. Second, there are investigations of a link between aggression and early androgen exposure among individuals known to have atypical androgen exposure.[2] We consider each in turn, with particular attention to issues of behavioral context and development. We also briefly consider a popular theory regarding how and why aggression has evolved in humans, and how and why this process might be sex-specific.

ASPECTS OF CONTEXT IN HUMAN AGGRESSION

There is an enormous body of literature examining correlates of human aggression at the individual level, especially sex differences. The literature has become more and more complex, as investigators have become more attentive to contextual factors in aggressive behavior, which create problems for specifying and distinguishing particular kinds of aggression. The research appears to obey "Young's Rule" (Young, 2000), which holds that the more sophisticated the research becomes, the more elusive are the sex differences.

Claims for the existence of specific patterns and directions of sex differences ultimately depend on contextual factors, including: the relationship between aggressor and target (especially whether or not they are a "pair" or dating); whether aggressor and target are same or other sex; the type of aggression (physical, verbal, indirect); whether

---

[2] Many other kinds of studies examine the relationship between hormones and aggression, including correlational studies and experiments that measure the effect of competitive interactions and hormone levels. Though these studies are extremely important for appreciating the iterative bidirectional relationship between behavior and hormones, we consider them to be less critical to a model for genetically based sex differences in aggression than the prenatal exposure studies, so we do not review such research here.

respondents are asked about real vs hypothetical events; and the social representation of aggression (Archer, 1997, 1999, 2000; Campbell, 1997a, 1997b; Harris, 1994, 1996). These contextual factors are not all independently related to aggression, or related insofar as they mediate the effect of gender, but they are involved in a more complex web of interactions.

For example, a fairly consistent sex difference has been demonstrated in the social representation of aggression in the abstract. It is claimed that men tend to hold "instrumental" representations (aggression is seen as a means to exert control over others), while women tend to hold "expressive" representations (aggression is seen as a loss of self-control due to anger) (Campbell, 1997b). However, John Archer's research has shown that "men and women indicate that they have different opponents in mind when thinking about aggression," and suggests that the social representation of aggression that one holds depends on the specific actor and target one has in mind, the type of aggression (physical, verbal, or indirect), whether one is referring to a real or a hypothetical event, as well as other contextual aspects of the situation (Archer, 1997, 1999, 2000).

Perhaps the ultimate contextual variable for human aggression is provided by one's culture. Classic definitions of culture have it comprising the complex whole of the language, the customs, beliefs, laws, morals, and so forth, which are intergenerationally transmitted within particular human groups (Tylor, 1891). Contemporary cultural theorists tend instead to view culture as shared meaning and interpretation (e.g., Geertz, 1973; White and Dillingham, 1973). Within either definition, culture comprises a sort of macro-context against which aggressive acts are enacted and understood. The presumption of cross-cultural stability in patterns of aggression is often used to assert the dominance of biology in the development of human aggression, and especially of sex differences in aggression.

Does cross-cultural stability indicate that an extraordinarily sweeping kind of context is in fact irrelevant in shaping aggression? Keeping in mind that cross-cultural studies have only begun to specify subtypes of aggression fairly recently, we nonetheless note six key patterns in these data (Daly and Wilson, 1988; Fry, 1998; Maccoby and Jacklin, 1974; Montagu, 1978; Munroe et al., 2000; Whiting and Edwards, 1973):

1. There is wide cultural variation in amount of average aggression.
2. There is wide variation in the degree of male–female difference in aggression.
3. There is wide variation in levels of male- and/or female-specific aggression.
4. Within specific cultures, male–female differences in aggression are not consistent across different types of aggression.
5. There is some (though not conclusive) evidence that females may employ indirect aggression more often than males.
6. Across cultures, males tend to be at a higher level of physical aggression than females.

These cross-cultural data are most often summarized with the single observation that males are more (physically) aggressive than females.

Yet, to interpret sex differences in human aggression according to some simple biological mechanism, there should be some basic agreement in the absolute levels of aggression among males or females, and this is clearly not the case. Not only do the levels of aggression among males and females vary quite widely by cultural groups, the average level of male aggression in nonviolent groups is often considerably lower than the average level of female aggression in violent groups. Robarchek notes, for example, that "the level of male violence in [relatively nonviolent] groups such as the Semai is much lower than the level of female violence in groups like the [relatively violent] Waorani" (Robarchek, personal communication). Additional evidence for this lack of agreement in absolute levels of male or female aggression can be found by comparing the men of peaceful groups as described by Fabbro (1978) or Howell and Willis (1989) and the women of aggressive groups as described by Greenberg (1989) or Cook (1992). The pattern of a higher average level of male than female aggression within any given culture is neither as strong nor as absolute as is often claimed. For example, in a recent study of aggression among children in four cultures, sex differences in aggression did not reach significance in three out of the four cultures, even using one-tailed tests (Munroe et al., 2000). The Six-Culture Study examined six other cultural groups. While boys engaged in assault more often than girls in four cultures, girls assaulted more often than boys in one culture, and the sexes were tied in the sixth (Whiting and Edwards, 1973). Studies of aggression among children of various ages often show relatively small sex differences or even no sex difference in the frequency of aggression (e.g., Felson and Russo, 1988; Jones, 1984; Keenan and Shaw, 1994; *see* also Underwood et al., 2001, for an excellent recent review of research on gender and aggression in children).

In terms of particular cultural features of gender that correlate with differences in patterns of aggression, there is conflicting evidence. On the one hand, Munroe and colleagues found that both the average level of aggression and the size of the sex difference showed a strong positive association with a patrilineal cultural pattern (Munroe et al., 2000). On the other hand, a study of cultural levels of violence found no evidence that gender differences in social role or rank could explain the extreme contrast between the Semai and the Waorani; there are no significant gender distinctions in rank among either group, no rigid gender division of labor in either society, and both groups are based on bilateral kinship structures (Robarchek and Robarchek, 1998; Robarchek, personal communication).

The important caveat to all cross-cultural data on sex differences is that of the broadly diffused social endorsement of male rather than female aggression. While it could be argued that this shared cultural pattern itself constitutes evidence that males are "biologically" more aggressive, such an argument teeters on teleology, since cross-cultural studies are meant precisely to discover what effect variation in socialization actually has on aggression. Experimental and cross-cultural studies indicate that cultural endorsement of aggression does have a potent effect on both the practice of aggression and on the interpretation of an act as aggressive.

As a number of classic studies of children in the 1970s demonstrated, gender bias strongly affects whether an observer identifies particular actions or emotions as "angry" as opposed to "fearful"; observers are more likely to attribute anger responses to children or babies whom they believe to be boys (Fausto-Sterling, 1985). A recent study of evaluative biases demonstrated that even young children (second and third graders) tend toward a "boys are bad" bias, more frequently attributing negative behaviors and attributions to boys than girls when presented with ambiguous information about eight unfamiliar children (Heyman, 2001). Given the somewhat small effect sizes for sex in most studies of aggression in children, there is the distinct possibility that observer bias might account for most or even all of the observed sex differences.

Research on gender role expectations in children indicates an additional dimension of context that must be considered to understand sex differences in aggression. As Diane Carlson Jones has noted, "When it comes to the use of coercive power, sex role expectations are clear: it is inappropriate for females to be aggressive, while it is acceptable and expected behavior for males" (Jones, 1984; *see* also Crick, 1997; Maccoby and Jacklin, 1974; Underwood et al., 2001). Jones' research on preschool children found no differences in the way that aggressive (or "dominance") behaviors are used in establishing and maintaining power structures in male vs female play groups and also found that girls who used dominance strategies (coercive influence) were as effective at controlling interaction sequences as were boys. However, the dominant girls "paid" for this achieved power by a loss of affective rank, that is, the dominant girls were strongly disliked, while the dominant boys were strongly liked by their peers (Jones, 1984). Jones' findings contrast with an earlier study that found dominance and acceptance to be positively associated in adolescent girls (Savin-Williams, 1979) and argues that the different findings for the two age groups most likely reflect a social learning process, whereby adolescent girls have learned to use "power strategies which are socially approved and to rely less on coercive or direct strategies" (Jones, 1984). This hypothesis is supported by evidence that the gender-dependent relationship between dominance or aggressive behavior and peer acceptance and/or rejection holds and perhaps even increases as children age (Crick, 1997; Lancelotta and Vaughn, 1989; Savin-Williams, 1976, 1979). Note, however, that at least one study has found no association between physical aggression scores and peer rejection for elementary school girls, presumably because the level of physical aggression in girls of this age group was so low (French, 1990).

In sum, sex role expectations affect the perception of the same set of behaviors as "aggressive" or not, create a sex-dependent condition on whether or not aggression will be a successful strategy for controlling interactions and resources, and create strong disincentives for girls to use aggression from a very young age. "Two- to three-year old girls who are able to consistently label gender engage in less physical aggression with peers than those less proficient at labeling" (Fagot et al., 1986; cited in Underwood et al., 2001). Clearly, if we are to succeed in making meaningful links between biology and sex differences in human aggressive behavior, we need to have a way of incorporating these pervasive kinds of contextual factors into forces acting on the brain that influence the way aggressive behaviors become linked to situational perceptions, and to sets of "favored" and "disfavored" behavioral alternatives. This seems to us very far removed from arguments about statistical correlations between single biological factors (be they genes, chromosomes, or hormones) and crime statistics.

HORMONES AND AGGRESSION

Androgens are directly or indirectly implicated in most discussions of sex differences in human aggression. Even theories of Y chromosome effects usually boil down to hormones, as androgen exposures are implicitly or explicitly identified as the mechanism through which sex chromosome differences affect development of aggression. Yet evidence that androgens play a causal role in human aggression is quite thin, though popular ideas to the contrary continue to be widely disseminated (*see*, e.g., Sullivan,

2000). While a number of studies have found correlations between aggression and circulating testosterone (*see*, e.g., Archer, 1994; Sanchez-Martin et al., 2000), other studies have failed to find such associations (Campbell et al., 1997a). Since circulating levels of testosterone are responsive to aggression (Archer, 1991; Gladue et al., 1989), some authors conclude that "current data supports a bidirectional model with androgens both influencing and being influenced by aggressive behavior" (Sanchez-Martin et al., 2000).

The theory that testosterone exerts the initial causal effect on aggression focuses on androgren exposures during critical periods of early development. This Organization Theory (OT) holds that androgens *in utero* "masculinize" the human brain in far-reaching ways, leading to male-typical traits, such as strong spatial relations skills, high energy expenditures, male heterosexuality, and aggression. While there is strong support for aspects of "male-type" behavior being related to early androgen exposures in many other mammalian species, the evidence that androgens "organize" the human brain is scanty, contradictory, and rests on studies that are deeply flawed methodologically (Birke, 1981; Bleier, 1984; Doell and Longino, 1988; Fausto-Sterling, 1985; Longino and Doell, 1983; Sloan, 1993; van den Wijngaard, 1997; Young, 2000).

Studies of girls with congenital adrenal hyperplasia (CAH) have been particularly influential and bear special attention here. CAH is a family of disorders associated with inborn errors of metabolism that cause various enzyme deficiencies, especially 21-hydroxylase deficiency (Dittmann et al., 1990). People with CAH do not properly synthesize cortisol and overproduce adrenocorticotrophic hormone (ACTH), androstenedione, and testosterone. In genetic females with CAH, the increase in "male" sex steroids during fetal development causes various degrees of masculinization of the genitalia. According to OT, high prenatal androgen levels would presumably also "masculinize" the brain and behavior.

From the earliest behavioral studies of girls with CAH, there has been interest in whether these children showed higher rates of aggressive or violent behavior, as well as traits that might be precursors to or correlates of aggression, such as "high energy expenditure," "rough-and-tumble play," or "tomboyism" (Dittmann, 1990; Money and Ehrhardt, 1972). At least three studies have directly investigated aggression among CAH children, with two teams reporting no increase (Ehrhardt and Baker, 1974; Money and Schwartz, 1976), and one recent study reporting higher levels of aggression in CAH females (but not males) relative to controls (Berenbaum and Resnick, 1997). The latter study purports to be methodologically more adequate than the earlier smaller reports, yet it is important to note how each of the CAH studies amplify the issues of gender stereotype, interpretation of observational data, and context that we raised earlier with regard to research on aggressive behavior in normal children. Key aspects of the CAH research have attracted a great deal of criticism (*see*, e.g., Birke, 1981; Bleier, 1984; Doell and Longino, 1988; Fausto-Sterling, 1985; Longino and Doell, 1983; Sloan, 1993; van den Wijngaard, 1997; Young, 2000), notably extremely small sample sizes; repeated reports on the same very small group of patients, suggesting much more independent data than actually exists; inadequate or no control groups; little or no attention to subjects' extremely unusual rearing experiences, or to the effect of chronic and sometimes life-threatening illnesses caused by the CAH; and extremely generous statistical treatments and interpretations of data, which systematically favor the hypothesis of brain organization.

In keeping with our overall focus on context, we will briefly consider the issue of comparability between androgen-exposed subjects' rearing experiences and the experiences of normal or control girls. Many of the subjects in CAH studies were born with ambiguous genitalia. For example, among children in the Berenbaum and Resnick (1997) study, the mean Prader value, which indicates genital virilization on a scale from 0 (normal female morphology) to 6 (normal male morphology) was 3. Scores ranged from 1 to 4, indicating that all subjects had some visible virilization of the genitalia. Late-treated patients, who show the clearest behavioral masculinization, "had lived for many years with the stigma of heavy [physical] virilization, sometimes uncorrected genital morphology, and lack of feminine secondary sex development" (Ehrhardt et al., 1968). The numerous social and clinical sequelae of having a child whose sex is either initially uncertain or is "reassigned" or "reannounced" in response to the diagnosis of CAH are quite serious and would surely affect parents' and clinicians' interaction with the children from the time such uncertainty begins. For the child him- or herself, the experience of genital surgeries, repeated clinical examinations, implicit and explicit messages about "unfinished" or "damaged" genitalia are likewise serious. The authors of OT studies of intersex patients tend to dramatically underemphasize these social and clinical experiences associated with ambiguous genitalia. They also tend to overestimate the degree to which such effects can be controlled by techniques such as asking parents how concerned or anxious they were about the child's condition. These indirect effects of hormone exposures on behavior, which operate through anomalous genitals and the sociomedical responses that they provoke, cannot be separated analytically from any putative direct effects of hormones on the brain.

From the late 1970s, investigators interested in brain organization in humans largely abandoned intersex studies in favor of other study designs (Young, 2000). Critics, and to some extent investigators themselves, had been troubled by the inability of intersex syndromes to separate direct effects of hormones on the brain from their indirect effects on behavior via anomalous genitalia and subsequent rearing experiences. Therefore, studies of hormone exposures that did not result in ambiguous genitalia offered a more attractive alternative for trying to establish that hormones affect human behavior by affecting the brain directly. In animal studies, critical periods for masculinizing the brain (or specific brain regions involved with different functions or traits) differ from the critical period for masculinizing the genitalia. Furthermore, different specific steroid hormones are involved in the development of specific sex-typed behaviors in animals. Diethylstilbestrol (DES) is an example of a synthetic steroid shown to increase male-type behavior in animal experiments, though it does not masculinize genitalia. Since DES was widely prescribed to prevent miscarriage in the 1960s and 1970s, offspring of DES-treated pregnancies have provided a nonintersex human population who would be expected to show behavioral masculinization. To date, only two published studies have reported on DES and aggression; neither found any increase in aggression among DES-exposed offspring (Lish et al., 1991, 1992; Yalom et al., 1973).

In sum, humans who are known to have been exposed to masculinizing hormones *in utero* have also almost universally been exposed to socialization as males. This stubborn fact leaves us with the perhaps unsatisfying inability to collect the requisite data to settle the question of sex differences in human aggression. This is probably good news for serious biological inquiry, because this recognition can steer research toward more precise, systematic, and ultimately answerable questions about the biology of aggression.

EVOLUTION AND HUMAN AGGRESSION

The evolutionary story of why one would expect sex differences in aggression has three main threads: (*i*) historical sex roles of "man the hunter" and "woman the gatherer," whose different subsistence activities required male and not female aggression; (*ii*) different reproductive strategies and parental investment, which both cause women to be more risk averse than men and reward male sexual aggression against females as a reproductive strategy; and (*iii*) a simple byproduct of the average difference in body size between the sexes. Each of these storylines has serious problems when viewed in light of contemporary archeological data and/or empirical research on human aggression.

The classic story of man the hunter is increasingly being questioned by both theoretical and archeological developments in anthropology (Adovasio, 1996; Hawkes, 2000; Tanner and Zihlman, 1978). Men, this argument goes, needed to develop aggressive traits in order to be successful hunters, while women's activities in gathering did not encourage the development of aggression. Recent evidence suggests, though, that a great deal of early hunting activity was not the mythological group of male warriors armed with spears, clubs, or arrows, stalking dangerous large prey in all-male groups. Instead, the discovery of very early net fragments (Soffer et al., 2001), as well as evidence that women do in fact hunt in some traditional societies, lends credence to the alternative hypothesis that entire communities (including children and adults of both sexes) may have been involved in hunting. Hunting parties of men with sharp weapons may have been less common than chasing prey into a net, off cliffs, or into traps. This sort of activity requires not aggression, but cooperation, and would not produce sex differences in any case.

Another evolutionary story proposes that females will avoid aggression because of their greater parental investment than men (Fry, 1998; Trivers, 1972; Washburn and Lancaster, 1968) and the resultant need to keep their own bodies intact in order to have reproductive success. Yet more recent data and theoretical work on the notion of parental investment have reopened the question of whether male parental investment is necessarily smaller than the females (Bleier, 1984; Fausto-Sterling, 1985, 1995). The calculation of parental investment is more assumption-laden and complicated than many evolutionary accounts let on. As Ruth Hubbard has asked, "Is it reasonable to count only the energy required to produce the few sperm that actually fertilize eggs, or should one not count the total energy males expend in producing and ejaculating semen?" (Hubbard, 1990). We would add that in many species, the male courtship that precedes mating is very demanding of time, energy, and physical resources, and should also be included in investment calculations.

The final thread of the evolutionary story about sex differences in human aggression is the simplest and in many ways the most plausible. Sex differences in aggression might simply have evolved along with the average size differences between males and females, reflecting a sex difference in the likelihood that aggression will be a successful strategy for access to resources. If this is the case, then patterns of aggression should have evolved that would mirror average relative size of aggressors and their targets. In other words, both males and females should be more likely to aggress against females, and females should be especially unlikely to aggress against males. Yet the experimental data do not support this theory: both males and females are more likely to aggress against males (Harris, 1994, 1996; Archer, 1999).

CONCLUSION

What lessons should people interested in linking biology and aggression take with them as a result of reading this chapter? We hope that three things have emerged from this discussion.

First, aggression is a problematic trait for biological studies, because it, of necessity, incorporates contextual features into its definition.

Second, to make headway in understanding biological correlates of aggression, one needs to separate those features of aggression that have to do with situational interpretations (perceptions of context) from those features that have to do with linking a particular contextual interpretation to aggressive behavior. We think that it is in the latter arena that biology is likely to make the most progress in the near future, provided that the studied contexts linked to aggression are common to most individuals in a species. It is in this domain of "representativeness" that we think animal research (which we view as fundamental to an understanding of biological correlates of aggression) has most seriously fallen short. There are two ways this situation could be improved. First, to shift the focus from highly artificial encounters between rodents housed in highly abnormal conditions to historically more typical situations (for instance, social conflict over stored food), and second, to develop animal analogs of nonphysical aggression, since it is commonly agreed that this is the most common form in humans.

Finally, we hope that the examination of material on sex differences in aggression has highlighted how ineffective an approach focusing on single causative factors will be in the face of a trait that will surely have many complicated situational dependencies. Perhaps aggression is an area where the biological strategy of starting with differences (for instance, between men and women, or between murderers on death row and medical students) fails because of the way that individual development intertwines context with physiology. The best understanding of the biological correlates of aggression may come from an examination of what makes us similar, not what makes us different.

REFERENCES

Adovasio, J. M., Soffer, O., and Klima, B. (1996) Upper Palaeolithic fibre technology: interlaced woven finds from Pavlov I, Czech Republic, c. 26,000 years ago. *Antiquity* **70**, 26–534.
Archer, J. (1991) The influence of testosterone on human aggression. *Br. J. Psychol.* **82**, 1–28.
Archer, J. (1994) Testosterone and aggression, in *The Psychobiology of Aggression.* (Hilbram, M. and Pallone, N. J., eds.), Haworth Press, New York, pp. 3–35.
Archer, J. (1997) Beliefs about aggression among male and female prisoners. *Aggress. Behav.* **23**, 405–415.
Archer, J. (1999) Sex differences in beliefs about aggression: opponent's sex and the form of aggression. *Br. J. Soc. Psychol.* **38**, 71–84.
Archer, J. (2000) Sex differences in aggression between heterosexual partners: a meta-analytic review. *Psychol. Bull.* **126**, 651–680.
Archer, J. (2001) A strategic approach to aggression. *Soc. Dev.* **10**, 267–271.

Balaban, E., Alper, J. S., and Kasamon, Y. L. (1996) Mean genes and the biology of aggression: a critical review of recent animal and human research. *J. Neurogenet.* **11,** 1–43.

Berenbaum, S. A. and Resnick, S. M. (1997) Early androgen effects on aggression in children and adults with congenital adrenal hyperplasia. *Psychoneuroendocrinology* **22,** 505–515.

Birke, L. (1981) Is homosexuality hormonally determined? *J. Homosexuality* **6,** 35–49.

Bjorkvist, K. (2001) Comments to top ten challenges for understanding gender and aggression in children: why can't we all just get along? Different names, same issue. *Soc. Dev.* **10,** 267–271.

Bleier, R. (1984) *Science and Gender.* Pergamon Press, New York.

Brodkin, E. S., Goforth, S. A., Keene, A. H., Fossella, J. A., and Silver, L. M. (2002) Identification of quantitative trait loci that affect aggressive behavior in mice. *J. Neurosci.* **22,** 1165–1170.

Campbell, A., Muncer, S., and Odber, J. (1997a) Aggression and testosterone: testing a biosocial model. *Aggress. Behav.* **23,** 229–238.

Campbell, A., Sapochnik, M., and Muncer, S. (1997b) Sex differences in aggression: does social representation mediate form of aggression? *Br. J. Soc. Psychol.* **36,** 161–171.

Carey, G. and Gottesman, I. (1996) Genetics and antisocial behavior: substance versus sound bytes. *Politics Life Sci.* **15,** 88–90.

Caspi, A., McClay, J., Moffitt, T. E., et al. (2002) Role of genotype in the cycle of violence in maltreated children. *Science* **297,** 851–854.

Cook, H. B. K. (1992) Matrifocality and female aggression in Margariteno society, in *Of Mice and Women: Aspects of Female Aggression.* (Bjorkvist, K. and Niemala, P., eds.), Academic Press, San Diego, pp. 149–162.

Crick, N. R. (1997) Engagement in gender normative versus gender nonnormative forms of aggrssion: links to social-psychological adjustment. *Dev. Psychol.* **33,** 610–617.

Daly, M. and Wilson, M. (1988) *Homicide.* Aldine de Gruyter, New York.

Dittmann, R. W., Kappes, M. H., Kappes, M. E., et al. (1990) Congenital adrenal hyperplasia I: gender-related behavior and attitudes in female patients and sisters. *Psychoneuroendocrinology* **15,** 401–420.

Doell, R. and Longino, H. (1988) Sex hormones and human behavior: a critique of the linear model. *J. Homosexuality* **15,** 55–78.

Ebert, P. D. (1983) Selection for aggression in a natural population, in *Aggressive Behavior: Genetic and Neural Approaches.* (Simmel, E. C., Hahn, M. E., and Walters, J. K., eds.), Lawrence Erlbaum Associates, Hillsdale, NJ, pp. 103–127.

Ehrhardt, A. A., Epstein, R., and Money, J. (1968) Influence of androgen and some aspects of sexually dimorphic behavior in women with the late-treated adrenogenital syndrome. *Johns Hopkins Med. J.* **123,** 115–122.

Ehrhardt, A. A. and Baker, S. W. (1974) Fetal androgens, human central nervous system differentiation and behavior sex differences, in *Sex Differences in Behavior.* (Friedman, R. C., Richart, R. M., and Vande Wiele, R. L., eds.), Wiley & Sons, New York, pp. 33–51.

Fabbro, D. (1978) Peaceful societies: an introduction. *J. Peace Res.* **15,** 67–83.

Fagot, B. I., Leinbach, M. D., and Hagan, R. (1986) Gender labeling and the adoption of sex-typed behaviors. *Dev. Psychol.* **22,** 440–443.

Fausto-Sterling, A. (1985) *Myths of Gender: Biological Theories about Women and Men.* Basic Books, New York.

Fausto-Sterling, A. (1995) Attacking feminism is no substitute for good scholarship. *Politics Life Sci.* **14,** 171–174.

Felson, R. B. and Russo, N. (1988) Parental punishment and sibling aggression. *Soc. Psychol. Q.* **51,** 11–18.

French, D. C. (1990) Heterogeneity of peer-rejected girls. *Child Dev.* **61,** 2028–2031.

Fry, D. P. (1998) Anthropological perspectives on aggression: sex differences and cultural variation. *Aggress. Behav.* **24,** 81–95.

Galen, B. R. and Underwood, M. K. (1997) A developmental investigation of social aggression among children. *Dev. Psychol.* **33,** 589–6000.

Geertz, C. (1973) *The Interpretation of Cultures.* Basic Books, New York.

Gladue, B. A., Boechler, M., and McCaul, K. D. (1989) Hormonal responses to competition in human males. *Aggress. Behav.* **15,** 409–422.

Greenberg, J. B. (1989) *Blood Ties: Life and Violence in Rural Mexico.* University of Arizona Press, Tucson, AZ.

Guillot, P. V., Carlier, M., Maxson, S. C., and Roubertoux, P. L. (1995) Intermale aggression tested in two procedures, using four inbred strains of mice and their reciprocal congenics: Y chromosomal implications. *Behav. Genet.* **25,** 357–360.

Harris, M. B. (1996) Aggression, gender, and ethnicity. *Aggress. Violent Behav.* **1,** 123–146.

Harris, M. B. (1994) Gender of subject and target as mediators of aggression. *J. Appl. Soc. Psychol.* **24,** 453–471.

Hawkes, K., O'Connell, J. F., Blurton Jones, N. G., Alvarez, H., and Charnov, E. L. (2000) *Grandmother Hypothesis and Human Evolution.* (Cronk, L., Chagnon, N., and Irons, W., eds.), Aldine de Gruyter, New York, pp. 237–258.

Heyman, G. D. (2001) Children's interpretation of ambiguous behavior: evidence for a 'boys are bad' bias. *Soc. Dev.* **10,** 230–247.

Hood, K. (1992) Female aggression in mice: developmental, genetic, and contextual factors, in *Of Mice and Women: Aspects of Female Aggression.* (Bjornkvist, K. and Niemela, P., eds.), Academic Press, San Diego, pp. 395–402.

Howell, S. and Willis, R. (Eds.) (1989) *Societies at Peace: Anthropological Perspectives.* Routledge, New York.

Hubbard, R. (1990) *The Politics of Women's Biology.* Rutgers University Press, New Brunswick.

Jones, D. C. (1984) Power structures and perceptions of power holders in same-sex groups of young children, in *Biopolitics and Gender.* (Watts, M., ed.), Haworth Press, Binghamton, pp. 147–163.

Keenan, K. and Shaw, D. S. (1994) The development of aggression in toddlers—a study of low-income families. *J. Abnorm. Child Psychol.* **22,** 53–77.

Lancelotta, G. X. and Vaughn, S. (1989) Relation between types of aggression and sociometric status: peer and teacher perceptions. *J. Educ. Psychol.* **81,** 86–90.

Lee, R. and Coccaro, E. (2001) The neuropsychopharmacology of criminality and aggression. *Can. J. Psychiatry* **46,** 35–44.

Lish, J. D., Ehrhardt, A. A., and Meyer-Bahlburg, H. F. L. (1991) Gender-related behavior development in females exposed to diethylstilbestrol (DES) in utero: an attempted replication. *J. Am. Acad. Child Adolesc. Psychiatry* **30,** 29–37.

Lish, J. D., Meyer-Bahlburg, H. F. L., Ehrhardt, A. A., Travis, B. G., and Veridiano, N. P. (1992) Prenatal exposure to diethylstilbestrol (DES): childhood play behavior and adult gender-role behavior in women. *Arch. Sexual Behav.* **21,** 423–441.

Longino, H. E. and Doell, R. (1983) Body, bias, and behavior: a comparative analysis of reasoning in two areas of biological science. *Signs J. Women Culture Society* **9,** 207–227.

Maccoby, E. and Jacklin, C. (1974) *The Psychology of Sex Differences.* Stanford University Press, Stanford, CA.

Maxson, S. C. (1992) Methodological issues in genetic analyses of agonistic behavior (offense) in male mice, in *Techniques for the Genetic Analysis of Brain and Behavior.* (Goldowitz, D., Wahlsten, D., and Wimer, R. E., eds.), Elsevier Science, New York, pp. 349–373.

Maxson, S. C. (1996) Searching for candidate genes with effects on an agonistic behavior, offense, in mice. *Behav. Genet.* **26,** 471–476.

Maxson, S. C. (2000) Genetic influences on aggressive behavior, in *Genetic Influences on Neural and Behavioral Functions.* (Pfaff, D. W., Berrettini, W. H., Joh, T. H., and Maxson, S. C., eds.), CRC Press, Boca Raton, pp. 405–416.

Miczek, K. A., Maxson, S. C., Fish, E. W., and Faccidomo, S. (2001) Aggressive behavioral phenotypes in mice. *Behav. Brain Res.* **125,** 167–181.

Money, J. and Ehrhardt, A. A. (1972) *Man and Woman, Boy and Girl.* Johns Hopkins University Press, Baltimore.

Money, J. and Schwartz, M. (1976) Fetal androgens in the early treated adrenogenital syndrome of 46 XX hermaphroditism: influence on assertive and aggressive types of behavior. *Aggress. Behav.* **2,** 19–30.

Montagu, A. (Ed.) (1978) *Learning Non-Aggression: The Experience of Non-Literate Societies.* Oxford University Press, New York.

Munroe, R. L., Hulefeld, R., Rodgers, J. M., Tomeo, D. L., and Yamazaki, S. K. (2000) Aggression among children in four cultures. *Cross-Cultural Res.* **34,** 3–25.

Österman, K., Björkqvist, K., Lagerspetz, K. M. J., et al. (1998) Cross-cultural evidence of female indirect aggression. *Aggress. Behav.* **24,** 1–8.

Pigliucci, M. (2001) *Phenotypic Plasticity.* The Johns Hopkins University Press, Baltimore.

Ramirez, J. M., Andreu, J. M., and Fujihara, T. (2001) Cultural and sex differences in aggression: a comparison between Japanese and Spanish students using two different inventories. *Aggress. Behav.* **27,** 313–322.

Robarchek, C. A. and Robarchek, C. J. (1998) Reciprocities and realities: world views, peacefulness, and violence among semai and waorani. *Aggress. Behav.* **24,** 123–133.

Robarchek, C.A. (2002) Personal communication with R.M. Young, email Friday July 19.

Roubertoux, P. L., Carlier, M., Degrelle, H., Haas-Dupertuis, M. C., Phillips, J., and Moutier, R. (1994) Co-segregation of the pseudoautosomal region of the Y chromosome with aggression in mice. *Genetics* **136,** 225–230.

Sanchez-Martin, J. R., Fano, E., Ahedo, L., Cardas, J., Brain, P. F., and Azpiroz, A. (2000) Relating testosterone levels and free play social behavior in male and female preschool children. *Psychoneuroendocrinology* **25,** 773–783.

Savin-Williams, R. (1976) An ethological study of dominance formation and maintenance in a group of human adolescents. *Child Dev.* **47,** 972–979.

Savin-Williams, R. (1979) Dominance hierarchies in groups of early adolescents. *Child Dev.* **50,** 923–935.

Schlichting, C. D. and Pigliucci, M. (1998) *Phenotypic Evolution: A Reaction Norm Perspective.* Sinauer, Sunderland, MA.

Sloan, E. (1993) *Biology of Women, 3rd ed.* Delmar Publishers, Albany, NY.

Sluyter, F., van der Vlugt, J. J., van Oortmerssen, G. A., Koolhaas, J. M., van der Hoeven, F., and de Boer, P. (1996) Studies on wild house mice. VII. Prenatal maternal environment and aggression. *Behav. Genet.* **26,** 513–518.

Sluyter, F., Bult, A., Lynch, C. B., Meeter, F., and van Oortmerssen, G. A. (1997) No evidence for a Y chromosomal effect on alternative behavioral strategies in mice. *Behav. Genet.* **27,** 477–482.

Soffer, O., Adovasio, J. M., Hyland, D. C., Illingworth, J. S., Klima, B., and Svoboda, J. (2001) Perishable industries from dolni vestonice I: new insights into the nature and origin of the gravettian. *Archaeology, Ethnology & Anthropology of Eurasia* **2,** 48–65.

Stowers, L., Holy, T. E., Meister, M., Dulac, C., and Koentges, G. (2002) Loss of sex discrimination and male-male aggression in mice deficient for TRP2. *Science* **295,** 1493–1500.

Sullivan, A. (2000) *Why Men are Different: the Defining Power of Testosterone.* The New York Times Magazine, **April 2,** 46.

Tanner, N. and Zihlman, A. (1976) Women in evolution, Part I: innovation and selection in human origins. *Signs* **1,** 585–608.

Tanner, N. and Zihlman, A. (1978) Women in evolution, Part II: subsistence and social organization in early hominids. *Signs* **4,** 4–20.

Trivers, R. (1972) Parental investment and sexual selection, in *Sexual Selection and the Descent of Man.* (Campbell, B., ed.), Aldine, Chicago, IL, pp. 136–179.

Tylor, E. B. (1891) *Anthropology: An Introduction to the Study of Man and Civilization.* D. Appleton and Co., New York.

Underwood, M. K., Galen, B. R., and Paquette, J. A. (2001) Top ten challenges for understanding gender and aggression in children: why can't we all just get along? *Soc. Dev.* **10,** 248–266.

van den Wijngaard, M. (1997) *Reinventing the Sexes: The Biomedical Construction of Femininity and Masculinity.* Indiana University Press, Bloomington, IN.

Van Oortmerssen, G. A. and Bakker, T. C. M. (1981) Artificial selection for short and long attack latencies in wild *Mus musculus domesticus. Behav. Genet.* **11,** 115–126.

Washburn, S. and Lancaster, C. (1968) The evolution of hunting, in *Man the Hunter.* (Lee, R. B. and De Vore, I, eds.), Aldine, Chicago, IL, pp. 293–303.

White, L. A. and Dillingham, B. (1973) *The Concept of Culture.* Burgess, Minneapolis.

Whiting, B. B. and Edwards, C. P. (1973) A cross-cultural analysis of sex differences in the behavior of children aged 3-11. *J. Soc. Psychol.* **91,** 171–188.

Yalom, I. D., Green, R., and Fisk, N. (1973) Prenatal exposure to female hormones: effect on psychosexual development in boys. *Arch. Gen. Psychiatry* **28,** 554–561.

Young, R. M. (2000) Sexing the Brain: Measurement and Meaning in Biological Research on Human Sexuality. Doctoral dissertation, Columbia University, Division of Sociomedical Sciences, New York.

Zihlman, A. (1978) Women in evolution, part 2: subsistence and social organization among early hominids. *Signs* **4,** 4–20.

12
The Family Environment in Early Life and Aggressive Behavior in Adolescents and Young Adults

Sven Barnow and Harald-J. Freyberger

INTRODUCTION

Aggressive behavior in adolescents is of great interest to mental health professions, as it causes disruptions in the family, school, and peer relations, and predicts higher risks for criminality, substance misuse, and personality disorders in adulthood (Coie et al., 1982; Elliott, 1994).

It is very doubtful, however, that "aggression" can be regarded as a coherent system of behavior and motivation. Indeed, it is reasonable to assume the existence of multiple forms of aggression. Petermann (1993), for example, identifies forms of expression of aggressive behavior in terms of such dimensions as overt vs covert, physical vs verbal, active vs passive, and direct vs indirect. One of the most commonly applied classifications relates to the differentiation between spontaneous and instrumental aggression. It should be noted, however, that it is often impossible to make clear distinctions between these forms of aggression, as aggressive adolescents frequently display mixed forms of aggressive behavior (Achenbach et al., 1989; Earls, 1994). Reflecting this complexity, factor analytic studies of clinical samples have combined many of these acting out behaviors to a broad behavioral problem scale called "externalizing" symptoms. One frequently used questionnaire, the Child Behavior Checklist (CBCL) (Achenbach, 1994) and the similar self-report version Youth Self Report (YSR) (Achenbach, 1991), divides, for example, externalizing symptoms into a broadband "behavioral problem scale" made up of aggressive and delinquent conduct problems (Achenbach et al., 1989; Achenbach and Edelbrock, 1978; Angold and Costello, 2000; Frick et al., 1993; Verhulst and Achenbach, 1995).

In view of the lack of clarity with respect to definitions, on the one hand, and the diversity of behavioral expressions of aggression, on the other hand, this chapter not only discusses studies focused on factors influencing aggressive behavior, but also analyzes the correlations between behavioral problems and family-related risks in general, whereby the two groups frequently overlap to a certain degree.

Factors that contribute to involvement with aggressive behavior are likely to be heterogeneous and may include biological characteristics of the individual and environmental conditions (Barber, 1996; Cloninger, 1994; Cohen et al., 1989; Kandel and Mednick, 1991;

Kumpfer and Turner, 1990; Pettit, 1997). For example, one study, which investigated the influence of 20 different risk factors presented at age 5 (e.g., child characteristics, peer-related factors, parenting factors, and sociocultural factors) on externalizing problems in middle childhood, found that child factors (e.g., temperament) explained 19%, peer-related factors up to 13%, parenting factors up to 6%, and sociocultural factors up to 4% of the variance in behavioral problems (Deater-Deckard et al., 1998).

Pellegrini (1990) distinguishes between indicators that point to biological and psychological traits in the individual (e.g., genetic vulnerability), or psychological features, and factors that relate to personal and environmental psychosocial characteristics (e.g., chronic disharmony in the family, sexual abuse, or specific personality traits). The biological risk factors include genetic influences, neurological factors, neurotransmitter changes, and psychophysiological conditions (Blanz, 1988; Edelbrock et al., 1995; O'Conner et al., 1995). Also classified within the group of psychosocial stress factors are conditions imposed by shared (e.g., family characteristics) and unshared (e.g., peer group characteristics) environments.

Keeping this complexity in mind, this chapter examines recent progress in understanding the influence of the early family environment on adolescent's aggressive behavior. We begin by discussing the question whether early family conditions play a role for aggressive behavior in adolescents.

FAMILY-RELATED RISK FACTORS FOR AGGRESSIVE BEHAVIOR AMONG ADOLESCENTS

Does the Family Environment Matter?

In 1995, Judith Rich Harris published a critical study of the influence of parenting behavior on personality traits in the *Psychological Review* (Harris, 1995). In her article, she comes to the conclusion that parenting behavior and the family environment have no significant influence on personality characteristics in children. She supports this conclusion with the following line of argumentation:

- Adopted siblings show no similarities as adults, i.e., they are no more alike than children who grow up in different families.
- Biological siblings show somewhat more similarity, but not a high degree of similarity.
- Twins raised in the same home are about 50% similar, but no more similar than those who have grown up in different homes.
- Studies of twins show that 50% of personality traits are genetically transmitted. Therefore, the remaining variance must be attributable to environmental factors. Harris thus concludes that monozygotic twins, whose genetic makeup is 100% identical and who have grown up in the same home, must therefore be 100% alike. Yet this is not the case, she points out, noting that matching personality traits (e.g., extraversion, neurotic behavior) account for approx 50%, which corresponds with the genetic contribution. She also contends that consideration of various other factors, e.g., that parents raise each child differently, that children interpret identical parenting behavior differently, does not help to explain why twins who grow up in the same family are no more or less alike than those who are raised in different households. She cites several different studies in support of her contention, including that of Ernst and Angst (1983), who found in their study of 1582 young adults surveyed for a range of different family characteristics and personality traits that such factors as sibling birth order and other family characteristics had no influence on later personality traits. She also mentions a

number of different studies showing that only 1–10% of variance is a product of the shared environment.

More recent findings suggest, however, some modifications of this position, and especially aggression has often been seen as exception to this rule (Rutter et al., 1999). Thus, a large number of studies provide evidence of a direct relationship between parenting behavior and negative family environment and behavioral problems that appear later in life (Brook et al., 1983; Campbell, 1995; Coleman and Frick, 1994; Robins, 1991). The findings of prospective studies suggest a causal link between parenting behavior and behavioral problems in adolescence and adulthood (Kendler et al., 2000a; 2000b). Several studies show that behavior training for parents led to a significant reduction of antisocial behavior in children (for review, *see* Lipson et al., 1981). Aside from the clinical impression, successes achieved through intervention can be seen as evidence that modifications in the family environment can produce behavioral changes, which would indicate that at least the two levels are related to one another.

It is important to realize, however, that biological factors, such as difficult temperament, for example, interact with the family environment and contribute to more or less adaptive adjustment. Belsky et al. (1989) distinguished between neurobiological risks in the child (e.g., difficult temperament) and social risks in the family environment, whereby both factors contribute to an insecure parent–child bond and favor the development of externalizing disorders (Greenberg et al., 1993). It is likely that psychosocial context variables (e.g., family poverty and disadvantage, large family size, parental psychopathology, divorce, and parental low education level) mediate such processes. This depends, however, on the extent to which parenting styles are affected. In other words, the more stable the (positive) parenting behavior and the less dependent it is on negative external factors, the less likely it is that childhood behavioral disorders will emerge. The following section is devoted to a description of the combinations of familial risk factors that appear early in life to facilitate the development of aggressive behavior later on.

Biosocial Theories of Aggression
Taking the Interaction Between Early Family Influences
and Biological Characteristics of the Child into Consideration

Many authors assume that aggressive behavior is at least partially determined by genetic makeup (Mednick and Kandel, 1988; Mednick et al., 1982; Pollock et al., 1990). Children who exhibit neuropsychological deficits resulting from prenatal, perinatal, or postnatal birth complications appear to be more susceptible to negative social environmental influences (Moffitt, 1993). Thus, for example, maternal alcohol abuse during pregnancy, which is known to cause damage during brain development, is associated with later behavior disorders (Streissguth, 1997).

Within the context of biosocial theory as applied to antisocial behavior disorders, Raine et al. (1997) demonstrated that deficits in the autonomic nervous system of children can lead to a reduced capacity for conditioning and conflict avoidance, and to an inability to adapt to social norms. Where a constellation of this kind encounters problematic parenting behavior, such as the tendency to promote aggressive behavior, failure to establish rules, rejection, or insufficient warmth and emotional support, the combination of factors may cause aggressive behavior in children at a later age. In a supportive family environment, however, the likelihood of the appearance of aggression would be lower (*see* also the section on protective factors). Buikhuisen's (1988) biosocial theory seeks to arrive at an understanding of aggressive behavior from the perspective of developmental psychology. Aggressive or criminal behavior is regarded as a product of the interaction of personal and environmental factors. Relevant personal factors include biological, cognitive, and psychiatric variants. Relevant environmental factors include family, neighborhood, and peer contact. Buikhuisen also characterizes the development of antisocial and/or aggressive behaviors as products of adverse socialization processes. He proposes that children who are unable to learn appropriate avoidance tactics exhibit a higher risk of delinquent or aggressive behavior at a later age. Buikhuisen also recognizes the relation between personal factors, (e.g., neuropsychological deficits and low verbal intelligence) and rejection of the child by the mother, which can prevent the child from developing sufficiently strong feelings of empathy and self-esteem. This, in turn, can lead to further failures in school or within the peer group. His theory relies on the assumption that children who exhibit both biological and social deficits run a particularly high risk of developing aggressive behavior at a later age. Moffitt (1993) goes a step further in developing his biosocial theory, contending that the biological roots of aggression are already present shortly before or after birth. Neuropsychological deficits can express themselves, for example, in a difficult temperament or cognitive deficits in the child. Such children are extremely susceptible to factors in the family environment. Moffit also recognizes the reverse possibility that deficits in the family environment, prenatal and postnatal complications, or abuse in early infancy can result in biological deficits in the child. This makes it difficult to distinguish specifically between causes and effects. One child may be biologically predisposed, while another healthy child later develops neuropsychological deficits due to neglect in early infancy. Using data from a long-term study conducted in New Zealand (Multidisciplinary Health and Development Study), Moffitt (1990) was able to evaluate his ideas relating to the association between neuropsychological functional capacity and aggressive behavior during adolescence as reported by subjects of the study. Findings based on a sample of more than 1000 children, examined every 2 yr from birth to age 13, revealed a significant interaction between negative family environment (e.g., low parental education levels, low income, divorced parents, poor health of the mother), neuropsychological deficits, and later aggressive behavior. The findings also confirmed his hypothesis that children with both neuropsychological deficits and a negative family environment have a significantly higher potential for aggressive behavior. These results were supported indirectly by the results of a long-term study conducted by Raine et al. (1994, 1997). The authors discovered that birth complications, in combination with maternal rejection, represented the best predictors of subsequent violent and/or criminal behavior in the respective children. Chronic aggressive and violence-prone behavior was defined on the basis of reports of arrests during youth or early adulthood.

In the Mannheim Risk Study, a prospective longitudinal study in which the development of a relatively large birth cohort, including high-risk children, was observed using a complex body of research instruments from early childhood to adolescence, a correlation between psychosocial stress and later behavioral problems was identified (Laucht, 2001; Laucht et al., 2001a, 2001b). The study also included children whose psychological development was endangered by early organic and psychosocial stress factors identifiable at the time of birth, such as premature birth or a parent with a mental disorder. The results showed that the proportion of 8-yr-olds showing evidence of retarded cognitive development in the high stress group was 2.95 times higher than that of the low stress group. In another study conducted by the same research group, the presence of a conduct disorder in children of elementary school age was correctly predicted by such early psychosocial risk factors as broken home, delinquency of the father, chronic family problems, or psychological problems of one parent in 64.8% of cases. A crucial factor in this context was the quality of interaction between mother and child. Thus, it was found that 8-yr-olds whose mothers communicated with them in a less loving manner (spoke little or used very little baby language) in early infancy (at 3 mo) exhibited behavioral problems more frequently than those in a control group of the same age whose mothers showed more affection. Another remarkable pattern of interaction, which appeared frequently in advance of subsequent mental disorders and particularly of behavioral problems among children and adolescents, was a reduced degree of maternal flexibility in response to the infant, especially in cases where this factor appeared in combination with negative affective behavior on the part of the infant.

Another study found an association between physical abnormalities due to prenatal or postnatal malformations and aggressive behavior at a later age. The findings of this study conducted by Arseneault et al. (2000), in which minor physical abnormalities were studied in a group of 170 adolescents in Montreal, showed that adolescents with a larger number of anomalies, particularly oral malformations, show an especially high propensity for later delinquent and aggressive behavior.

In summary, the empirical findings provide evidence of the significance of interactive behavior between mother and child in early infancy. The adolescents who showed more behavioral problems were those, who as infants, were frequently dysphoric, exhibited difficult temperaments (high degree of impulsiveness, negative affective behavior) (*see* Chess and Thomas, 1977; Thomas et al., 1968), had experienced complications during birth, and had also experienced that their mothers responded to their moods with little sensitivity (Laucht, 2001; Laucht et al., 2001a, 2001b).

The Influence of Child Gender on Parenting Behavior and Adolescent Aggression

Gender has proven to be a robust predictor of later aggressive behavior. Many studies provide evidence for the biological origin of gender differences. Thus, it has been shown, for example, that boys are more susceptible to alcoholism than girls and, in particular, to type II alcoholism, which is closely associated with aggression. However, it is likely that parental expectations interact with basic genetic factors in boys and girls and that other group influences modify the manner in which aggression is expressed in behavior. We know, for example, that boys tend chiefly toward overt forms of aggression, even in groups, whereas less overt forms of aggression, such as tattling and conspiracy, are more prevalent among girls. Garnefski and Diekstra (1996) found that psychosocial risks play a more significant role with respect to later behavioral problems in boys than in girls. According to the authors, girls are more successful in learning to adjust; they are generally taught not to attract attention and to place less emphasis on their own needs. Accordingly, they argue, the superior social competencies of female adolescents enable them to experience a greater measure of social support from parents and significant adults or peers than that experienced by boys.

Early Traumatic Events in the Family

Sexual and Physical Abuse

Generally speaking, traumatic and frustrating experiences tend to reduce the stimulus threshold for aggressive behavior. Consequently, the child tends to react more quickly with anger to minor incidents and to exhibit a higher level of aggression (anger [expressive] aggression). Exacerbated by reinforcement processes, this predisposition can develop into a marked behavioral tendency. Furthermore, many studies have shown that maltreatment leads to a tendency to overreact to hostile cues and to interpret ambiguous provocation stimuli as threatening (*see* section on social processing theory).

Domestic child abuse and sexual abuse are psychotraumatic experiences that have lasting effects on child development and can facilitate the development of aggressive behavior. They are indicators of severely disturbed family dynamics, although the accompanying symptoms are largely independent of a given family's social status. Child abuse occurs in all social classes, although substandard social conditions, confined living space, and poor financial situations tend to bring intrafamily problems to the surface more quickly.

In his survey article, Richter-Appelt (1997) came to the conclusion that physical and sexual abuse are closely related to problems in adolescence and adulthood. This clearly shows how important it is to avoid viewing sexual abuse in isolation and to give equal consideration, above all, to physical abuse, even where it is not directly associated with sexual abuse.

Browne and Finkelhor (1986) distinguish between two types of effects of sexual abuse: initial effects, i.e., immediate reactions on the part of the child to sexual abuse, which become apparent during the first 2 yr after abuse, and long-term effects, i.e., the long-term consequences of sexual abuse expressed in adulthood. The authors also divide the immediate consequences (initial effects) into four symptom groups: emotional reactions, inappropriate sexual behavior, physical and psychosomatic effects, and anomalies in social behavior (delinquency, aggressive behavior, such as deliberate property destruction, physical attacks, and so forth). Empirical findings on long-term consequences are widely distributed. The results suggest that sustained incest relationships, in which sexual intimacy (and particularly sexual intercourse) is achieved through force, produce more pronounced initial effects and persistent long-term consequences. The findings also show a strong correlation, quite apart from specific characteristics of the sexual abuse of a child, between a pattern comprised of "incest relationship," the use of force, frequent and persistent child sexual abuse and sexual intercourse, with a significantly more pronounced tendency toward depression and aggression, a negative self-image, a greater incidence of interpersonal disorders, and more sexual problems.

In a survey study by Kendall-Tackett et al. (1993), a number of studies devoted to comparison of sexually abused children and children who had not experienced sexual abuse were subjected to meta-analysis. The authors were able to calculate effect values for 7 of the 13 symptoms identified, for which a sufficient number of studies provided information relevant to a comparison between abused and nonabused children. The highest effect values (Etas) were found for such overt behaviors as sexualized behavior and aggression. Sexual abuse explained 43% of variance for these two behavioral patterns. However, the findings with respect to aggression must be interpreted with caution, as this symptom can be the result of a wide range of underlying causes possibly associated with domestic abuse (e.g., lower-class social status, low parental education level, experience of violence, negative parenting styles).

One of the aspects examined by Lauscher and Schulze (1998) in their study is the association between the severity of sexual abuse and the effects of abuse. One means of classifying the different degrees of severity of abuse cited in their study is the division of subjects into the following categories: (1) no abuse during childhood; (2) physical abuse during childhood; (3) sexual abuse during childhood; and (4) physical and sexual abuse during childhood. The authors found that the types and frequency of behavioral problems increased progressively from Group 1 to Group 4.

In our own retrospective study of 99 adolescents, of which 33 came from social welfare institutions (referred to as "Crisis Centers") and exhibited extremely aggressive behavior, 78.8% of the aggressive group (AG) and 42.4% of the nonaggressive control group (CG, $n = 66$) reported that their parents had separated by the time they (the respondents) had reached the age of 10 (Barnow et al., 2001). Roughly twice as many aggressive adolescents as nonaggressive subjects (69.7 vs 33.3%) stated that they had been physically abused (hard blows administered by parents or family members), while 42.4% of adolescents in the AG (compared to 21.2% of those in the CG) cited sexual abuse (whereby all unwanted sexual advances were classified as sexual abuse). The majority of incidents of sexual abuse occurred within the family. An especially remarkable finding was that reports of simultaneous occurrences of physical and sexual abuse were eight times as frequent among adolescents from the AG as compared to those of the CG.

Parental Violence

Simply witnessing parental violence can also be a traumatizing experience. Children of abused women, for example, reported incidents of aggressive behavior similar to those exhibited by children who had suffered abuse personally. It should be noted, however, that these children are often not only witnesses to parental violence, but frequently the targets of parental violence themselves, as a number of studies have shown.

Broken Home

Other causes of aggressive behavior in children and adolescents included family disharmony and chronic conflicts between adult partners. Boys from families rendered fatherless by divorce are especially susceptible to the effects of such circumstances. The importance of fathers in childhood development and identify formation has been sufficiently documented in extensive research on fathers (Fthenakis, 1992). Male adolescents in particular frequently react to the broken home situation with aggressive and dissocial behavior, and boys show a tendency to compensate for identity problems by acting out aggressive impulses.

In summary, the majority of studies found that chronically maltreated children run a higher risk of abuse, neglect, or both, across multiple developmental phases, which may impair their capacity to master developmental functions such as controlling behavior and forming social relationships.

PARENTAL CHARACTERISTICS, THE QUALITY OF RELATIONSHIPS BETWEEN PARENTS AND CHILDREN AND BEHAVIORAL PROBLEMS IN ADOLESCENTS

Social Learning and Attachment Theory

Exponents of the social learning theory (Bandura, 1977) propose that aggressive behaviors represent a direct reaction by the child to his or her environment. Bandura proposes that learning from success reinforces willingness to repeat the behavior in question, while the relinquishment of a demand following conflict results in negative reinforcement. In other words, negative reinforcement sustains aggressive behavior when a child succeeds in averting or eliminating a perceived threat or an unpleasant situation through corresponding aggressive behavior. Furthermore, modeling and observational learning experiments demonstrated that reinforcement is not necessary for learning. Instead, it has been found repeatedly that individuals can learn by observing others being reinforced. Banduras' social learning theory provides one explanation for the importance of early family environment for the development of social behavior in children. Patterson and Bank (1992), for example, demonstrated that parents who provide inappropriate reinforcement (e.g., too many contradictory instructions to the child) are likely to exacerbate the child's behavior problems and reduce his or her willingness to cooperate.

The attachment theory (Bowlby, 1969, 1980) assumes that children develop mental representations of themselves, which reflect the influences of significant others. As conceptual assessments of the relationship between child and mother, such internal working models have a lasting impact on the child's interactive behavior and his or her behavior in general. Accordingly, the authors argue that dysfunctional parent–child interaction produces insecure working models, and that a low level of parental responsibility and/or attention coupled with more or less irresponsible parenting behavior contribute to the formation of these insecure working models. Hazan and Shaver (1987) postulate three types of relationships between parents and children: secure, avoidant, and anxious–ambivalent. Accordingly, maltreated children often have an insecure or anxious, mistrustful internal working model, which can be expressed in the perception of a hostile environment and in aggressive behavior. The fact that aggressive children often have extremely hostile attribution patterns (Crick and Dodge, 1994) also has its place within the context of attachment theory. It is assumed, for example, that insecure attachment behavior prompts children to interpret neutral or even positive behaviors as hostile (Shaw and Bell, 1993). Shaw et al. (1996), who investigated 100 infants regarding the relationship between attachment style and subsequent aggressive behavior, found, for example, that infant insecurity predicted behavioral problems at ages 3 and 5, whereas 6 out of 10 children who were insecure-attached showed elevated levels of aggression compared to 17% of secure-attached infants.

Social cognitive models can serve as aids to a better understanding of why and how perception of self and others is influenced by the quality of attachment. Combinations of threatful or confusing environmental experiences may contribute to inaccurate information processing and to extreme activation or frustration of motivational systems leading to behavioral disturbances. By filtering incoming information, for instance, hostile attribution scripts can produce situations in which only that information is registered that corresponds with individual internal concepts and views. Thus, even essentially neutral

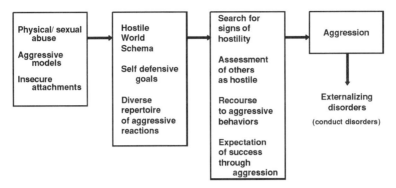

Fig. 1. Development of a hostile scheme as a result of early family influences. Modified from Dodge (1993).

cues can be interpreted as aggressive when the prevailing internal scripts are hostile. This provides one explanation for why bonding behavior between children and parents mediated by internal working models and expectations can produce aggressive behavior later on (*see* Fig. 1).

Empirical Findings Regarding the Association Between Parental Characteristics and Aggressive Behavior in Children

A number of studies have identified parental characteristics, such as low education level, young age of the mother at birth, low socioeconomic status, and separation or divorce before the child reaches age 6, as predictors for later aggressive behavior disorders (Cohen et al., 1998; Fagot et al., 1998; Fergusson et al., 1994; Nagin et al., 1997). The results of a prospective study of 1037 boys between the ages of 6 and 15 showed, for example, that maternal factors, such as low education level and low birth age, resulted in a significant increase in the risk that children would exhibit aggressive behavior at age 15 (Nagin and Tremblay, 2001). In explaining their findings, the authors state that the factors cited increase the risk of problematic behavior on the part of the mother toward the child. It was found, for example, that aggressive younger women were more likely to bear a child while still in their teens, to be discharged from school, and to be less reactive parents. Findings also suggest that such mothers tend to exhibit rather more strict punishment-oriented rigid parenting styles.

Mental Disorders of Parents and their Influence on Parenting Behavior and Behavior Problems in Adolescence

Many studies have reported that adults who are raised in alcoholic households had higher scores on tests of aggression and delinquency (Alterman et al., 1998; Barnow et al., in press; Dimileo, 1989; Giancola et al., 1996; Hesselbrock et al., 1992; Jacob et al., 1999; Martin and Sher, 1994; Stowell and Estroff, 1992; Windle, 1996). In a large community sample, it was demonstrated that, after controlling for other risk factors, childhood exposure to parental problem drinking was associated with greater risk of antisocial behavior (Greenfield et al., 1993). However, the great extent of co-morbidity of antisocial personality disorders (ASPD) in alcoholic fathers (Hesselbrock, 1992) appears to be one variable that may mediate between parental problem drinking and the development of behavioral problems among children of alcoholics (COAs) (Schuckit et al., 2000). For example, antisocial children are disproportionately drawn from families in which one or both parents also show antisocial tendencies (Farrington et al., 1996; Frick et al., 1992). Moreover, higher levels of delinquency and aggression, as measured by the Achenbach Child Behavioral Checklist, are found in children whose fathers have a co-morbid ASPD (Helzer and Pryzbeck, 1988). It is likely that the correlation between child's aggressive behavior and the father's and/or mother's ASPD is at least partially mediated through the negative parenting styles (e.g., rejection, harsh punishment, and coercive parenting) of antisocial parents (Rutter et al., 1998).

Other findings indicate that general parental psychopathology (e.g., depression of the mother) is associated with an increased risk of a variety of behavioral problems in their offspring (Johnson et al., 1999). It is reasonable to assume, in this context, that the reduced reactive flexibility and negative affective behavior of the mother have an unfavorable impact on the psychosocial development of the child (*see* the discussion further below in this chapter).

The Association Between Perceived Parenting Styles and Subsequent Behavioral Problems of Offspring

Parenting styles and the parental environment are significant predictors for aggressive behavior and delinquency in adolescence (Barnow, 2001). It is important to note, however, that many aspects of parenting behavior are determined by parental behavior that is not independent of parental genes and experience. Moreover, parenting styles are influenced at least to a certain degree by the behavior of the child. Thus, it is reasonable to assume that genetic and environmental factors interact in childhood development.

A number of studies have identified an association between aggression and parental neglect and/or the experience of negative parenting styles. Both a strict punishing parenting style (Dodge et al., 1990) and a lack of emotional warmth and support (Loeber and Dishion, 1983; Loeber and Stouthammer-Loeber, 1998) have been cited as significant factors contributing to later aggressive and antisocial behavior. Patterson (1995) showed that less contingent or insufficiently supportive parental behavior in combination with strict punishing tendencies increased the probability that children would later develop antisocial or aggressive behavior. Bolger and Patterson (2001) emphasize that chronic neglect, as opposed to neglect over a brief period of time, can contribute substantially to severe disturbances in childhood development processes, where by chronically neglected children have a higher probability of experiencing abuse and neglect over several successive development phases, which in turn diminishes the capacity of these children to solve development tasks adequately and manage their behavior appropriately. Lack of emotional warmth and support by parents may also be associated with aggressive behavior. The lack of affective ties between the adolescent and his or her parents entails the absence of ties with conventional society, which may in turn generate the development of deviant behavior. Gardner (1989) emphasized the role of inconsistent parenting for subsequent behavioral problems. He found that mothers of conduct problem children handled 43% of their conflict episodes inconsistently, compared with only 5% for mothers of children without conduct disorders. Finally, overprotection has also been found to be associated with behavioral problems in offspring (Barber, 1996; Steinberg, 1990). It is likely that extensive use of psychological controlling behaviors (e.g., guilt induction, overprotection) may increase behavioral problems of offspring by suppressing adolescents' autonomy and providing insufficient space to assert an independent sense of identity.

In a study of our own, which is currently still in progress, we have thus far surveyed 192 families in Pomerania (Germany), focusing on a variety of family factors and personality or individual traits of children in an attempt to identify correlates with behavioral problems and alcohol abuse in adolescence (Barnow, 2001; Barnow et al., 2002a, 2000b). In this study, we are not only examining individual characteristics, behavioral parameters, peer-group characteristics and attributes of the family environment, but also surveying parents directly with respect to such aspects as personality, parenting styles, and so forth. Initial analyses of the findings for 180 adolescents have shown that aspects of parenting, especially perceived parental rejection, correlate significantly with aggression and delinquency in adolescence. In subsequent phases of analysis, we investigated the relationship between psychological disorders in parents, birth complications, parenting behavior, and adolescent behavioral problems in a subsample of 128 adolescents, for whom all data from the first follow-up survey (after 1 yr) were available. Only feelings of rejection by the parents, however, contributed to about 14–15% of cleared variance in the YSR broadband behavioral problem scale of aggression–delinquency (Barnow et al., in press). Similar findings were obtained through an analysis of a structural equation model for a subsample of 168 adolescents, which showed that perceived parental rejection correlated significantly with behavioral problems (path coefficient was $\beta = 0.38$) (Barnow et al., submitted). However, owing to the specific design of the study, it was not possible to demonstrate causal relationships, and thus, the alternative hypothesis that adolescents with behavioral problems perceive a stronger sense of rejection in their parent's behavior cannot be ruled out. An interesting secondary finding was that the effect for girls was markedly stronger ($\beta = 0.57$) than for boys ($\beta = 0.19$). This may indicate that parenting behavior is highly significant for female behavioral problems. The finding on gender effects discussed above support this interpretation to the extent that the effects of socialization appear to be more significant for girls. On the other hand, the possibility that boys and girls assess parenting behavior differently cannot be ruled out, as girls may tend to describe parental behavior more openly and critically.

It should be noted, however, that the above-mentioned parental characteristics have also been found to be associated with other forms of psychopathology in the child: phobias and social anxiety (Arrindell et al., 1988), obsessive–compulsive disorders, and Type A behavior. It is likely that such negative parenting practices may make children vulnerable to psychopathology in general, rather than to one or a few specific forms of psychopathology.

Therefore, further research will be required in developing behavioral (rather than self-reporting) measures to assess the parenting practices. When such measures have been evaluated, prospective studies will be needed for the purpose of determining whether children of "high-risk" families, in which parenting practices are characterized by rejection, lack of emotional warmth, and overprotection, are particularly vulnerable to becoming aggressive.

FAMILY PROTECTIVE FACTORS AND LATER AGGRESSIVE BEHAVIOR

Not all children who come from problematic families or exhibit a higher psychosocial risk profile develop aggressive behavior patterns at a later age. It is important to consider the interaction between genetic disposition, biological risks, and environmental factors, on the one hand, and protective factors that contribute to positive development of the individual and/or help to prevent negative development, on the other hand. Distinction must be made between characteristics of the children themselves (personal resources) and those of their family environment (social resources). Findings obtained in the study conducted by Laucht et al. (1998), on risk and protective factors, clearly showed that both aspects together explain more of the variance than either of the two groups of variables alone. One frequently cited protective factor is the supportive mother–child interaction, which, owing primarily to the experience of emotional security and dependability it provides the child, is regarded as a particularly significant basis for the development of self-confidence and self-reliance in the child. This is supported by our own study results using cross-sectional data on 168 adolescents. Here, we found that perceived feelings of emotional warmth of parents were positively related to self-esteem and negatively related to measures on aggression and delinquency (Barnow et al., submitted). Thus, reflecting the extensive literature on parenting, it is relatively safe to assume that parental behavior, which combines emotional warmth and firm but fair control, is associated with well-being in children and can protect children from developing behavioral problems (Barnow, 2001; Bornstein, 1995).

SUMMARY

The literature provides some evidence for the causal status of parenting skills in childhood behavior. Previous research has indicated, however, that parenting behavior can be adversely affected by parental psychopathology and offspring temperament, both of which seem to be determined in part by genetic factors. Only rarely is a single risk factor (e.g., neuropsychological deficits) sufficiently significant to justify equating it with the cause. In most cases, independent risks factors are added together or form complex patterns of interaction (e.g., biologically induced brain damage, difficult temperament, and parental rejection). Thus, for example, the probability of postnatal birth complications is higher when the mother (or the father, if he is the primary reference person) neglects the child, is divorced or separated, and communicates feelings of rejection to the child. Other negative psychosocial factors within the family included primarily recurring traumatic experiences and negative parenting behavior, which opposes the development of a positive self-image in the child. Whether or not an aggressive behavior disorder emerges ultimately depends upon which neurobiological, environmental and protective factors appear during the course of development. Figure 2 represents an attempt to illustrate the patterns of interaction described above.

In evaluating the findings described above, it should be noted, however, that methodological problems have been found in many of the studies cited. First of all, most of the studies did not use multiple methods and agents when describing parental practices and child outcomes. Second, responses to self-ratings and questions about childhood trauma, parenting styles, and occupational achievement may be biased by the degree of willingness of specific individuals to disclose their feelings about and interpretations of their

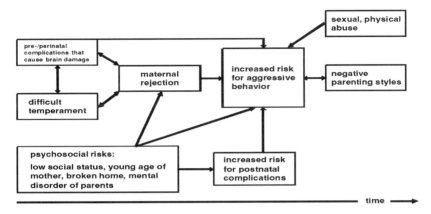

Fig. 2. The role of family environment in early life for later aggressive behavior.

experiences, by selective recall, and by their desire to present themselves in a socially desirable way. Third, negative parenting behavior is associated relatively nonspecifically with child psychopathology and, thus, does not represent a specific risk factor for aggression. Fourth, psychiatric co-morbidity in parents influences both behavioral problems in children and parenting behavior, making it very difficult to establish causal relationships.

REFERENCES

Achenbach, T. M. (1991) *Manual for the Youth Self Report and 1991 Profile.* University of Vermont, Burlington.

Achenbach, T. M. and Edelbrock, C. S. (1978) The classification of child psychopathology: a review and analysis of empirical efforts. *Psychol. Bull.* **85,** 1275–1301.

Achenbach, T. M., Conners, C. K., Quay, H. C., Verhulst, F. C., and Howell, C. T. (1989) Replication of empirically derived syndromes as a basis for taxonomy of child/adolescent psychopathology. *J. Abnorm. Child Psychol.* **17,** 299–323.

Achenbach, T. M. (1994) Child Behavior Checklist and related instruments, in *The Use of Psychological Testing for Treatment Planning and Outcome Assessment.* (Maruish, M. E., ed.), Lawrence Erlbaum Associates, Hillsdale, NJ, pp. 517–549.

Alterman, A. I., Bedrick, J., Cacciola, J. S., et al. (1998) Personality pathology and drinking in young men at high and low familial risk for alcoholism. *J. Studies Alcohol* **59,** 495–502.

Angold, A. and Costello, E. J. (2000) The epidemiology of disorders of conduct: nosological issues and comorbidity, in *Conduct Disorders in Childhood and Adolescence.* (Hill, J. and Maughan, B., eds.), Cambridge University Press, Cambridge, pp. 126–168.

Arrindell, W. A., Perris, H., Denia, M., et al. (1988) The constancy of structure of perceived parental rearing style in greek and spanish subjects as compared with the dutch. *Int. J. Psychol.* **23,** 3–23.

Arseneault, L., Tremblay, R. E., Boulenice, B., Séguin, J. R., and Saucier, J. F. (2000) Minor physical anomalies and family adversity as risk factors for adolescent violent delinquency in adolescence. *Am. J. Psychiatry* **157,** 917–923.

Bandura, A. (1977) *Social Learning Theory.* Englewood Cliffs, Prentice Hall, NJ.

Barber, B. K. (1996) Parental psychological control: revisiting a neglected construct. *Child Dev.* **67,** 3296–3319.

Barnow, S., Lucht, M., and Freyberger, H. J. (2001) Influence of punishment, emotional rejection, child abuse, and broken home on aggression in adolescence: an examination of aggressive adolescents in Germany. *Psychopathology* **34,** 167–173.

Barnow, S. (2001) Aggression in adolescence: empircal findings with regard to family influences, in *Prevention in Psychiatry and Psychology.* Wauthe, J. H., ed.), Axept, Konigscutter, Germany, pp. 51–87.

Barnow, S., Schuckit, M., Lucht, M., John, U., and Freyberger, H. J. (2002a) The importance of a positive family history of alcoholism, parental rejection and emotional warmth, behavioral problems and peer substance use for alcohol problems in teenagers: a path analysis. *J. Studies Alcohol* **63**, 305–315.

Barnow, S., Schuckit, M. A., Smith, T., Preuss, U., and Danko, G. (2002b) The relationship between the family density of alcoholism and externalizing symptoms among 146 children. *Alcohol Alcohol.* **37**, 383–387.

Barnow, S., Lucht, M., Hamm, A., John, U., and Freyberger, H. J. The relation of a family history of alcoholism, obstetric complications and family environment to externalizing symptoms among 154 adolescents in Germany: results from the Children of Alcohols Study in Pomerania. *Eur. Addict. Res.,* in press.

Barnow, S. and Freyberger, H. J. Correlates of behavioral problems among 168 adolescents in Germany: a path analysis. *J. Abnorm. Psychol.,* submitted.

Belsky, J., Rovine, M., and Fish, M. (1989) The developing family system, in *Systems and Development.* (Gunnar, M. R. and Thelen, E., eds.), Erlbaum, Hillsdale, pp. 119–186.

Blanz, B. (1988) Biologische Korrelate aggressiven Verhaltens. *Zeitschrift für Kinder und Jugendpsychiatrie* **26**, 43–52.

Bolger, K. E. and Patterson, C. J. (2001) Developmental pathways from child maltreatment to peer rejection. *Child Dev.* **72**, 549–568.

Bornstein, M. (1995) *Handbook of Parenting.* Mahwah, Lawrence Erlbaum, NJ.

Bowlby, J. (1969) *Attachment and Loss, Vol. I: Attachment.* Basic Books, New York.

Bowlby, J. (1980) *Attachment and Loss, Vol. III: Loss.* Basic Books, New York.

Brook, J., Brook, D., Whiteman, M., and Gordon, A. (1983) Depressive mood in male college students father-son interactional pattern. *Arch. Gen. Psychiatry* **40**, 665–669.

Browne, A. and Finkelhor, D. (1986) Impact of child sexual abuse: a review of the research. *Psychol. Bull.* **99**, 6–77.

Buikhuisen, W. (1988) Chronic juvenile delinquency a theory, in *Explaining Criminal Behavior.* (Buikhuisen, W. and Mednick, S. A., eds.), E. J. Brill, Leiden, Netherlands, pp. 27–50.

Campbell, S. B. (1995) Behavior problems in preschool children: a review of recent research. *J. Child Psychol. Psychiatry* **36**, 113–149.

Chess, S. and Thomas, A. (1977) Temperament and the parent-child interaction. *Pediatr. Ann.* **6**, 574–582.

Cloninger, C. R. (1994) Temperament and personality. *Curr. Opin. Neurobiol.* **4**, 266–273.

Cohen, P., Kasen, S., Brook, J. S., and Hartmark, C. (1998) Behavior patterns of young children and their offspring: a two-generation study. *Dev. Psychol.* **34**, 1202–1208.

Cohen, P., Velez, C. N., Brook, J., and Smith, J. (1989) Mechanisms of the relation between perinatal problems, early childhood illness, and psychopathology in late childhood and adolescence. *Child Dev.* **60**, 701–709.

Coie, J., Dodge, K., and Coppotelli, H. (1982) Dimensions and types of social status: a cross-age perspective. *Dev. Psychol.* **18**, 557–570.

Coleman, F. L. and Frick, P. J. (1994) MMPI-2 profiles of adult children of alcoholics. *J. Clin. Psychol.* **50**, 446–454.

Crick, N. R. and Dodge, K. A. (1994) A review and reformulation of social information-processing mechanismus in children's social adjustment. *Psychol. Bull.* **11**, 74–101.

Deater-Deckard, K., Dodge, K. A., Bates, J. E., and Pettit, G. S. (1998) Multiple risk factors in the development of externalizing behavior problems: group and individual differences. *Dev. Psychopathol.* **10**, 469–493.

Dimileo, L. (1989) Psychiatric symptoms in adolescent substance abusers. *Am. J. Psychiatry* **146**, 1212–1214.

Dodge, K. A., Bates, J. E., and Pettit, G. S. (1990) Mechanisms in the cycle of violence. *Science* **250**, 1678–1683.

Dodge, K. A. (1993) Social-cognitive mechanisms in the development of conduct disorder and depression, in *Annual Review of Psychology*. (Porter, L. W. and Rosenzweig, M. R., eds.), Palo Alto, CA, pp. 559–584.

Earls, F. J. (1994) Violence and today's youth. *Future Child* **4**, 4–23.

Edelbrock, G., Rende, R., Plomin, R., and Thompson, I. A. (1995) A twin study of competence and problem behavior in childhood and early adolescence. *J. Child Psychol. Psychiatry* **36**, 775–785.

Elliott, D. (1994) Serious violent offenders: onset, developmental course and termination. *Criminology* **32**, 1–21.

Ernst, C. and Angst, J. (1983) *Birth Order: Its Influence on Personality*. Springer Verlag, Berlin.

Fagot, B. I., Pears, K. C., Capaldi, D. M., Crosby, L., and Leve, C. S. (1998) Becoming an adolescent father: presursors and parenting. *Dev. Psychol.* **34**, 1209–1219.

Farrington, D., Loeber, R., Stouthamer-Loeber, M., van Kammen, W., and Schmidt, I. (1996) Self-reported delinquency and a combined delinquency seriousness scale based on boys, mothers, and teachers: concurrent and predictive validity for African-Americans and Caucasians. *Criminology* **34**, 520–525.

Fergusson, D. M., Horwood, L. J., and Lynskey, M. T. (1994) Parental separation, adolescent psychopathology, and problem behaviors. *J. Am. Acad. Child Adolesc. Psychiatry* **33**, 1122–1131.

Frick, P. J., Lahey, B. B., Loeber, R., Stouthamer-Loeber, M., Christ, M. A., and Hanson, K. (1992) Familial risk factors to oppositional defiant disorder and conduct disorder: parental psychopathology and maternal parenting. *J. Consult Clin. Psychology* **60**, 49–55.

Frick, P., Horn, Y., Lahey, B., et al. (1993) Oppositional defiant disorder and conduct disorder: a meta analytic review of factor analyses and cross-validation in clinic sample. *Clin. Psychol. Rev.* **13**, 319–340.

Fthenakis, W. E. (1992) Zur Rolle des Vaters in der Entwicklung des Kindes. *Prax. Psychother. Psychosom.* **37**, 179–189.

Gardner, F. E. M. (1989) Inconsistent parenting: is there evidence for a link with children's conduct problems? *J. Abnorm. Child Psychol.* **17**, 223–233.

Garnefski, N. and Diekstra, R. F. (1996) Perceived social support from family, school, and peers: relationship with emotional and behavioral problems among adolescents. *J. Am. Acad. Child Adolesc. Psychiatry* **35**, 1657–1664.

Giancola, P., Moss, H., Martin, C., Kirisci, L., and Tarter, R. (1996) Executive cognitive functioning predicts reactive aggression in boys at high risk for substance abuse: a prospective study. *Alcohol. Clin. Exp. Res.* **20**, 740–744.

Greenberg, M. T., Speltz, M., and DeKlyen, M. (1993) The role of attachment in the early development of disruptive behavior problems. *Dev. Psychopathol.* **5**, 191–214.

Greenfield, S. F., Swartz, M. S., Landerman, L. R., and George, L. K. (1993) Long-term psychosocial effects of childhood exposure to parental problem drinking. *Am. J. Psychiatry* **150**, 608–613.

Harris, J. (1995) Where is the child's environment? A group socialization theory of development. *Psychol. Rev.* **102**, 458–489.

Hazan, C. and Shaver, P. (1987) Romantic love conceptualizid as an attachment process. *J. Pers. Soc. Psychol.* **52**, 511–524.

Helzer, J. E. and Pryzbeck, T. R. (1988) The co-occurrence of alcoholism with other psychiatric disorders in the general population and its impact on treatment. *J. Studies Alcohol* **49**, 219–224.

Hesselbrock, V., Meyer, R., and Hesselbrock, M. (1992) Psychopathology and addictive disorders, in *Addictive States*. (O'Brien, C. and Jaffe, J., eds.), Raven Press, New York, pp. 179–191.

Jacob, T., Windle, M., Seilhamer, R. A., and Bost, J. (1999) Adult children of alcoholics: drinking, psychiatric and psychosocial status. *Psychol. Addic. Behav.* **13**, 3–21.

Johnson, J., Cohen, P., Skodol, A., Oldham, J., Kasen, S., and Brook, J. (1999) Personality disorders in adolescence and risk of major mental disorders and suicidality during adulthood. *Arch. Gen. Psychiatry* **56,** 805–811.

Kandel, E. and Mednick, S. (1991) Perinatal complications predict violent offending. *Criminology* **29,** 519–530.

Kendall-Tackett, K. A., Meyer-Williams, I. M., and Finkelhor, D. (1993) Impact of sexual abuse on children: a review and synthesis of recent empirical studies. *Psychol. Bull.* **113,** 164–180.

Kendler, K. S., Bulik, C. M., Silberg, J., and Hettema, J. M. (2000a) Childhood sexual abuse and adult psychiatric and substance use disorders in women. *Arch. Gen. Psychiatry* **57,** 953–959.

Kendler, K. S., Myers, J., and Prescott, C. A. (2000b) Parenting and adult mood, anxiety and substance use disorders in female twins: an epidemiological, multi-informant, retrospective study. *Psychol. Med.* **30,** 281–294.

Kumpfer, K. L. and Turner, C. W. (1990) The social ecology model of adolescent substance abuse: implications for prevention. *Int. J. Addict.* **25,** 435–463.

Laucht, M., Esser, G., and Schmidt, M. H. (1998) Risiko- und Schutzfaktoren frühkindlicher Entwicklung. *Zeitschrift für Kinder und Jugendpsychiatrie Psychotherapie* **26,** 6–20.

Laucht, M. (2001) Antisocial behavior in adolescence: risk factors and developmental types. *Zeitschrift für Kinder Jugendpsychiatrie Psychotherapie* **29,** 297–311.

Laucht, M., Esser, G., and Schmidt, M. H. (2001a) Längsschnittforschung zur Entwicklungsepidemiologie psychischer Störungen: Zielsetzung, Konzeption und zentrale Befunde der Mannheimer Risikokinderstudie. *Zeitschrift für Klinische Psychologie und Psychotherapie* **29,** 246–262.

Laucht, M., Esser, G., and Schmidt, M. H. (2001b) Differential development of infants at risk for psychopathology: the moderating role of early maternal responsivity. *Dev. Med. Child Neurol.* **43,** 292–300.

Lauscher, S. and Schulze, C. (1998) Schweregrad von sexuellem Mißbrauch und Langzeitfolgen. *Zeitschrift für klinische Psychologie* **27,** 181–188.

Lipson, A., Yu, J., O'Halloran, M., and Williams, R. (1981) Alcohol and phenylketonuria. *Lancet* **1,** 717–718.

Loeber, R. and Dishion, T. (1983) Early predictors of male delinquency: a review. *Psychol. Bull.* **94,** 68–99.

Loeber, R. and Stouthammer-Loeber, M. (1998) Development of juvenile aggression and violence: some common misconceptions and controversies. *Am. Psychol.* **53,** 242–259.

Martin, E. D. and Sher, K. J. (1994) Family history of alcoholism, alcohol use disorders and the five-factor model of personality. *J. Studies Alcohol* **55,** 81–90.

Mednick, S. A. and Kandel, E. (1988) Congenital determinants of violence. *Bull. Am. Acad. Psychiatry Law* **16,** 101–109.

Mednick, S. A., Pollock, V., Volavka, J., and Gabrielli, W. F. (1982) Biology and violence, in *Criminal Violence.* (Wolfgang, M. E. and Weiner, N. A., eds.), Sage, Beverly Hills, pp. 21–80.

Moffitt, T. E. (1990) The neuropsycholgy of juvenile delinquency, in *Crime and Justice: A Review of Research.* (Tonry, N. and Morris, N., eds.), University of Chicago Press, Chicago, pp. 99–169.

Moffitt, T. E. (1993) The neuropsychology of conduct disorder. *Dev. Psychopathol.* **5,** 135–151.

Nagin, D. S. and Tremblay, R. E. (2001) Parental and early childhood predictors of persistent physical aggression in boys from kindergarten to high school. *Arch. Gen. Psychiatry* **58,** 389–394.

Nagin, D. S., Pogorsky, G., and Farrington, D. P. (1997) Adolescent mothers and the criminal behavior of their children. *Law Soc. Rev.* **31,** 137–162.

O'Connor, T. G., Hetherington, E. M., Reiss, D., and Plomin, R. (1995) A twin-sibling study of observed parent-adolescent interactions. *Child Dev.* **66,** 812–829.

Patterson, G. R. (1995) Coercion as a basis for early age of onset for arrest, in *Coersion and Punishment in Long-Term Perspectives.* (McCord, J., ed.), Cambridge University Press, New York, pp. 81–105.

Patterson, G. R. and Bank, L. (1992) Some amplifying mechanism for pathologic processes in families, in *System and Development. The Minnesota Symposium on Child Psychology.* (Gunnar, M. R. and Thelen, E., eds.), Erlbaum, Hillsdale, pp. 167–209.

Pellegrini, D. S. (1990) Psychosocial risk and protective factors in childhood. *Dev. Behav. Pediatry* **11**, 201–209.

Petermann, F. P. U. (1993) *Training mit Aggressiven Kindern (6. Auflage ed.).* Psychologie Verlags Union, Weinheim.

Pettit, G. S. (1997) The developmental course of violence and aggression. Mechanisms of family and peer influence. *Psychiatry Clin. N. Am.* **20**, 283–299.

Pollock, V., Briere, J., Schneider, L., Knop, J., Mednick, S., and Goodwin, D. (1990) Childhood antecedents of antisocial behavior. Parental alcoholism and physical abusiveness. *Am. J. Psychiatry* **147**, 1290–1293.

Raine, A., Brennan, P., and Mednick, S. (1994) Birth complications combined with early maternal rejection at age 1 year predispose to violent crime at age 18 years. *Arch. Gen. Psychiatry* **51**, 984–988.

Raine, A., Brennan, P., and Mednick, S. A. (1997) Interaction between birth complications and early maternal rejection in predisposing individuals to adult violence: specificity to serious, early-onset violence. *Am. J. Psychiatry* **154**, 1265–1271.

Richter-Appelt, H. (1997) Differentielle Folgen von sexuellem Missbrauch und körperlicher Misshandlung, in *Sexueller Missbrauch: Überblick zu Forschung, Beratung und Therapie: ein Handbuch.* (Amann, G., Wipplinger, R., eds.), Deutsche Gesellschaft für Verhaltenstherapie, DGVT Verlag, Tübingen, pp. 201–216.

Robins, L. N. (1991) Conduct disorder. *J. Child Psychol. Psychiatry* **32**, 193–212.

Rutter, M., Giller, H., and Hagell, A. (1998) *Antisocial Behavior by Young People.* Cambridge University Press, Cambridge.

Rutter, M., O'Connor, T., and Simonoff, E. (1999) genetics and child psychiatry I. Advances in quantitative and molecular genetics. *J. Child Psychol. Psychiatry* **40**, 3–18.

Schuckit, M. A., Smith, T. L., Radziminski, S., and Heyneman, E. K. (2000) Behavioral symptoms and psychiatric diagnoses among 162 children in nonalcoholic or alcoholic families. *Am. J. Psychiatry* **157**, 1881–1883.

Shaw, D. S. and Bell R. Q. (1993) Developmental theories of parental contributions to antisocial behavior. *J. Abnorm. Child Psychol.* **21**, 493–518.

Shaw, D. S., Owens, E. B., Vondra, J. I., Keenan, K., and Winslow, E. B. (1996) Early risk factors and pathways in the development of early disruptive behaviour problems. *Dev. Psychopathol.* **8**, 679–700.

Steinberg, L. (1990) Interdependence in the family: autonomy, conflict, and harmony in the parent-adolescent relationship, in *At the Threshold: The Developing Adolescent.* (Feldmann, S. and Elliott, G., eds.), Harvard University Press, Cambridge, MA, pp. 225–276.

Stowell, R. J. and Estroff, T. W. (1992) Psychiatric disorders in substance-abusing adolescent inpatients: a pilot study. *J. Am. Acad. Child Adolesc. Psychiatry* **31**, 1036–1040.

Streissguth, A. (1997) *Fetal Alcohol Syndrome.* Paul H. Brookes, Baltimore, MD.

Thomas, A., Chess, S., and Birch, H. (1968) *Temperament and Behavior Disorders in Children.* New York University Press, New York.

Verhulst, F. C. and Achenbach, T. M. (1995) Empirically based assessment and taxonomy of psychopathology: cross-cultural applications. A review. *Eur. Child Adolesc. Psychiatry* **4**, 61–76.

Windle, M. (1996) On the discriminative validity of a family history of problem drinking index with a national sample of young adults. *J. Studies Alcohol* **57**, 378–386.

13
Television and Movies, Rock Music and Music Videos, and Computer and Video Games

Understanding and Preventing Learned Violence in the Information Age

Susan Villani and Nandita Joshi

INTRODUCTION

Advances in technology over the past several decades have thrust the global population into what is now called the Information Age. While the United States leads in this area with more homes with televisions, videocassette recorders (VCRs), video games, and computers, the rapid spread of technology, coupled with entertainment industry products, has created a global village where news and entertainment circle the globe reaching millions of adults and children in seconds. The pace of research regarding the effects of media, particularly violent media and its relationship to aggression, has been steady since the 1950s, with thousands of research studies examining the association between media violence and violent behavior.

This chapter will review the research on violent media and its effects on children and adults, with emphasis on the developing brain and children. The culmination of the research field to the point of defining violent media as a public health concern, presented by physicians to the U.S. Congress in the form of a Congressional Public Health Summit in July 2000, points to the growing consensus about the role violent media plays in understanding the evolution of violent behavior (Joint Statement on the Impact of Entertainment Violence on Children, 2000). Prevention strategies and programs will be discussed with emphasis on the need for a multipronged public health approach implemented through health care and education systems, grass roots nonprofit groups for public education, and increased governmental oversight of the public airways, entertainment industry, and electronic media.

THEORETICAL BASIS OF RESEARCH: SOCIAL LEARNING THEORY AND NEUROBIOLOGICAL RESEARCH

Early media research begins with the basic tenets of social learning theory, which state that behavior is learned by repeated exposure to the desired (or undesired) behavior (Bandura, 1977). The more familiar the social context and the more attractive the protagonists, the more likely the viewer will identify and mimic the behavior. The vulnerability to mimic may be mediated by a variety of factors, including but not limited to: (*i*) developmental age; (*ii*) inherited temperament; (*iii*) cultural sanction or rejection; and (*iv*) cognitive capacity. Therefore, all viewers who are exposed will not be similarly affected, much like the link between smoking and lung cancer, where it is widely accepted that all who smoke will not get cancer, but that smoking does significantly increase one's risk of lung cancer. The degree to which individuals exposed to media violence will be affected is variable and dependent on many factors, factors which are also not constant and change over time. Thus the more accepted way of looking at the interaction between violent media and violent behavior is to look at cumulative risk over time with exposure to media violence, amount and content being risk factors.

Social learning theory, while hard to quantify, is very powerful in terms of its ability to shape values and attitudes and modify behavior. The twentieth century multibillion dollar advertising industry is predicated on social learning theory tenets as it seeks to influence viewers to purchase items not based on their objective qualities, but instead on the subjective identification of the viewer with the behaviors and lifestyles depicted in the advertisements and the desire for the viewer to mimic both.

The link between learning theory and current neuroscience research is emerging from the field of neuroradiology. Neuroimaging studies, positive emission tomography (PET) scans, and functional magnetic resonance imaging (fMRIs), now allow the activity of the brain to be observed. One can actually see how learning affects the brain in neurochemical and structural ways. A dramatic example of this is the study that compared the neuroimaging effects of fluoxetine hydrochloride, a serotonin-reuptake inhibitor, on the brain compared to the effects of cognitive behavior therapy in patients being treated for obsessive–compulsive disorder. The changes in the PET scans observed with cognitive behavioral therapy were almost identical to the changes the occurred with fluoxetine (Baxter et al., 1992).

Other neuroscience research has recently looked at early trauma and its effects on brain structures and endocrinologic axes. Largely in animal models, but with clear applicability to humans, animals abandoned early in life show evidence of anatomical brain changes. Similarly, those who were stressed early in life showed changes in cortisol regulation that remained through the adult life of the animal (Coplan et al., 1996; Ladd et al., 1996).

With this kind of knowledge as background, one understands that the linkages between exposure to violent media and later violent behavior, while difficult to prove using a one-to-one causality model, are well founded in the increasingly abundant research that demonstrates the interactions between learning and neuroscience research.

TELEVISION AND MOVIE VIOLENCE: EXPOSURE, CONTENT, AND EFFECT

The exposure to television and movie violence has grown steadily since the advent of television in the 1950s. In fact, movies are now lumped with television, because within a relatively short time span, movies become television fare. Bandura's original studies from the early 1960s (Bandura et al., 1961, 1963) showed that young children imitate what they see on television, particularly if the behavior is performed by attractive role models and is either rewarded or goes unpunished. This was followed by longitudinal correlational studies, which tracked children into adulthood. These studies showed that viewing media violence at young ages, 8 yr and younger, is a significant risk factor for adolescent or adult aggressive behavior or criminal violence (Lefkowitz et al., 1972). The 1972 Surgeon General's Report and the 1982 National Institute on Mental Health report both drew on voluminous research (Andison, 1977; Malamuth and Check, 1981; Robinson and Bachman, 1972; Singer and Singer, 1981) Malamuth and Check (1981) concluding:

> After 10 more years of research, the consensus among most of the research community is that violence on television does lead to aggressive behavior by children and teenagers who watch the programs. This conclusion is based on laboratory experiments and on field studies. Not all children become aggressive, of course, but the correlations between violence and aggression are positive. In magnitude, television violence is as strongly correlated with aggressive behavior as any other behavioral variable that has been measured.

Even though this consensus was reached in 1982, research and compilations of research continued. Studies regarding stability of aggression over time and generations (Huesman et al., 1984), naturalistic experiments (Williams, 1986), and cross-national comparisons (Huesman and Eron, 1986) were published. Given this research activity, the American Psychological Association appointed a committee in 1986 to review the effects of media on youth and to further the debate about how to solve the problems created. The committee's work culminated in a 195-page book, *Big World, Small Screen: The Role of Television in American Society* (Huston et al., 1992), with 420 references of predominantly research-based studies done in the 1970s and 1980s in behavioral psychology and communications journals, as well as government surveys and studies. The book provided a comprehensive review and discussion of both the positive, negative, and potential positive effects of media on our culture and our children.

The early 1990s brought the publication of two additional scientific books, *Media, Children, and the Family: Social Scientific, Psychodynamic, and Clinical Perspectives* (Zillman et al., 1994) and *Media Effects Advances in Theory and Research* (Zillmann and Bryant, 1994). The first book provides five chapters describing how media have affected the family and highlights the reality that children's use of media is socialized primarily in the family. Subsequent sections examine the developmental and educational implications; the effects of violence and horror; sexual content and family context; the effects of erotica and pornography; and social awareness and public policy. The second book, which is not exclusively oriented toward the effects on children, provides important chapters on the effects of media on personal and public health, as well as consideration of minorities and the media.

In addition, two highly significant meta-analytic reviews were published, Wood et al. (1991) and Paik and Comstock (1994). Wood et al. (1991) examined 28 research reports on children and adolescents exposed to media violence and subsequently observed in unconstrained social interactions. They concluded that exposure to media violence increases aggressive interactions with strangers, classmates, and friends. Paik and Comstock (1994) reviewed 217 studies conducted between 1957 and 1990, which examined the effects of television violence on antisocial behavior. The age range in the studies was 3–70 yr with 85% of the sample ages 6–21 yr. Their analysis revealed a positive and significant correlation between television violence and aggressive behavior, regardless of age. From the public health perspective, it is noteworthy that the correlation between media violence and aggressive behavior (0.3) is greater than that of calcium intake and bone mass

(0.1), lead ingestion and lower intelligence quotient (IQ) (0.15), condom nonuse and sexually acquired human immunodeficiency virus (HIV) infection (0.2), or environmental tobacco smoke and lung cancer (0.15), all of which are associations widely accepted and upon which preventative programs are based (Bushman and Huesmann, 2000).

Another significant finding from the Paik and Comstock (1994) meta-analysis is that the greatest effect size demonstrated was for preschool children. While by nature of size and age, the aggression exhibited was not particularly problematic, the authors cautioned about the possible long-term effects of cumulative viewing, given the large effect size demonstrated at the younger ages. Researchers have, in fact, shown that children learn attitudes about violence at a very young age and that once learned, the attitudes become life-long (Eron, 1995). Other research is consistent with this pointing to the fact that developmentally, children younger than 8 yr cannot discriminate between fantasy and reality and are, therefore, uniquely vulnerable to learning directly from what they see on television (Flavell, 1986; Morrison and Garnes, 1978; Wright et al. 1994).

In addition to the studies that examine violence, there are numerous studies conducted on consenting young adult males regarding the viewing of erotica and violent erotica. Zillman and Weaver (1989) and Mullin and Linz (1995) described the development of "sexual callousness." This demonstrated that erotica, with or without violence, which depicts women as promiscuous, encourages the subsequent development of callous attitudes or behaviors toward women among males so exposed. Geen and Donnerstein (1998) in *Human Aggression: Theories, Research, and Implications for Social Policy,* provide a chapter, "Harmful Effects of Exposure to Media Violence: Learning of Aggression, Emotional Desensitization, and Fear," which examines the issue of cognitive development and aggression citing a multitude of basic research studies from the 1980s and original later works. Sissela Bok's book *"Mayhem: Violence as Public Entertainment"* (Bok, 1998) historically traces the fall of societies as they have gravitated toward violence as entertainment as opposed to violence for the purpose of survival or even for the advancement of religious or social beliefs.

In a recent article, which raises concerns about the influence of the unrestrained access to erotica on adolescents and young adults, Zillmann (2000) describes how prolonged exposure leads to habituation of excitatory reactions. He further states that this excitatory habituation cycle of diminished enjoyment and the need for increasingly novel materials in order to achieve sustained enjoyment, is the first phase of the habituation paradigms of sexual deviancy (Zillmann and Bryant, in press).

With the firmly established causal links between viewing violence and aggressive behavior created from the 1960s to 1990, the research of the 1990s has focused almost entirely on the content of media and viewing habits. Modern research ethics preclude experiments that would subject children and adolescents to violent material and then observe their behavior, even though millions of children and adolescents are exposed on a daily basis. Nielsen Media Research indicate (Nielsen, 1998) that the average American child spends more than 21 h/wk viewing television. By age 18 yr, youngsters will have spent more time in front of the television, from 16,000–20,000 h, than in classroom education, 14,000 h. Given the yearly average of 10,000 acts of violence viewed, by age 20 one will have seen 200,000 acts of violence on television alone.

To further examine the content of television and thus understand its potential to be imitated, the National Television Violence Study was undertaken from 1996 to 1998.

Three volumes of data were gathered examining more than 10,000 h of programming across a variety of channels, cable and noncable (Federman, 1996, 1997, 1998). There was a striking consistency from year to year, with 61% of all shows containing violence of some type. This overall percentage is, however, somewhat misleading when one considers that only 15% of public television shows have violent content, while 85% of shows on the premium movie channels have violent content.

Given Bandura's original social learning theory tenets, concern is further heightened when one examines the context in which the violence occurs. Fully 26% of violent interactions involved the use of a weapon, and 38% were committed by attractive perpetrators. There was no pain associated with the violence in 50% of the incidents and almost 75% of violent acts involved no evidence of remorse, criticism, or penalty. Humor accompanied the violence in 41% of the incidents. The three volumes of the National Television Violence Study concluded: (*i*) television violence contributes to antisocial effects on viewers; (*ii*) the three primary effects of viewing television violence are learning of aggressive behaviors and attitude, desensitization to violence, and fear of being victimized by violence; and (*iii*) not all violence poses the same degree of risk of these harmful effects. The third volume of the study examined more closely the contextual fac-tors and their predictive impact. Attractive perpetrators, presence of weapons, and humor were associated with learned aggression; graphic violence and realistic violence were associated with generating fear; and humor and graphic violence associated with predicting desensitization.

Television is also the media medium that is introduced to children earlier in life than other media forms, beginning before age 2 yr. Until very recently, shows were designed to begin to appeal to preschoolers and older children. Since 1998, a particular genre of shows emerged designed specifically to appeal to infants and toddlers. While these shows do not contain violence, they do train the child from an early age to be a viewer consuming television and the power of its messages. Epidemiological research published by Centerwall (Centerwall, 1992) postulated a 10- to 15-yr lag between the time of childhood exposure to television and the primary behavior-modifying effects becoming operant in adulthood. His research examined the change in homicide rates following the introduction of television into South Africa in 1975. The homicide rate in 1974 was 2.5 per 100,000 and rose to 5.8 per 100,000 in 1987, an increase of 130%. This is consistent with the results of a 1986 Canadian study that took place in a remote town that did not have television available until 1973. The behaviors of first and second graders were studied and objectively measured with results of an increase of hitting, shoving, and biting by 160% after 2 yr of having television in their homes (Williams, 1986).

From the standpoint of amount of exposure alone, the concerns about television have even increased over the last 5 yr. A recent survey of 2065 U.S. third through twelfth graders showed that two-thirds have a television in their bedroom, more than one-third have their own VCR, and 15% receive premium cable channels in their bedrooms (Roberts, 2000). This survey also discovered important socioeconomic and race–ethnicity differences: the likelihood of finding televisions, VCRs, and video games in children's bedrooms was inversely related to household income; and African-American youth report 10 h/d of total media exposure, Hispanic youths 9 h/d, and whites 7/d.

An additional, highly significant longitudinal study reports on 707 families with a child between 1 and 10 studied successively in 1975, 1983, 1985–86, and 1991–1993 regarding the amount of time spent watching television and subsequent aggressive behavior in 2000 (Johnson et al., 2002). What is striking and highly significant is that there was a statistically significant association between the amount of time spent watching television during adolescence and early adulthood and subsequent aggressive acts, even when previous aggressive behavior, childhood neglect, family income, neighborhood violence, parental education, and psychiatric disorders were controlled statistically. The authors acknowledged that aggressive individuals spent somewhat more time watching television than others, but this tendency did not explain the association between television viewing and aggressive behavior.

A perspectives article published in the same issue of *Science* (Anderson and Bushman, 2002) adds further to this topic by providing a table that summarizes correlation data from 46 longitudinal studies, 86 cross-sectional samples, 28 field experiments, and 124 laboratory experiments (Fig. 1). The authors point out that all studies show a positive correlation between media violence and aggression and that the effects are not trivial in magnitude.

ROCK MUSIC AND MUSIC VIDEOS: EXPOSURE, CONTENT, AND EFFECT

Rock music lyrics became more explicit in their references to sex and drugs (Fedler et al., 1982) during the 1970s and 1980s. By the 1980s, the first studies of music videos showed that frank violent content was present in more than 50%. Durant et al. (1997), building on these data, analyzed the content of over 518 music videos from Music Television (MTV), Video Hits One (VH1), Country Music Television (CMT), and Black Entertainment Network (BET). The data revealed that 22% of MTV videos portrayed overt violence, with the other three networks airing 11% of programs with violence. When examined by music genre, 20% of rap videos portrayed violence, and the carrying of weapons was similarly highest in rap and rock videos, at approx 19%. Most alarmingly, a child was seen carrying a weapon 15% of the time and an adolescent 8%, and children (11%) and adolescents (8%) were actually engaged in violence. Rich et al. (1998) looked further at the videos for gender and contextual factors, which might influence identification and learning. In the videos showing violence, white females were most frequently the victims, in more than 80% the aggressor was an attractive role model, with males more than three times as likely to be the aggressor, and black males and females were overrepresented compared with U.S. demographics as both aggressors and victims.

In a seminal work, Waite et al. (1992) examined the potential causal effect of music television on violent behavior. The study was prompted by the observation that the television sets on six forensic wards were often tuned to MTV, where there was a high level of sexual and violent themes and a seeming deterioration of behavior after prolonged viewing. Data were collected over a 55-wk period on 222 patients on the six wards, ages 18–67 yr, mean age 28.65 ± 9.45 yr, before and after the removal of MTV from the wards. The results showed a statistically significant reduction in violent incidents, from 44 to 27, after the removal of MTV, which was further supported by time-series analysis. The authors acknowledged that the population indeed might be more vulnerable, and postulated a mechanism of kindling as possible etiology.

The genre of rap music, increasingly popular with the African-American population, has created particular controversy in the last decade with limited empirical research

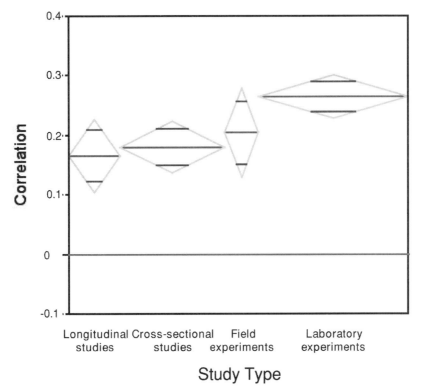

Fig. 1. Effects of media violence on aggression for different types of studies. Diamond widths are proportional to the number of independent samples. There were 46 longitudinal samples involving 4975 participants, 86 cross-sectional samples involving 37,341 participants, 28 field experiment samples involving 1976 participants, and 124 laboratory experiment samples involving 7305 participants. Red lines indicate the mean effect sizes. Blue lines indicate 95% confidence interval. Note that zero (dashed line, indicating no effect) is excluded from all confidence intervals. Reprinted with permission from Anderson, C. A., Bushman, B. J., "The Effects of Media Violence on Society," vol. 295, 29 March 2002, p. 2377, American Association for the Advancement of Science.

examining its effects. Wingood et al. (in press) designed a longitudinal study to investigate the association between exposure to rap music videos and a spectrum of health status indicators among African-American girls. Between December 1996 and April 1999, project recruiters screened teens in school health classes and adolescent health clinics as part of a study to prevent HIV/sexually transmitted diseases (STD). Teens were eligible if they were African-American females ages 14–18 yr, sexually active in the previous 6 mo, and provided written informed consent. At baseline, 522 females completed a self-administered survey identifying sociodemographic characteristics and media exposure, followed by face-to-face private interviews to assess alcohol use, drug use, sexual behaviors, and violence. An overwhelming majority, 95.4% reported viewing rap music videos, for on average 3.7 h/d for an average of 4.9 d/wk. Logistic regression analyses at 1 yr follow-up showed greater exposure to rap music videos at baseline was associated over the course of the year with being 3.0 times more likely to hit a teacher, 2.6 times more likely to have been arrested, 1.6 times more likely to have an incident of STD, and 1.6 times more likely to use drugs. The authors concluded that "Given the pervasiveness of rap music videos, its popularity, and its controversial content, future research needs to examine both the short and long-term effects of exposure to rap music videos."

COMPUTER AND VIDEO GAMES: EXPOSURE, CONTENT, AND EFFECT

In the past decade, new forms of media and entertainment, such as computer and video games, have entered the lives of most American youngsters. The annual industry growth of video and computer games has grown leaps and bounds in the United States, from $100 million in 1985 to $7 billion in 1994 (Provenzo, 1991) to 2001 estimates of $14 billion. Each year brings forth newer and more sophisticated material along with improved marketing strategies, which attract the public's fancy and imagination, leading to increased use of video and computer games. The average child 2–17 yr of age plays video games for 1 h each day with varying degrees of supervision, depending on the product (Gentile and Walsh, 1999). Another study, which evaluated 357 seventh and eighth grade students for frequency and location of video game use revealed that the average time spent by children playing video games was 4.2 h/wk. Boys were found to play more frequently in video arcades, 50% compared with 20% of girls playing in arcades (Funk, 1993). In addition to arcades, video games and computers have found their way into the homes of a large majority of American families, with 70% of American families owning video game players and more than two-thirds owning home computers. Thus, the access to cyberspace and the availability of digital media are burgeoning.

Video games are not entirely new to today's youth. The first appearance of video games in the United States was as early as 1970s with the arrival of "Pong," a video game by Atari, which was a basic version of the game ping-pong. "Pac-Man" was a popular game in the 1980s, where a yellow orb moved around the screen and ate ghosts. Some people considered this "eating of ghosts" as aggressive, and raised many objections to its use in children. The content of the video games has since taken new forms, with the 1990s seeing the emergence of explicit violence portrayed through video games. "Mortal Kombat," a game introduced in the early 1990s gained widespread popularity over the ensuing years. The ultimate goal of the game is for the player to kill any opponent he or she encounters.

Sophisticated special effects with increasingly graphic descriptions have made virtual violence more believable and appealing. The packaging of some games virtually spells out the violence with descriptions and wording, such as that on "Bionic Commando Elite Forces," which states "charging through enemies with barrels blazing and leaves only destruction in his wake." Another game "Metal Gear Solid," invites children "to build up an arsenal of weaponry" where "blood decorates all the bullet-ridden walls." However, many packagings do not mention that these video and computer games are based on adult-rated games. Dietz (1998) reported that violent video games dominate the market; 80% of 33 popular Sega and Nintendo games were found to depict violence, and 21% of the games featured violence against women.

The rating systems intended to provide information about the content and appropriate audiences for different games have been an attempt by the industry itself to provide consumer information. A study conducted by Walsh and Gentile (2001) found that when an entertainment industry rates a product as inappropriate for children, parents agree that it is inappropriate for children. However, parents disagree with many of the industry labels that are classified as "suitable" for the usage of children of different ages. Products rated as appropriate for adolescents are of greatest concern. The study further revealed that 90% of teenagers say that their parents never check the video game ratings before allowing them to rent or buy computer or video games. The amount of violence and depictions of violence are the primary markers for disagreement between parent raters and industry raters. In addition, many parents find the entertainment industry's media ratings confusing to use owing to the dissimilarity in ratings for each medium, such as television, movies, music, and video games. Walsh and Gentile further mention that the guidelines for rating media products, such as video games, are neither specific nor explicit. They add that the ratings are established and approved from within the industry and not from outside and that the systems do not seem to be subject to any kind of testing to ensure the appropriateness of labeling the products.

Images and behaviors that are modeled in the video games can have significant impact on children, especially with the increasing realism and graphic content that is portrayed in the video games. Portrayals of aggression, sex, and violence can shape the outlook and the values that children learn as they grow up. With the more active participation that is required to play the game, as contrasted to the passive absorption of material with watching television or movies, video games can pose a serious influence on the youth. While some researchers have found that video games are useful in teaching spatial performance, particularly for children with relatively poor skills in this area (Subramanyam and Greenfield, 1994) and some have proposed that "video game playing may be a useful means of coping with pent-up and aggressive energies" (Emes, 1997), many studies have outlined the deleterious effects of excessive video game playing, such as "nintenditis," a pathological preoccupation with video games, aggression, and an effect on pro-social behavior. Ferrie et al. (1994) documented 50 cases of video game-induced seizures reported worldwide. One-third of the cases had documented previous seizures, not related to video games. Keepers (1990) documented a single case study of video game preoccupation.

Studies have examined the relationships between video games and aggression vs pro-social effects. Scott (1995) examined college students and found little support for the theory of playing games as inducing aggressive behavior. However, a study of 278 children in the seventh and eighth grades by Wiegman and van Schie (1998), revealed a significant negative relationship with video game use and pro-social behavior. More consistent results were found for the preference for aggressive video games. Children, especially boys, who preferred aggressive video games were more aggressive and showed less pro-social behavior than those with a low preference for these games. Further analysis showed that children who preferred playing aggressive video games tended to be less intelligent.

Anderson and Dill conducted two studies on consenting college students, one analyzing violent video game exposure in real life, and the other being a laboratory exposure to graphically violent video games. Both studies examined violent video game effects on aggression-related variables. The studies use a theoretical construct called: General Affective Aggression Model (GAAM) (Anderson, 1995, 1996, 1997; Dill et al., 1998), which integrates theory and data concerning learning, development, instigation, and expression of human aggression. GAAM describes a multistage process by which personological (aggressive personality) and situational (video game playing and provocation) input variables lead to aggressive behavior. A significant finding from one of the studies is that the effect of violent video games appears to be cognitive in nature, such that in the short term, playing a violent video game appears to affect aggression by priming aggressive thoughts, including aggressive scripts. The findings do not rule out the possibility that under some circumstances violent video game effects on subsequent aggressive behavior may be mediated by increased feelings of hostility or general arousal effects. The researchers espouse the belief that violent video games can provide a paradigm for learning and practicing solutions to conflict situations.

The Anderson study further examines the long-term media violence effects on aggression and posits that longer-term effects result from the development, overlearning, and reinforcement of aggression-related knowledge structures. The longer-term effects are likely to be more enduring over time as well. The study showed that both in the real world set-up and in an experimental design using a standard objective laboratory measure of aggression, violent video game playing was positively related to increases in aggressive behavior. An interesting finding from the same study is that poor academic achievement was not related to prior violent video game play in particular, but was related to long-term exposure to video games in general. There are reasons to believe that exposure to violent video games could have more deleterious consequences than exposure to violent television or violent movies. The Anderson group posits three concerns regarding the unique dangers of violent video game exposure, namely, identification with the aggressor (Leyens and Picus, 1973), active participation in playing the games, and the addictive potential of the games.

A review of the literature by Benseley and Van Eenwyk (2001) was conducted to determine the validity of a major public health concern that violent video games contribute to real life aggression. The review included studies that examined an association between video game playing or violent video game playing and measures of aggressive, hostile, or antisocial behavior; personality; ideation; or mood. The results were evaluated separately for each age group. In the preschool children and elementary school students, three of four studies using behavioral observations of aggression during free-play found that violent video game play causes increased aggression or aggressive play immediately after the video game. Among middle and high school students, one study found that boys who reported more video game play in arcades, but not at home, reported more aggression and delinquency (Dominick, 1984). A study by Kestenbaum and Weinstein (1985) found an association of high levels of video game play with problems with police and low frustration tolerance. The same study, however, also found an association with feeling more relaxed after play, thus suggesting some help for youths dealing with developmental conflicts. The review further examined the effects on college students and young adults, which yielded mixed results, with two studies showing an increase in hostility after video game play, and four studies not finding any such association.

SUMMARY OF MEDIA VIOLENCE: EXPOSURE, CONTENT, AND EFFECT

The data regarding the average child's exposure to media violence presented in the previous three sections dramatically demonstrates effects of prolonged exposure over time. The causality research is compelling in supporting the position that exposure to violent media is a contributory risk factor to the occurrence of violent behavior, based on learning theory and neuroscience research, coupled with the accumulation-of-risk model. Such media exposure is likely to continue to grow as new forms of media are created and marketed to children and adults. The vast majority of products that are available now were not even on the market 10 yr ago. One can also predict, with confidence, that as the current products become less expensive, new ones with more sophisticated technology will be created to be sold at higher prices. The trendy computer game cafes with high-tech virtual games played to the early morning hours in darkened rooms, which have recently sprouted up in metropolitan areas, are just one example of how the culture of exposure can quickly change.

PREVENTION: A MULTIPRONGED PUBLIC HEALTH APPROACH

Basic public health tenets need to be applied to the prevention of learned violence secondary to media exposure. The goals of the public health approach should be to decrease exposure, specifically the exposure of the most vulnerable population, children; to influence the modification of content that is less violent; and to expand the research base in ways using new technology that elucidate more clearly the interaction between behavior and brain neurobiology. This requires, first, dissemination of the conclusive research data to healthcare professionals, professionals in the education field, and public policy experts. As this is occurring, information flows out to the general public, and parents become increasingly aware of the dangers posed by exposure to violent media. Grass roots efforts subsequently emerge (Table 1). While these tracks occur in parallel fashion simultaneously, for the purposes of discussion, they will described separately.

Medical Profession

The medical profession became increasingly aware of the research done by psychologists and medical epidemiologists in the early 1990s. This culminated in 1996 with the American Medical Association (AMA) publishing and distributing free of charge a 26-page brochure, *Physician Guide to Media Violence* (Walsh et al., 1996). The opening summary states that the purpose of the guidebook is to provide suggestions and options for dealing with violence in the media and to protect children from "its notorious and insidious effect." To add credence to the AMA's stance, they listed other organizations that support the link between violent entertainment and violent behavior:

- American Academy of Child and Adolescent Psychiatry
- American Academy of Pediatrics
- American Medical Association
- American Psychological Association
- Centers for Disease Control and Prevention
- National Institute of Mental Health
- Surgeon General's Office

Table 1
Trends in Media Research and the Prevention of Learned Violence

1960s and 1970s:	Beginning Basic Research
	Bandura's Social Learning Theory.
1980s:	Research date mounts:
	Professional organizations become involved:
	American Academy of Pediatrics Policy Statements.
	American Psychological Association's Policy Statements and *Big World, Small Screen: The Role of Television in American Society.*
1990s:	Content analysis in the information age:
	Professional organizations update policy statements.
	Education of practicing professionals and trainees.
	Grass roots nonprofit organizations develop.
2000–2010:	Professional organizations and nonprofits join efforts:
	Research using new technology advances the field.
	Media literacy promoted through educational system.
	Parent education at well-child health visits.
	General public educated via print media and Web-based information sites.
	Public pressure mounts due to failure of the industry to self-regulate violent content and continued marketing to children.
	Government involvement increases due to public concern about the need to protect children from media violence and other untoward effects of media exposure.

Each of these organizations has approached the task of educating its membership and the public in different ways. The American Psychological Association published and disseminated to its membership and the press, *Policy Statement on Social Practices that Induce Violence* (August 1996), which includes statements regarding the concern about "content for amusement for all ages in films, television, toys and games, music, and general media resulting in multiple daily exposures to violent content."

The American Academy of Pediatrics (AAP) has been in the forefront of these efforts, publishing in its academic journal *Pediatrics* well-researched policy statements relating to the media dealing with media violence. These include: "Media Violence" (1995a); "Media Education" (1999); "Impact of Music Lyrics and Rock Music Videos on Children and Youth" (1996); and "Children, Adolescents, and Television" (1995b). From these efforts, the Academy designed a 5-yr public education campaign, "Media Matters," with the goal of increasing public awareness and advocating for media education as a way to mitigate against the negative impact of media. The list of nine detailed recommendations to pediatricians in the policy statement "Media Education" includes a multi-pronged recommendation, which states that "pediatricians should begin incorporating questions about media use into their routine visits, including the use of the AAP's Media History form." The recommendation goes on to state that pediatricians should give advice to parents suggesting eight specific points to be made, ranging from encouraging selective viewing, to limiting time spent with media, to teaching critical viewing skills. The AAP, throughout their policy statements and in other articles published within their journal (Dietz and Strasburger, 1991; Strasburger and Donnerstein, 1999), has maintained the stance that exposure to violent media is a risk factor for violent behavior in children and adolescents and, therefore, is a legitimate public health concern. As a professional organization, the AAP repeatedly urges its membership to become involved at multiple levels of intervention, from direct patient education as mentioned above, to involvement with schools, to working with other health activists to monitor media, to advocating for pro-social programming, and to encouraging government to become more active in the field of media education and media education research.

Simultaneous to the efforts of educating practicing pediatricians has been the effort to incorporate information about the effects of media on children into pediatric training programs. To emphasize the importance of this issue, Rich and Bar-on (2001) surveyed the 209 accredited pediatric training programs in the United States to determine what training programs are teaching trainees about the media and its influence on the physical and mental health of children. The results revealed that, in spite of increasing awareness of media influence on the health of children, less than one-third of pediatric training programs teach about media exposure. The authors recommended the development of a pediatric media curriculum and urged its use throughout all training programs.

While the results of this survey might be interpreted as discouraging given the overwhelming evidence about media exposure, it can also be seen as another step in the logical progression of education. The one-third of training programs will likely grow to two-thirds over the next 5 yr and to fully a 100% over the next decade. The pediatricians who complete their training 10 yr from now will not question whether media education should be a part of the counseling task with new parents. Future pediatricians will naturally begin the process of cautioning parents about the dangers of the television and other media, as being the all too easy, all too readily available, electronic babysitter. The recommendation to begin these discussions at the well-child check-up at age 2 yr, will likely be adjusted downward to 1 yr, since television shows are now developed to appeal to infants and toddlers. Physicians will, thus, be training parents to raise children much differently than a mere decade ago. Whether is it bicycle helmets, car seats, or media education, the inculcation of such public safety messages occurs slowly over time through routine healthcare visits and public education campaigns.

Education Profession

Schools of education have also begun to incorporate media literacy into the training of teachers and into school curricula design. Some states, New Mexico and North Carolina being two, have mandated media literacy as a core element of school curricula. The New Mexico Media Literacy Project includes the development and distribution of a CD-ROM and video, *Understanding Media* (McCannon, 1999), which can be used by teachers to instill critical analysis skills through media literacy. The debate within the education field at the present time centers mostly on how to teach media literacy, not whether or not it should be taught. The methods being debated are whether to use separate curriculum materials with media literacy as a distinct topic, or to utilize the infusions method, where it is incorporated into existing areas such as math and English. Graduate schools of education are adding lectures, and in some instances full courses, devoted to educating teachers on how to incorporate media literacy into the instruction of today's youth.

In addition to schools providing education about media and its effects, one research study has shown that a school-based intervention, which promoted decreasing television and video game exposure, also decreased aggressive behavior. Robinson et al. (2001), in a randomized controlled school-based trial in two sociodemographically and scholastically matched public elementary schools, measured pre-intervention levels of aggressive behavior in both schools. One school then received an 18-lesson, 6-mo classroom curriculum to reduce television, videotape, and video game use. Postintervention measures showed that the students in the school receiving the intervention had a statistically significant decrease in aggression. While this study has yet to be replicated, it provides a prototype of intervention at the school level, which has potential widespread applicability toward the goal of decreasing aggressive behavior in school age children.

Educational institutions at the university level also serve to continue to expand the base of knowledge and understanding about the role of media in shaping and changing values and behavior, in addition to serving as honest brokers in the area of public debate about what we know, what we need to know, and what can and should be done. Examples include the University of Pennsylvania's Annenberg School of Communication and the Freedom Forum Media Studies Center at Columbia University, both of which organize regular conferences with published proceedings regarding trends in media and their impact on culture. Another noteworthy recent example is The University of Chicago, Third Annual Arts and Humanities in Public Life Conference, October 2001, which addressed what it described as the "pressing concern to the public interest," in the topic, "Playing by the Rules: Video Games and Cultural Policy." To answer its own question as to why a cultural policy research center should hold a conference on computer games, the conference organizers noted that as a "new cultural form that has begun to rival film in popularity, revenues, and formal complexity, computer games raise a host of questions about their impact on American society." Specifically, citing the Columbine School massacre as raising the question of the influence of violent games on the young shooters, the University noted the need for the review of current research into the effects of violent games on player behavior, specifically aggression. The conference was self-described as the first public presentation of the evidence and arguments for and against the thesis that "violent video games cause aggression." Conference papers are available on the Web site (http://culturalpolicy.uchicago.edu/conf2001).

Grass Roots Efforts

As parental concern about violent media has grown, grass roots efforts to combat the problem have also taken hold. The Center for Media Education, the Center for Media Literacy, the Coalition for Quality Children's Media, and the Lion and the Lamb (Table 2) are leading examples of nonprofit organizations whose goals are to gather and disseminate information to parents and to lobby government and schools to be more active. The Lion and the Lamb has chosen the more narrow focus of combating violent entertainment under the premise that it teaches violence to children. It publishes a regular online newsletter available free at www.lionlamb.org and covers such topics as the Federal Trade Commission's findings in June 2002 that the entertainment industry is still marketing adult-rated violent movies, video games, and music to children in spite of government's request for self-regulation; as well as the report of a case of a 5-yr old injuring his 22-mo-old cousin by imitating a violent wrestling move that he had seen

Table 2
Media Education Nonprofit Groups and Content Web Sites

Nonprofit groups	Content Web sites
Center for Media Education (202) 331-7833 www.cme.org	www.kids-in-mind.com www.gradingthemovies.com www.moviemom.com www.screenit.com
Center for Media Literacy (800) 226-9494 www.medialit.org	
Coalition for Quality Children's Media (505) 989-8076 www.kidsfirstinternet.org	
Lion and the Lamb Project (301) 654-3091 www.lionlamb@lionlamb.org	
Children Now (510) 763-2444 www.childrennow.org	

on television. In addition, the Lion and the Lamb organizes an annual press conference to alert parents about the most violent toys being marketed to children during the holiday season, many of which are linked to media entertainment. All of the nonprofits play a large role in providing information and testimony to federal agencies and to Congress regarding potential regulations and laws that will decrease violent entertainment, provide parents with more accurate information about media content, and educate the public about the untoward effects of children being exposed to violent media.

Other grass roots efforts include the development of Web sites to provide parents with information about media content. The naturally occurring movie release cycle during the holiday season and the summer often gives rise to public debate about how media has become much more graphic in terms of both sex and violence. In these forums, the movie rating system is regularly described and debated as an arbitrary system that has evolved over time to be more accepting of sex and violence, such that PG-13 in the year 2002 is markedly different than it would have been in previous years. Exposes about the rating systems, descriptions of current movies, and Web site sources are often published in newspapers and magazines. Thus, the print media has increasingly taken the role of educating its readership about the negative effects of television, movie, and video game media and providing parents with alternative ways to further educate themselves.

Public Policy Officials

Professional organizations and nonprofit grass roots groups continue to join together to lobby government officials to change public policy regarding the freedom of the airwaves. Given the U.S.' historical grounding in freedom of speech, this debate has far-reaching and serious implications. The Federal Communications Commission (FCC) is regularly lobbied to be more active in its rules and regulations, and Congress has been urged similarly to pass legislation that protects children. To address this topic, Newton Minow and Craig LaMay published, *Abandoned in the Wasteland: Children, Television and the First Amendment* (1995). In the acknowledgement, Minow, a former FCC chair, begins, "This book is the result of thirty-four years of work and contemplation of a medium, television, that since mid-century has transformed the world, the nation, the community, the family." The book persuasively argues against the false application of the First Amendment defense by broadcasters, citing the Supreme Court and Congress as requiring that public airways be used to serve public good and further recognizing that children are indeed a special audience that needs to be nurtured, not harmed. Although The Children's Television Act, requiring commercial broadcasters to provide "educational and informational" programs for children had been passed in 1990, it was largely ignored until after Minow and Lamay's book and the hue and cry of the mid-1990s following the outbreak of school shootings in the United States at Paducah, Jonesboro, and Columbine. Congress was lobbied to make the requirement more specific, with advocates urging for 1 h/d of educational and informational broadcasting. The media industry, however, prevailed, and the relatively meager requirement of 3 h/wk was added to the regulation.

During the same time period of the late 1990s, two other public policy debates raged, the television ratings debate and the V-chip technology debate. These two debates culminated in the passage of the Telecommunications Act of 1996. This mandated the development of a rating system for television similar to that developed by the Motion Picture Industry of America for the purpose of informing parents, or adult viewers, if the show about to be aired would have either violent content, noted with a V, or sexual content, noted by an S. The rating system could then be combined with the V-chip technology, wherein a computer chip inserted into televisions can be programmed to recognize all shows with a V and, thereby, prevent such shows from being viewed. The V-chip became available in 1999 and is a required part of all new television sets made to be sold in the United States after January 2000. While there will be many televisions without V-chip technology, over time, these sets will be replaced and parents raising future generations of children will have been given an additional tool to help them protect their children from being exposed to media violence.

Public awareness about the poorly enforced movie rating system has also occurred over the last 5 yr. Parental outrage at violent and sexually explicit movies, coupled with stories of their reenactment by young teens who had viewed them, led to parents influencing movie theaters to ask for age identification for teens purchasing tickets for R-rated movies. The movie industry countered quickly to make more movies PG-13, realizing that by reaching into the younger age group, the market base is expanded, and the potential dollars made by the film increased. Whereas only a few years ago, if one read the local paper for first run movies, the majority would have been rated R, now an increasing number are rated PG-13. While a PG-13 film is much more violent and sexual than the PG-13 film 5 yr ago, it is still less so than the typical R-rated film that teens were routinely watching in the early 1990s.

The Federal Trade Commission (FTC) has recently entered the fray of this debate owing to parents' complaints that violent movies, video games, and music are actually being actively marketed to children. In the year 2000, the entertainment industry was urged to impose self-regulation or face Congressional action that would impose regulation, much like what happened to the tobacco and alcohol industries. When the industry was monitored 2 yr later, the FTC found "little change" in marketing practices. Whether this will lead to further legislation or whether parents will be on their own in terms of monitoring the violent media marketed to their children remains in the realm of public debate. The full report of the FTC is available at http://wwwftc.gov/reports/violence/mvecrpt0206.pdf.

THE FUTURE: RESEARCH, REGULATIONS, AND PUBLIC EDUCATION

Research

The United States, as the largest manufacturer and exporter of technology and entertainment media in the world, the leader of the free world, and the promoter of free speech, has an obligation to the "global village" to provide more research regarding the effects of media on the developing brain and on the development of cultural values and behaviors. The National Institutes of Health should actively encourage and fund research that applies current technology, such as fMRIs and PET scans, to studies involving exposure to violent media. While research ethics will naturally preclude children from exposure to media with clearly violent content, milder and/or briefer forms of violent media, could be used and data extrapolated to the more violent forms and to prolonged exposure. Also, as is currently done, research on young adults who are able to give consent should also be pursued, because it provides information that can be extrapolated downward to younger ages.

Given the rapidity of technologic advances and the special status of children as a more vulnerable population, the National Institutes of Mental Health should establish a division devoted to monitoring and researching the trends in media and their effects on children. A biennial report from such a division would serve as an update for healthcare professionals and for watchdog nonprofit groups seeking to provide information to parents. It would also send a strong message to the entertainment industry that its behavior and its effects on youth are being monitored in a regular fashion consistent with the best science available in the country.

Regulations

The First Amendment of the United States guaranteeing free speech will likely continue to guide our country in terms of limiting regulation of the airwaves. Regulations, however, which seek to give children and adolescents special status, due to the developing-brain issues, which warrant protection from violent media, much like those that exist to protected them from tobacco and alcohol, could and should be forthcoming. A clearer rating system, with consistency of rating warnings on all entertainment media, should be developed at the federal level and safeguards developed, such as enforced fines against marketing violence to youth, are but two ways that government can and should become more active in the future.

Government should also encourage the development of educational and pro-social programming through either enhanced funding for public broadcasting or incentives to media to develop such shows. The comparison of what the United States spends for public broadcasting, $1.09 per capita, as opposed to other industrialized countries, United Kingdom $38.56, Canada $32.15, and Japan $17.71 (Kubey, 1996; Palmer, 1998), is but another example of how much more can be done to offer alternatives, which use the media to educate in positive social and health-promoting ways, vs the teaching of violence through violent entertainment.

Requiring broadcasters to increase their amount of educational programming from 3 h/wk to 1 h/d, and placing this hour in prime viewing time, vs in early morning hours when viewing is less, is an additional way that media can be shaped. The FCC and the FTC should increase their efforts to monitor airwaves practices and to match policy with areas of demonstrated concern. With increased Internet usage, this will obviously have to be extended to the development of technology and regulations regarding youth's access to violent and pornographic Web sites.

Public Education

Many professionals who study child development have expressed concern about the pervasively present "toxic pollutants" created by the Information Age. With parents working more hours and spending less time directly with their children, the entertainment media, with its pervasive violent content, has become the electronic babysitter. For children born after 1994, who will have grown-up their entire lives with the Internet, even more toxic elements will be available at an earlier age. Educating parents and helping them to monitor media, spend more time with their children in positive interactive ways, and to talk with them about what they are seeing on media, has to be a primary intervention of violence prevention. Whether through well-child check-ups with their pediatricians, public health messages on television, print media education about media ratings systems, regularly published governmental scientific proceedings from the National Institutes of Health, or school-based media literacy intervention programs, the message to parents should be clear: be aware of the media's impact on your children and be active to guard against it.

REFERENCES

Andison, P. S. (1977) TV violence and viewer aggressiveness: a cumulation of study results. *Public Opin. Q.* **41**, 324–331.

Anderson, C. A., Anderson, K. B., and Deuser, W. E. (1996) Examining an affective aggression framework: weapon and temperature effects on aggressive thoughts, affect, and attitudes. *Pers. Soc. Psychol. Bull.* **22**, 366–376.

Anderson, C. A. and Bushman, B. J. (2002) The effects of media violence on society. *Science* **295**, 2377–2378.

Anderson, C. A. and Dill, K. E. (2000) Video games and aggressive thoughts, feelings, and behavior in the laboratory and in life. *J. Pers. Soc. Psychol.* **78**, 772–790.

Anderson, C. A., Deuser, W. E., and DeNeve, K. M. (1995) Hot temperatures, hostile affect, hostile cognition and arousal: test of a general model of affective aggression. *Pers. Soc. Psychol. Bull.* **21**, 434–448.

Anderson, K. B., Anderson, C. A., Dill, K. E., and Deuser, W. E. (1998) The interactive relations between trait hostility, pain, and aggressive thoughts. *Aggress. Behav.* **24**, 161–171.

Baker, W. and Dessart, G. (1998) *Down the Tube: An Inside Account of the Failure of American Television.* Basic Books, New York.

Bandura, A., Ross, D., and Ross, S. A. (1961) Transmission of aggression through imitation of aggressive models. *J. Abnorm. Soc. Psychol.* **63**, 575–582.

Bandura, A., Ross, D., and Ross, S. A. (1963) Imitation of film-mediated aggressive models. *J. Abnorm. Soc. Psychol.* **66**, 3–11.

Bandura, A. (1977) *Social Learning Theory.* Prentice-Hall, Englewood Cliffs.
Baxter, L. R., Jr., Schwartz, J. M., Bergman, K. S., et al. (1992) Caudate glucose metabolic rate changes with both drug and behavior therapy for obsessive-compulsive disorder. *Arch. Gen. Psychiatry* **49**, 681–689.
Bensley, L. and Van Eenwyk, J. (2001) Video games and real-life aggression: review of the literature. *J. Adolesc. Health* **29**, 244–257.
Bok, S. (1998) Mayhem, *Violence as Public Entertainment.* Addison Wesley.
Bushman, B. J. and Huesmann, L. R. (2001) Effects of televised violence on aggression, in *Handbook of Children and the Media.* (Singer, D. and Singer, J. L., eds.), Sage Publications, Thousand Oaks, CA, pp. 223–254.
Centerwall, B. S. (1992) Television and violence: the scale of the problem and where to go from here. *JAMA* **267**, 3059–3063.
Committee on Communications, American Academy of Pediatrics (1999) Media education. *Pediatrics* **104**, 341–343.
Committee on Communications, American Academy of Pediatrics (1995a) Media violence. *Pediatrics* **95**, 949–951.
Committee on Communications, American Academy of Pediatrics (1996) Impact of music lyrics and rock music videos on children and youth. *Pediatrics* **98**, 1219–1221.
Committee on Communications, American Academy of Pediatrics (1995b) Children, adolescents, and television. *Pediatrics* **96**, 786–787.
Coplan, J. D., Andrews, M. W., Rosenblum, L. A., et al. (1996) Persistent elevations of cerebrospinal fluid concentrations of corticotropin-releasing factor in adult non-human primates exposed to early life stressors: implications for the pathophysiology of mood and anxiety disorder. *Proc. Natl. Acad. Sci. USA* **93**, 1619–1623.
Dietz, T. L. (1998) An examination of violence and gender role portrayals in video games: implications for gender socialization and aggressive behavior. *Sex Roles* **38**, 425–442.
Dietz, W. H. and Strasburger, V. C. (1991) Children, adolescents, and television. *Curr. Probl. Pediatr.* **21**, 8–31.
Dill, K. E., Anderson, C. A., Anderson, K. B., and Deuser, W. E. (1997) Effects of aggressive personality on social expectations and social perceptions. *J. Res. Pers.* **31**, 272–292.
Dominick, J. R. (1984) Video games, television, and aggression in teenagers. *J. Comm.* **34**, 136–147.
Durant, R. H., Rich, M., Emans, S. J., Rome, E. S., Allred, E., and Woods, E. R. (1997a) Violence and weapon carrying music videos. *Arch. Pediatr. Adolesc. Med.* **151**, 443–448.
Emes, C. E. (1997) Is Mr. Pac-Man eating our children? A review of the effect of video games on children. *Can. J. Psychiatry* **42**, 409–414.
Eron, L. R. (1995) Media violence. *Pediatr. Ann.* **24**, 84–87.
Federman, J. (1996) *National Television Violence Study I.* Sage, Thousand Oaks, CA.
Federman, J. (1997) *National Television Violence Study II.* Sage, Thousand Oaks, CA.
Federman, J. (1998) *National Television Violence Study III.* Sage, Thousand Oaks, CA.
Fedler, R., Fall, J., and Tanzi, L. (1982) Popular songs emphasizes sex, deemphasize romance. *Mass. Commun. Rev.* **9**, 10–15.
Ferrie, C. D., De Marco, P., Grunenwald, R. A., Giannakodimos, S., and Panyiotopoulos, C. P. (1994) Video game induced seizures. *J. Neurol. Neurosurg. Psychiatry* **57**, 925–931.
Flavell, J. H. (1986) The development of children's knowledge about the appearance-reality distinction. *Am. Psychol.* **41**, 418–425.
Funk, J. B. (1993) Reevaluating the impact of video games. *Clin. Pediatr. (Phila)* **32**, 86–90.
Gentile, D. A. and Walsh, D. A. (1999) Media Quotient: National Survey of Family Media Habits, Knowledge, and Attitudes. National Institute on Media and the Family, Minneapolis, MN.
Geen, R. and Donnerstein, E. (eds.). (1998) *Human Aggression: Theories, Research, and Implications for Social Policy.* Academic Press, New York.
Huesmann, L. R. and Eron, L. D. (eds.). (1986) *Television and the Aggressive Child: A Cross-National Comparison.* Lawrence Erlbaum, Hillsdale, NJ.

Huesmann, L. R., Eron, L. D., Lefkowitz, N. M., and Walder, L. O. (1984) Stability of aggression over time and generations. *Dev. Psychol.* **20,** 1120–1134.

Huston, A. C., Donnerstein, E., Fairchild, H., et al. (1992) *Big World, Small Screen: The Tole of Television in American Society.* University of Nebraska Press, Lincoln, NE.

Johnson, J. G., Cohen, P., Smailes, E. M., Kasen, S., and Brook, J. S. (2002) Television viewing and aggressive behavior during adolescence and adulthood. *Science* **295,** 2468–2471.

Joint Statement on the Impact of Entertainment Violence on Children: Congressional Public Health Summit. (2000) July 26 WWW (http://www.aap.org/advocacy/releases/jstmtevc.htm).

Keepers, G. A. (1990) Pathological preoccupation with video games. *J. Am. Acad. Child Adolesc. Psychiatry* **29,** 49–50.

Kestenbau, G. I. and Weinstein, L. (1985) Personality, psychopathology, and developmental issues in male adolescent video game use. *J. Am. Acad. Child Psychiatry* **24,** 329–333.

Kubey, R. W. (1996) Television dependence, diagnosis, and prevention, in *Tuning in to Young Viewers: Social Science Perspectives on Television.* (MacBeth, T. M., ed.), Sage, Thousand Oaks, CA, pp. 221–260.

Ladd, C. O., Owens, M. J., and Nemeroff, C. B. (1996) Persistent changes in corticotropin-releasing factor neuronal systems induced by maternal deprivation. *Endocrinology* **137,** 1212–1218.

Leflowitz, M. M., Eron, L. D., Walder, L. O., and Huesmann, L. R. (1972) Television violence and child aggression: a follow-up study, in *Television and Social Behavior, III. Television and Adolescent Aggressivness.* (Comstock, G. A. and Rubinstein, E. A., eds.), US Government Printing Office, Washington, DC, pp. 35–135.

Leyens, J. P. and Picus, S. (1973) Identification with the winner of a fight and name mediation: their differential effects upon subsequent aggressive behavior. *Br. J. Soc. Clin. Psychol.* **12,** 374–377.

Malamuth, N. M. and Briere, J. (1986) Sexual violence in the media: indirect effects on aggression against women. *J. Soc. Issues* **42,** 75–92.

Malamuth, N. M. and Check, J. V. (1981) The effects of mass media exposure on acceptance of violence against women: a field experiment. *J. Res. Personal.* **15,** 436–446.

McCannon, B. (1999) Understanding Media (CD-ROM). New Mexico Media Literacy Project, Alburquerque Academy, Alburquerque.

Minow, N. and Lamay, C. (1995) *Abandoned in the Wasteland: Children, Television and the First Amendment.* Hilland Want, New York.

Morison, P. and Garner, H. (1978) Dragons and dinosaurs: the child's capacity to differentiate fantasy from reality. *Child Dev.* **49,** 642–648.

Mullin, C. R. and Linz, D. (1995) Desensitization and resensitization to violence against women: effects of exposure to sexually violent films on judgments of domestic violence victims. *J. Pers. Soc. Psychol.* **60,** 449–459.

National Institute of Mental Health. (1982) *Television and Behavior: Ten Years of Scientific Progress and Implications for the Eighties, I.* US Government Printing Office, Washington, DC.

Nielsen Media Research (1998) *1998 Report on Television.* Nielsen Media Research, New York.

Paik, H. and Comstock, G. (1994) The effects of television violence on antisocial behavior: a meta-analysis. *Commun. Res.* **21,** 516–546.

Palmer, E. L. (1988) *Television and America's Children.* Oxford University Press, New York.

Peterson, D. L. and Pfost, K. S. (1989) Influence of rock videos on attitudes of violence against women. *Psychol. Rep.* **64,** 319–322.

Policy Statement on Social Practices that Induce Violence (1996) American Psychological Association, Washington, DC.

Provenzo, E. F. (1991) *Video Kids: Making Sense of Nintendo.* Harvard University Press, Cambridge, MA.

Rich, M. and Bar-on, M. (2001) Child health in the information age: media education of pediatricians. *Pediatrics* **107,** 156–162.

Rich, M., Woods, E. R., Goodman, E., Emans, S. J., and DuRant, R. H. (1998) Aggressors or victims: gender and race in music video violence. *Pediatrics* **101,** 669–674.

Roberts, D. F. (2000) Media and youth: access, exposure, and privitazation. *J. Adolesc. Health* **27S,** 8–14.

Robinson, J. P. and Bachman, J. G. (1972) Television viewing habits and aggression, in *Television and Adolescent Aggressiveness.* (Comstock, G. A. and Rubinstein, E. A., eds.), US Government Printing Office, Washington, DC, pp. 372–382.

Robinson, T. N., Wilde, M. L., Navracruz, L. C., Haydel, K. F., and Varady, A. (2001) Effects of reducing children's television and video game use on aggressive behavior: a randomized controlled trial. *Arch. Pediatr. Adoles. Med.* **155,** 17–23.

Scott, D. (1995) The effect of video games on feelings of aggression. *J. Psychol.* **129,** 121–132.

Singer, J. L. and Singer, D. G. (1981) *Television, Imagination and Aggression: A Study of Preschoolers.* Lawrence Erlbaum, Hillsdale, NJ.

Strasburger, V. C. and Donnerstein, E. (1999) Children, adolescents, and the media: issues and solutions, *Pediatrics* **103,** 129–139.

Subrahmanyam, K. and Greenfield, P. (1994) Effect of video game practice on spatial skills in girls and boys. *J. Appl. Dev. Psychol.* **115,** 13–32.

Surgeon General's Scientific Advisory Committee on Television and Social Behavior. (1972) *Television and Growing Up: The Impact of Televised Violence.* US Government Printing Office, Washington, DC.

Waite, B. M., Hillbrand, M., and Foster, H. G. (1992) Reduction of aggressive behavior after removal of music television. *Hosp. Community Psychiatry* **43,** 173–171.

Walsh, D. A. and Gentile, D. A. (2001) A validity test of movie, television, and video-game ratings. *Pediatrics* **107,** 1302–1308.

Walsh, D., Goldman, I. S., and Brown, R. (1996) *Physician Guide to Media Violence.* American Medical Association, Chicago.

Wiegman, O. and van Schie, E. G. (1998) Video game playing and its relations with aggressive and prosocial behaviour. *Br. J. Soc. Psychol.* **37,** 367–378.

Williams, T. B. (1986) *The Impact of Television: A Natural Experiment in Three Communities.* Academic Press, New York.

Wingood, G. M., DiClemente, R. J., Bernhardt, J. M., et al. A prospective study of exposure to rap music videos and African-American female adolescents' health. *Am. J. Public Health*, in press.

Wood, W., Wong, F., and Chachere, J. (1991) Effects of media violence on viewers' aggression in unconstrained social interaction. *Psychol. Bull.* **109,** 371–383.

Wright, J. C., Huston, A. C., Reitz, A. L., and Pieymat, S. (1994) Young children's perceptions of television reality: development differences. *Dev. Psychol.* **30,** 229–239.

Zillmann, D. and Bryant, J., eds. (1994) *Media Effects, Advances in Theory and Research.* Lawrence Erlbaum Associates, Mahwah, NJ.

Zillman, D., Bryant, J., and Huston, A., eds. (1994) *Media, Children, and the Family: Social Scientific, Psychodynamic, and Clinical Perspectives.* Lawrence Erlbaum Associates, Mahwah, NJ.

Zillmann, D. and Bryant, J. Pornography: models of effects on sexual deviancy, in *Encyclopedia of Criminology and Deviant Behavior.* (Davis, N. and Geis, G., eds.), Taylor & Francis, New York, in press.

Zillman, D. and Weaver, J. B. (1989) Pornography and men's sexual callousness toward women, in *Pornography: Research Advances and Policy Considerations.* (Zillman, D. and Bryant, J., eds.), Erlbaum, Hillsdale, NJ, pp. 95–125.

Zillman, D. (2000) Influence of unrestrained access to erotica on adolescents and young adults. Dispositions toward sexuality. *J. Adolesc. Health* **27S,** 41.

14
Social Drinking and Aggression

Kathryn Graham

INTRODUCTION

Alcohol has effects that people find positive or desirable. In fact, it is the number one drug of choice in many societies (Heather, 2001). However, alcohol also has effects that are seen as negative or undesirable by individuals and by society in general. One such effect is the role alcohol plays in aggressive behavior. For example, in the United States, 30–40% of violent crimes committed in the years 1993–1998 involved an offender who had consumed alcohol prior to the crime (Greenfield and Henneberg, 2000). Although this does not mean that alcohol "caused" the crime, cross-cultural analyses show a greater than chance involvement of alcohol in homicide and violent crime (Murdoch et al., 1990). Moreover, there is converging evidence that alcohol contributes to aggression and violence in a causal way. Time series analyses have found that alcohol consumption and violent crime tend to covary over time (Cook and Moore, 1993; Norström, 1998), and meta-analysis of experimental research on alcohol and aggression suggests a general increase in aggression after consuming alcohol, especially for men (Bushman, 1997; Bushman and Cooper, 1990). Despite the strong links between alcohol and aggression, however, it is clear that alcohol is neither necessary nor sufficient to cause aggression; that is, aggression occurs without alcohol, and alcohol consumption occurs without aggression. In particular, whether drinking leads to aggression is a function of the combination of the effects of alcohol, the characteristics of the drinker, and the environmental constraints and pressures relating to aggression, including cultural factors such as drinking patterns (Rossow, 2001; *see* also Room and Rossow, 2000) and norms regarding appropriate drinking behavior (MacAndrew and Edgerton, 1969).

Not only is alcohol associated with an increase in aggressive behavior, there is some evidence to suggest that the effects of alcohol consumption may contribute to the escalation of conflict and increase risk of injury when aggression does occur. Martin and Bachman (1997) analyzed the U.S. National Crime Victimization Survey data and found that men's conflicts with strangers were more likely to escalate from threat to physical violence if the assailant had been drinking; similarly, attacks on women by an intimate partner were more likely to result in the woman being injured if the partner had been drinking. Another indication of the potential role of alcohol in the escalation of violence is the finding that male perpetrators who killed their intimate female partners were significantly

From: *Neurobiology of Aggression: Understanding and Preventing Violence*
Edited by: M. Mattson © Humana Press Inc., Totowa, NJ

more likely to have consumed alcohol and/or drugs compared to perpetrators who had physically abused their partners but not killed them (Sharps et al., 2001). There is also evidence that alcohol use by a perpetrator predicts rape completion vs attempted rape (Brecklin and Ullman, 2002). Consistent with this, women receiving emergency services for injury from domestic violence were significantly more likely to have a male partner who abused alcohol or drugs compared with women receiving emergency services for other reasons (Kyriacou et al., 1999).

Many reviews of the literature on the relationship between alcohol and aggression have been published.[1] Because it is not possible to cover all the relevant literature in the space of a brief chapter, and in order not to duplicate existing reviews, the present chapter will focus on aggression that occurs in social drinking situations, highlighting the interaction of alcohol, the person, and the environment, and describing prevention interventions that focus on the social drinking context.

The first section of the chapter provides a brief overview of the main findings from the research on the relationship between alcohol and aggression, including the pharmacological effects of alcohol that increase the likelihood of aggression, as well as the characteristics of the drinker and the social context of drinking, which may moderate this relationship. The second part of the chapter will apply these explanations of alcohol-related aggression to examples of incidents in one specific drinking context, the licensed premise (i.e., bars, nightclubs, pubs, and taverns), which has been shown to be high risk for aggression and other problems. The third part of the chapter summarizes existing approaches to prevention and describes an example of a prevention intervention designed to minimize aggression and injury in licensed premises by changing the social environment for drinking and the way that drinking behavior is managed in this environment.

AN OVERVIEW OF THE RELATIONSHIP BETWEEN ALCOHOL AND AGGRESSION

Pharmacological Effects of Alcohol

The pharmacological effects of alcohol are many and varied, including paradoxical effects, such as excitation but also sedation (Grupp, 1980) and emotional lability (Pliner and Cappell, 1974), but also reduced emotional responding (Stritzke et al., 1995). It also appears that the link between alcohol and aggression is related to several different pharmacological effects of alcohol rather than a single particular effect (*see* Giancola, 2000; Graham et al., 1997; Miczek et al., 1997; Pernanen, 1976; Pihl et al., 1993; Yudko et al., 1997). The following briefly summarizes the effects of alcohol for which there is empirical evidence of a relationship with aggression.

---

[1]Relevant reviews include entire books [Galanter, M. (Ed.), *Recent Developments in Alcoholism, Volume 13, Alcohol and Violence* (1997); Martin, S. E. (Ed.), *Alcohol and Interpersonal Violence: Fostering Multidisciplinary Perspectives* (1993)] and journal issues [*Alcohol Research and Health* 25(1) (2001); *Alcohol Health and Research World*, 17(2) (1993); *Contemporary Drug Problems*, 24(4) (1997); *Journal of Studies on Alcohol*, 11(Supple.) (1993)] plus a number of recent journal articles and chapters (e.g., Chermack and Giancola, 1997; Giancola, 2000; Graham et al., 1996; 1998; Graham and West, 2001; Ito et al., 1996; Leonard, 2000; Roberts et al., 1999; Room and Rossow, 2000) as well as some older particularly thorough or classic reviews (Fagan, 1990; Pernanen, 1976).

Increased Risk Taking

Alcohol appears to increase the likelihood [...] people less able to accurately assess risks or identify inh[...] 1979), less likely to give thought to or expect possible [...] ne et al., 1997; Graham and Wells, in press; Parker, 1[...] risks even when these risks have been accurately assess[...] eased risk taking has been linked to anxiolytic effects of alcohol that are thought to be mediated by alcohol's actions on γ-amino butyric acid (GABA) neurotransmitter signaling (Miczek et al., 1993; Pihl et al., 1993). Alcohol's effects on serotonin may also account for some of alcohol-induced risk taking, especially in terms of aggression that occurs later in a drinking session. In particular, alcohol consumption results in an initial increase and then a decrease in serotonin (Virkkunen and Linnoila, 1993). Because decreased level of serotonin has been associated with reduced impulse control for some individuals (*see* Higley, 2001; Linnoila and Virkkunen, 1992; Pihl and Peterso[...] ssion may be more likely when serotonin level is lower in the later stages [...]

Focus on the Present and Increased Emotional Labilit[y]

Based on anthropological research, Washburne (1[...] le are drinking, they have reduced awareness and a ten[...] ne aspect of the present situation. Other research has s[...] on leads to a state of emotional plasticity or high em[...] ell, 1974). The combination of these effects (i.e., narro[...] mo-tionality or emotional liability), also referred to as alcohol-induced "myopia" ([...] and Josephs, 1990), has been observed to be a major contributing factor in naturally occurring aggression (Graham et al., 2000).

Impaired Cognitive Functioning Leading to Reduced Ability to Appraise the Situation and the Motives or Roles of Others as well as Leading to Poorer Problem Solving

Alcohol impairs cognitive abilities associated with the prefrontal and temporal lobes of the brai[...] ency, and memory (Peterson et al., 1990). It has been h[...] ontribute to aggression by leading to inadequate appr[...] g less able to take another person's perspective, and [...] la, 2000; Gibbs, 1986; Hoaken et al., 1998; Pernanen, [...] ing evidence for this theory was found in an experimer[...] ed ability to problem solve in conflict situations after [...] nd in a study that demonstrated reduced abilities to as[...] 980) when intoxicated. Related to cognitive effects, a number of reviews (Pernanen, 1976; Pihl and Ross, 1987) have suggested that alcohol makes a person less self-reflective and may make them more likely to attribute blame to others rather than recognizing their own contribution to conflict. Consistent with this proposed cognitive bias, a recent general population survey found that people reported that alcohol played a greater role for their opponents than for themselves in incidents of aggression, even when the respondent reported that he or she was as intoxicated or even more intoxicated than the opponent (Graham and Wells, 2001a). Similarly, experimental studies have also found that alcohol increases perceived aggressiveness of a bogus or hypothetical opponent (Pihl et al., 1981; Sayette et al., 1993; Schmutte et al., 1979).

Increased Power Concerns

Using a variety of methods, McLelland and his colleagues (1972) found that concerns with personal power were a contributing factor in male aggression, both as an effect of alcohol and as a contributing personality factor. As an extension of this, it has been suggested that alcohol's effects may make a person feel a greater sense of mastery (Pernanen, 1991) and make the person more likely to view another's behavior as a challenge (Gibbs, 1986). Consistent with this hypothesized increased concern with power when drinking, research on naturally occurring aggression has found that power concerns in the form of "macho" posturing and behavior play an important role in barroom aggression among young males (Graham and Wells, in press; Graham et al., 2000; Tomsen, 1997; Tuck, 1989). These observations, together with the evidence from McLelland's studies, suggest a possible interaction of alcohol and macho concerns.

The Pharmacological Effects of Alcohol on More than One Person in a Drinking Group

The link between alcohol and aggression is likely to be potentiated in natural drinking situations in which there are a number of people drinking, that is, when the effects of alcohol that increase the probability of aggression are experienced by more than one person. For example, violence may be the result of conflict in a drinking situation because both people involved are more focused on the present, impaired at problem solving, and overly concerned with personal power owing to the effects of alcohol. Empirical support for this hypothetical scenario was found in an experimental study of the effects of alcohol on aggression (Leonard, 1984) using real opponents in a randomized design that included: (*i*) both opponents drinking; (*ii*) one drinking; and (*iii*) neither drinking. This study found that alcohol increased aggression for all subjects who consumed alcohol, but aggression tended to escalate only when both subjects had consumed alcohol. Field studies of barroom aggression have also found a relationship between the level and severity of aggressive incidents and the overall intoxication level of the patrons in the bar (Graham et al., 1980; Homel and Clarke, 1994), as well as a contribution of intoxication of more than one person to specific incidents of aggression (Graham et al., 2000)

The Possible Relationship Between the Positive and Negative Effects of Alcohol

It is interesting to note that the effects of alcohol most likely to be related to aggression are often the same effects perceived as desirable outcomes of drinking. For example, the same anxiolytic effect that leads to increased risk taking and associated aggression could also account for alcohol's desirable quality of acting as a "social lubricant" by freeing people to take more social risks. Similarly, increased feelings of power and competence are often desirable, and increased emotionality and present-orientation with drinking can have positive results, such as feelings of greater social bonding and connectedness (Washburne, 1956). Even cognitive impairment due to alcohol may be perceived as positive, for example, if it contributes to the sense of the drinking occasion as a "time out" period from normal responsibilities and accountabilities (*see* MacAndrew and Edgerton, 1969). In sum, alcohol often leads to positive experiences rather than aggression, and, as will be described in the following sections, whether alcohol's effects lead to aggression depends on the environmental context of drinking and the characteristics of the drinker.

Personal Characteristics of the Drinker Associated with Increased Risk of Aggression

Gender

There is a large literature demonstrating that males are more likely to commit violent crimes (Chilton and Jarvis, 1999a; 1999b; Daly and Wilson, 1998; Kellerman and Mercy, 1992) and generally more likely to engage in physical aggression than females (Eagley and Steffen, 1986), although this difference is moderated by circumstances such as provocation and other factors (Bettencourt and Miller, 1996). Only marital aggression shows equal or greater rates of physical aggression by women, but even in this context, men are more likely to engage in severe aggression that causes injury (Archer, 2000). It has also been found across a wide range of countries that men drink more alcohol than women (Wilsnack et al., 2000). The high rates of both alcohol consumption and aggression among men suggest that the link between alcohol and aggression is likely to be particularly relevant for men. Moreover, although some studies have found equal effects of alcohol on aggression for men and women, the overall trend seems to be for alcohol to have a greater impact on aggressive behavior for men than for women (*see* Giancola et al., 2002, and meta-analysis by Bushman, 1997). The gender of the opponent may also moderate the extent that alcohol is associated with an increase in aggression with alcohol most likely to be involved in physical aggression when the incident involves conflict between males (Graham and Wells, 2001b).

Age

Younger adults are more likely to engage in aggression and violent crime generally (Beck et al., 1993; Wells et al., 2000), and alcohol-related aggression is particularly likely to involve young adults (Dawson, 1997; Rossow, 1996). However, at present, there is no evidence to suggest that alcohol is more likely to lead to aggression in younger vs older adults.

Heavy Drinking, Alcoholism, or Alcohol Abuse

There is a longstanding literature linking heavy or problem alcohol consumption with aggression. At least some of this relationship appears to be due to an increased risk of developing drinking problems among persons exhibiting childhood or adolescent aggression (Adalbjarnardottir and Rafnsson, 2002; Harford and Muthén, 2000; White et al., 1993). Similarly, antisocial or aggressive personality has been found to be associated with a greater likelihood of early-onset alcoholism in men (*see* Jaffe et al., 1988; Moeller and Dougherty, 2001; Virkkunen and Linnoila, 1997) and with a greater likelihood of becoming aggressive after consuming alcohol (*see* Moeller and Dougherty, 2001). A link between a preexisting deficit in serotonin, high impulsivity, heavy drinking, and violence has been proposed to account for the higher aggressivity among problem drinkers (*see* Heinz et al., 2001; Higley, 2001; Virkkunen and Linnoila, 1997), although research on nonhuman primates suggests that this link may be moderated by the early childhood environment (*see* Heinz et al., 2001; Higley and Linnoila, 1997). Finally, effects of chronic alcohol abuse such as dementia (Miller and Potter-Efron,

1989) and temporal lobe dysfunction (Bradford et al., 1992) may also increase the likelihood of these individuals being aggressive, especially when drinking.

A Pattern of Drinking Large Amounts per Occasion

A recent study of physical aggression in the general population found that high quantity drinking was associated with increased risk of alcohol-related aggression, but not with increased risk of aggression generally (Wells and Graham, 2003). Similarly, in their emergency room study, Borges et al. (1998) found that drinking prior to the event was a better predictor than usual drinking pattern of violent vs nonviolent injury. Finally, epidemiological studies have found that the relationship between usual drinking pattern and crime disappears when drinking at the time of the event is controlled for (Collins and Schlenger, 1988; Wiley and Weisner, 1995). Thus, it seems likely that at least part of the relationship between heavy or alcoholic drinking and aggression is due to heavy drinkers being more likely than light drinkers to experience the acute effects of alcohol (*see* Jaffe et al., 1988). In particular, there is growing evidence that the aspect of heavy drinking, which is the most relevant to aggression, is the acute exposure to the effects of alcohol (i.e., becoming intoxicated). Consistent with this conclusion, a number of studies have found that drinking to intoxication or drinking large amounts per occasion predicts aggression (Dawson, 1997; Rossow, 1996; 2001; Wells et al., 2000).

Environmental Factors that Moderate the Relationship between Alcohol and Aggression

Experimental studies of aggression and field studies of high risk drinking environments have identified a number of aspects of the drinking environment that appear to moderate the relationship between alcohol and aggression. These include: (*i*) provocation, threat, or other triggers; (*ii*) standards and expectations for behavior; (*iii*) rewards and punishment for aggression and for nonaggression; and (*iv*) the role of third parties and (*v*) the characteristics of persons present at the drinking occasion.

Provocation, Threat, or Triggers

At least some provocation appears to be necessary for alcohol consumption to increase aggression in laboratory research settings (*see* Graham et al., 1996; Gustafson, 1993). On the other hand, one meta-analysis, comparing high and low provocation, concluded that alcohol's effects on aggression are less when provocation is high, possibly due to the large general effect of high provocation on increasing aggression (Ito et al., 1996). This finding suggests that the effect of alcohol on aggression may be limited to nonextreme levels of provocation. Provocation also seems to be an important environmental determinant of alcohol-related aggression in real-life settings. For example, studies of marital violence have found that alcohol is associated with aggression primarily among hostile or discordant couples (*see* Leonard, 2000). Similarly, barroom environments at high risk for aggression tend to have activities or other aspects of the environment that trigger aggression. For example, crowding and bumping can be a trigger in bars and nightclubs (Graham et al., 1980; Graham and Wells, 2001c; MacIntyre and Homel, 1997). Pool and other competitive games and provocation in the form of perceived insults or offensive behavior can also serve as triggers for aggression in barroom settings (Felson et al., 1986; Graham et al., 1980; Graham and Wells, 2001c; Graves et al., 1981).

Standards and Expectations for Behavior

An experimental study of alcohol's effects on aggression found that nonaggressive norms can eliminate the increase in aggression caused by alcohol (Jeavons and Taylor, 1985). In terms of standards that encourage aggression, a number of field studies have found that permissive barroom norms are highly correlated with aggression, especially severe aggression (Graham et al., 1980; Homel and Clark, 1994) and appear to play a role in the majority of barroom incidents (Graham et al., 2000). Changes in norms and expectations in drinking environments have also been associated with changes in behavior. For example, community intervention research in Australia (Hauritz et al., 1998; Homel et al., 1997) found decreased aggression and crime when bars adopted policies and practices that limited over consumption and other problem-causing behaviors by patrons.

Rewards and Punishments for Aggression

Rewards and punishments may also moderate the alcohol–aggression relationship. For example, an experimental study of the effects of alcohol on aggression (Hoaken et al., 1998) found that monetary rewards for not aggressing were effective in preventing the increase in aggression normally found in experimental studies when subjects have been given alcohol. In terms of rewards and punishments in naturalistic environments, an observational study of aggression in bars frequented by young people (Graham et al., 2000) found that bars with a lot of aggression not only did not punish aggressors, but often actually encouraged aggressive behavior. Related to the lack of punishments for barroom aggression, alcohol-related aggression tends to have a lower emotional impact on persons involved in a conflict compared with aggression that does not involve alcohol, and this impact is especially low for incidents occurring in a bar (Graham et al., 2002).

The Role of Third Parties

In general, alcohol-related incidents tend to involve more people than incidents in which no one has been drinking (Wells and Graham, 2003), possibly because alcohol-related aggression is more likely to occur in a social context, which typically involves more than two people. Thus, the role of third parties is an especially important consideration in alcohol-related aggression. In experimental research on aggression, Taylor and Gammon (1976) found that having an observer encourage research subjects to set lower shocks reduced shock setting in both the alcohol and no-alcohol condition. In barroom settings, barstaff are the main agents of social control, and their behavior is critical to preventing and managing aggression. Bars that have high levels of aggression typically have staff who are either ineffectual at controlling problem behavior or who are violent themselves and actually escalate conflicts between patrons (Graham et al., 1980; Graves et al., 1981; Homel and Clark, 1994; Wells et al., 1998)

The Type of People Present in the Drinking Environment

Some drinking settings attract aggressive people, which may account for some of the aggression in that context, aside from any aggression that may be due to the effects of alcohol. For example, some incidents of aggression in bars appear to be due to the presence of groups of young males who are looking for trouble or who take offense easily (Burns, 1980; Homel et al., 1992; Pernanen, 1991; Tomsen, 1997). These incidents often involve angry reactions to accidental bumps, racist or other offensive remarks, taking offense at another person's behavior (when no offense was intended), sexual harassment or dominance, and just generally looking for a fight (Graham et al., 2000; Graham and Wells, in press). In terms of other at-risk populations, some studies have suggested that aggression is more likely in drinking settings frequented by marginalized (e.g., skid row) (Graham et al., 1980) or poorer persons (Parker and Rehbun, 1995).

EXAMPLES OF BARROOM AGGRESSION: COMBINING THE EFFECTS OF ALCOHOL, ENVIRONMENTAL FACTORS THAT INCREASE RISK OF AGGRESSION, AND DRINKERS WHO ARE PRONE TO AGGRESSION

Licensed premises (i.e., bars, nightclubs, pubs) are high-risk settings for aggression and violence (Hobbs et al., 2000; Ireland and Thommeny, 1993; Stockwell et al., 1993), especially licensed premises that combine environmental risk factors with intoxicated patrons (Graham et al., 1980; Homel and Clark, 1994; Lang et al., 1995; Stockwell et al., 1991). The following incidents have been selected from our various studies of aggression in licensed premises to illustrate different types and contexts of aggression and to link the aggression in these incidents to proposed explanatory factors described in the first part of this chapter. Although causal attributions regarding these incidents are speculative, these examples of naturally occurring aggression help to identify the dynamics of alcohol–person–environment at play in real-life aggression.

Example 1. Brawling males (interview) (Graham and Wells, in press). The respondent, a South Asian man in his early 20s, was leaving a bar with some Caucasian friends when another group of young men began making racist comments and jokes. One of the respondent's friends confronted the group. One man from the group said they were just joking, but the friend remained angry and eventually there was a scuffle between the two groups, and both groups were pushed out of the bar by bar staff. A fight then started with the respondent's friend being pushed and jumped on by one of the opposing men. At this point, the respondent grabbed another of the opposing men and started punching him. He described his emotions at the time as "once I saw that there was no turning back, it was just, you know, instant anger, aggression." The respondent noted that his friend who initially confronted the men had "a few ounces of liquid courage" and that alcohol had also affected his own judgment. "Rather than assessing the situation, I just jumped in head first without realizing that there were more of them than us...if I had been sober, I probably would just have grabbed my friend and said 'Let's just go.'"

As shown in Table 1, a number of the explanations for alcohol-related aggression described in the earlier part of this chapter appear relevant as potential contributing factors in this incident. First, the incident reflects the highly scripted "macho" exchanges between young men that has been found in barroom research in the United States (Burns, 1980), Canada (Graham and Wells, in press), the United Kingdom (Tuck, 1989), and Australia (Homel et al., 1997; Tomsen, 1997), with the term "macho" reflecting concerns with masculinity, male dominance, sexual prowess, physical strength and honor (Neff et al., 1991). Macho aggression in the context of conflict between groups of young males typically involves giving and taking offense easily, sexist heroics, and supporting a buddy in a fight, even if the buddy is in the wrong (Graham and Wells, in press; Hunt and Laidler, 2001).

Table 1
Probable Contributing Factors in Each Example of Barroom Aggression

	Brawling males—intergroup conflict	Skid row culture	Inappropriate drunken sexual overtures
Effects of alcohol			
Risk taking and not considering risks.	The respondent reported that he did not notice the large number of opponents (i.e., underestimated risks) due to alcohol.	Increased risk taking appeared to be a factor for the woman who was the target of most of the aggression in that she persisted in taunting the other women.	P1 was reported to be quite intoxicated so it seems likely that the social risk-taking of trying to pick up every woman he saw was partly caused by alcohol.
Focus on the present and increased emotional lability.	An emotional response appeared to be a factor especially once the respondent decided to fight.	It appeared that the women were totally focused on the situation without consideration of risks or injury.	P4 was very emotional and angry, possibly partly due to alcohol.
Impaired cognitive functioning.	Respondent said he would have noticed that they were outnumbered by opponents and would have pulled friend away if he had been sober.	Unknown whether problem solving would have been less aggressive without alcohol.	P1 seemed unaware that his overtures were generally not wanted by the women he approached, possibly due to poorer cognitive functioning and problem solving from alcohol.
Increased power concerns.	"Liquid courage" of friend noted by respondent.	Unknown.	P1's pestering women may have had an element of alcohol-increased power and dominance.
Effects of alcohol on more than one person.	No information about opponents, but according to the respondent if either he or his friend had been sober, the brawl would have been less likely.	Very likely a factor in the continuing escalation of aggression.	The effects of alcohol on both P1 and P4 seemed to be a factor in the initiation and escalation of the incident.
Characteristics of the drinker			
Male.	Groups of young males.	No.	Yes.
Young.		No.	Yes.
Heavy drinker and/or alcoholic.	The respondent's self-reported drinking pattern did not indicate alcohol abuse or problems.	Given the context, it seems likely that all were chronic heavy drinkers.	Usual drinking pattern of participants not known.

(continued)

Table 1 (Continued)

	Brawling males—intergroup conflict	Skid row culture	Inappropriate drunken sexual overtures
Characteristics of the drinker			
A pattern of drinking large amounts per occasion.	Respondent reported being quite intoxicated and that he drank more than 5 drinks per occasion fairly frequently.	The high level of drunkenness of the women (that was taken for granted by staff and other patrons) suggests a usual pattern of drinking to intoxication.	Unknown.
Environmental factors			
Provocation, threat, or triggers.	Racist remarks, "macho" posturing.	Revenge over drug money seemed to be the provocation.	Unknown. P1 was not reacting to provocation, and it was unclear what P1 did to provoke P4.
Standards and expectations for behavior.	That one group felt comfortable making racist remarks suggests a permissive environment.	Definitely an environment where aggression accepted as evident from the length of time before staff intervened and the fact that neither of the two very aggressive women were ejected from the bar.	Incident suggests that harassment of women was normative in the bar; even physical aggression may have been acceptable as demonstrated by the fact that no staff noticed or became involved when P4 was shouting angrily at P1.
Rewards and punishment for aggression.	No punishments were evident other than being beaten by the opponents, and the incident was probably rewarding, at least for the friend who stood up for the respondent.	Aggression was not punished and may have been rewarded in terms of status in this subculture.	It is possible that P4 felt chivalrous about protecting the women with him against the advances of P1 (and that therefore, his aggression was rewarding).
The role of third parties.	Friend was a third party who escalated incident; bar staff were third parties to the incident and escalated it by pushing the two groups out the door to fight rather than intervening in a preventive way.	Third parties increased and decreased aggression—the women who did the most severe aggressive acts came into the incident as a third party; on the other hand, the incident was stopped by a peaceful third party patron and a staff member.	Third parties increased and decreased aggression—P4 was an initial third party when he became angry with P1, apparently on behalf of the female patrons involved; the women became third parties in preventing P4 and P1 from fighting.
Type of people present in the drinking environment.	Groups of young males.	Skid row subculture known to be very violent.	Unknown.

The respondent described alcohol as a contributing factor both for himself and his friend ("liquid courage," jumping in without accurately assessing the odds). There may well have been effects of alcohol on the other group as well, in that the incident evolved over a period of time with lots of opportunity for a sober nonaggressive member of either group to encourage his friends to back away. The fact that this did not occur suggests that increased risk taking and other effects of alcohol may have been operating for both groups of men.

In terms of personal characteristics of those involved in aggression in macho milieus, this context may reflect at least partly the propensity to aggression of young males generally. However, a number of environmental factors also appeared to play a role, including rewards for fighting and the permissive poorly-controlled environment. The rewarding aspect was most apparent for the respondent's friend who intervened on behalf of the respondent in response to the racist remarks. The respondent reported that he was grateful to his friend, even though they lost the fight. Thus, the rewards for the friend would have included feeling that he had done the right thing as well as being seen by the respondent as having done the right thing. The environment also contributed to the incident by (*i*) the apparent tolerance of racist remarks by staff and others in that setting, and (*ii*) the bar staff allowing the incident to escalate by pushing the two groups outside.

Example 2. Skid row culture (observed incident) (ongoing study, Graham et al., 2000–2003). In a fairly run-down bar that could be classified as skid row, both by its rough location and the type of clientele, two female patrons (P1 and P2), in their mid-thirties, who appeared to know each other, were squaring off and exchanging angry words. Suddenly P1 slapped P2 hard across the face (twice). As P1 walked away, the two women yelled threats at each other and made slashing motions across the throat to emphasis the threat. Both appeared very intoxicated, slurring their words, and having trouble keeping their balance. A third woman (P3) who was about the same age stepped in and started shaking P2 by the shoulders. P3 then slapped P2's head and body and pushed her to the floor. P3 then walked around P2, screaming obscenities, and then kicked P2 hard in the head three times. At this point, a male staff member approached, sent P3 back to her seat, helped P2 up and talked to her for a while. P3 then started walking toward P2 and the male staff member with a steel-legged chair over her head. The staff member turned and told her to put it down and to go sit down, which she did. The staff member talked to P2 for a while and then went back to work. P2 sat down and started taunting P3 (they were seated about 10 feet apart). They began shouting at each other and, again, P3 approached P2 and shook her violently. Just as they both started fighting (both locked in struggle, grabbing each other with hands on their arms), a male patron came over to break it up, soon joined by the male staff member. The male staff member told P2 to leave. (The observers later learned that the cause of the incident was that P2 had spent all the money on drinking that P1, P2, and P3 had been planning to use for buying crack.)

This incident appears to be dominated by three main contributing factors, clearly aggressive persons who probably had chronic alcohol (and possibly) drug problems, a cultural context that is extremely permissive and accepting of aggression, and probably at least some acute effects of alcohol, at least on the target of the violence (P2), who engaged in the highly risky behavior of continuing to taunt the women who beat her. This incident reflects the highly aggressive climate of skid row bars described in previous research (Graham et al., 1980).

Example 3. Inappropriate drunken sexual overtures (Graham et al., 2000–2003).
A male patron (P1) tried to pick up every women who stood near him, walked by him, or was getting a drink at the bar—even if she was clearly with other people. Sometimes, P1 touched, kissed, or invaded the space of those females near him. Most of the women would politely push him away, say something so that he would go away, or just walked away from him. At about 1:15, P1 was standing at the bar when a female patron (P2) came up to get a drink, already showing signs of intoxication (glassy eyes, swaying slightly). P2 conversed with P1 at the bar for about 10 minutes, picked up her drinks and walked over to her friends. P2 gave one drink to her male friend (P4) and took her female friend (P3) over to meet P1 at the bar, while P4 remained at the edge of the dance floor watching. P4 was protective of both P2 and P3, but did not seem to be concerned about their conversation with P1 at this time. At about 1:30, however, P4 was observed being very angry, moving about aggressively (he paced around with fists clenched and walked fast), and talking loudly. P2 and P3 talked to P4 and tried to calm him down by both holding on to his arms. Later, P4 disappeared and could be heard yelling from the area near the men's restroom. The observers followed the sound immediately, because it was so repetitive and loud. P4 was yelling "What," "What," "What" very loudly, repeating it to P1 over and over. P4 was aggressively invading P1's space by talking down to him in a humiliating way. The observers could not hear the reaction of P1, but he did not seem afraid or seem to care. P2 and P3 tried to calm P4 down by telling him to leave P1 alone. When P1 entered the restroom, P4 wanted to follow, but was held back by P2 and P3. P4 was saying "He's dead. I'm going to kill him." P1 came out of the restroom about 5–10 min later and walked to the other side of the bar. For the rest of the night, P1 remained on the opposite side of the bar from P4, where he continued making sexual overtures toward women and bothering other patrons generally. Although his targets continued to push him, no further aggression took place.

One social function of some licensed premises is meeting sexual or romantic partners. In this environment, unwanted overtures occur frequently (de Crespigny et al., 1998); however, our ongoing observations (Graham et al., 2000–2003) suggest that persistent and/or indiscriminate unwanted overtures, such as those made by P1 in this incident, tend to be done by highly intoxicated individuals, possibly due to intoxication affecting the person's judgement and awareness regarding the feelings of others (Giancola, 2000; Pernanen, 1976). Aggression in response to these behaviors is usually mild (angry words or gestures by the female target), but sometimes aggression becomes more serious, either by the target or, in this case, by a protective male third party.

DEVELOPMENT OF POLICES AND INTERVENTIONS FOCUSED ON REDUCING ALCOHOL-RELATED VIOLENCE: THE EXAMPLE OF THE *SAFER BARS* PROGRAM

Alcohol, the characteristics of the drinker, and the drinking context all contribute to alcohol-related aggression; therefore, interventions to prevent alcohol-related aggression can focus on any or all of this triad, as well as on broader cultural aspects of drinking (Pernanen, 1998). Most often, prevention strategies have focused on alcohol and have not been specific to aggression. For example, a Swedish policy forcing liquor stores to close on Saturdays was associated with a decrease in the overall rate of domestic violence (Olsson and Wikström, 1982). This finding suggests that policy measures focused on reducing alcohol consumption or on reducing consumption in certain settings or circumstances can reduce violence.

Prevention interventions can also focus on people who become violent when they drink. For example, treatment focused on alcohol problems (but not specifically on violence) has been shown to reduce domestic violence by males who have alcohol problems, and this reduction in violence is correlated with reduced drinking or abstinence (*see* review by O'Farrell and Murphy, 2002). Thus far, a reduction in violence among alcohol abusers has been either a side effect of successful treatment for alcohol problems (in which violence was not addressed directly) or a result of Behavioral Couples Therapy for men who have alcohol problems in which marital violence was addressed as a component of the alcohol treatment program (O'Farrell et al., 1999). In recent years, however, increasing attention has been paid to developing specialized treatments that address both alcohol problems and violent behavior, although evidence of the effectiveness of this strategy is not yet available (Collins et al., 1997).

As yet, preventing alcohol-related violence has received little attention in policy and prevention programming. For example, although cross-cultural variability in the link between alcohol and violence is well known (MacAndrew and Edgerton, 1969; Rossow, 2001), strategies for changing drinking patterns of high risk cultures in order to reduce the incidence of alcohol-related aggression have not been developed. Similarly, although considerable attention has been paid to the role of expectancies in the link between alcohol and aggression (*see* e.g., Graham et al., 1998; Lipsey et al., 1997), there has been no development of prevention efforts addressing this issue in order to change societal norms and expectations regarding the acceptability of intoxicated aggression.

The most promising target for prevention of alcohol-related violence to date has been the public drinking context. As noted previously in this chapter, licensed premises are high risk drinking contexts and, because they are licensed and public, these are contexts that are more easily targeted for prevention interventions than private drinking contexts (Stockwell, 1997). Prevention interventions directed toward licensed premises have generally focused on preventing intoxication by training alcohol servers in "responsible beverage service," increasing enforcement of alcohol serving laws and regulations, and passing legislation that holds alcohol servers responsible for injury and damage done by the persons to whom they have sold alcohol (*see* Graham, 2000). Although these interventions have focused on preventing intoxication and not violence, *per se*, some interventions have demonstrated effects on preventing violence and injury. For example, a community alcohol abuse–injury prevention project in Rhode Island, which combined serving policies and server training with enhanced police enforcement, resulted in more alcohol-related arrests and fewer emergency room visits for injury related to assault in the intervention site, with no change in the comparison sites (Putnam et al., 1993). In terms of other alcohol-focused interventions directed toward licensed premises, Sloan et al. (1994) found that states, which held servers of alcohol legally liable for actions done by someone who has been served to intoxication, had lower homicide rates compared with states that did not have this legislation.

More recently, several intervention projects in Australia have focused specifically on reducing aggression in and around licensed premises. These projects have successfully demonstrated reduced violence using a community mobilization approach involving community committees, coordinated policies and monitoring, a nightclub code of practice, risk assessment policy checklist, training of bar staff in managing problem behavior, and enhanced police enforcement (Hauritz et al., 1998; Homel et al., 1997).

Safer Bars: *An Example of Research-Based Prevention Programming*

The *Safer Bars* program builds on the work of Homel and his colleagues, but instead of being community focused, the *Safer Bars* program was developed to provide a stand-alone standardized systematic approach to preventing and better managing aggression. It consists of two main interventions, a structured risk assessment for bar owners and managers to identify and change aspects of the bar environment that may be contributing to increased risk of aggression, and a 3-hour training program for bar staff and managers regarding preventing the escalation of aggression or problem behavior. The program also includes a pamphlet outlining the legal obligations of bar owners and staff to prevent aggression. As described in the following, the content of the *Safer Bars* program was drawn directly from the published research literature, as well as descriptive studies of aggression in bars undertaken as part of developing the *Safer Bars* program (Graham and Wells, in press, 2001c; Graham et al., 2000; Wells and Graham, 1999; Wells et al., 1998).

The Safer Bars *Risk Assessment Workbook*

Both the physical and social environment may create circumstances that increase the likelihood of aggression. For example, environmental irritants generally have been found to increase aggression (*see* Geen, 1990), and irritants, such as smoke, noise, and general discomfort, have been associated with more aggressive bar environments (*see* Graham and Homel, 1997). Crowding generally has been found to be associated with aggression (Anderson, 1982), possibly because crowding increases arousal and involves frustration. Similarly, crowding in bars (Homel and Clark, 1994), especially congestion in specific areas (MacIntyre and Homel, 1997), has also been found to be related to aggression. Frustration has long been thought to be associated with aggression (*see* Geen, 1990) especially if this frustration is seen as arbitrary or unfair; moreover, intoxication has been found to increase the effects of frustration on aggression (Ito et al., 1996). Therefore, frustration in bars may be a factor in aggressive behavior. Some of the more common sources of frustration in bars are line-ups, games, and delays or refusal of service. Other general environmental risks identified in observational studies include the presence of other intoxicated persons, the characteristics of patrons who frequent the bar, pool playing, patrons milling about, and bored patrons (*see* Graham and Homel, 1997).

The general social environment that sets the standards for acceptable behavior is possibly the most important environmental factor in the amount and severity of aggression in bars (*see* Graham and Homel, 1997). Permissive drinking environments have been associated with higher levels of aggression across cultures (MacAndrew and Edgerton, 1969) and bars (Graham et al., 1980; Homel and Clark, 1994). In addition, certain aspects of drinking environments, such as the presence of violent symbols (*see* Berkowitz, 1993), formality of drinking setting (Radlow, 1995), care and maintenance of the bar (Graham et al., 1980), physical characteristics of staff (Lawrence and Leather, 1999), and provision of nonaggressive norms (Jeavons and Taylor, 1985) have been found to influence expectations regarding aggression.

Using findings described above and in the first part of this chapter, the risk assessment workbook (Graham, 1999) was developed to help the bar owner or manager evaluate and address potential environmental risk factors that increase the likelihood of aggression without reducing the patronage and profitability of the business. The workbook provides questions for the owner and/or manager to rate the bar on 92 environmental risk factors (e.g., drinking by staff, bottlenecks where crowding occurs, high risk activities such as "slam" dancing), as well as five areas related to hiring, training, and supervising staff. Explanations of the nature of risks for each factor are also provided. Finally, the workbook includes a section for the owners to write down specific plans for change relating to each risk area.

Safer Bars *Training*

Social control, including the behavior of third parties (Taylor and Gammon, 1976), is a major factor in alcohol-related aggression (Homel and Clark, 1994). Therefore, the importance of staff behavior in providing effective social control in bars is discussed in several sections of the risk assessment workbook and is the main focus of the *Safer Bars* training. Observational studies of naturally occurring aggression in bars have identified both effective and ineffective strategies for handling problem behavior shown by bar staff. For example, Gibbs' (1986) description of maintaining order around pool tables identified methods of control such as keeping observers out of disputes. Wells et al. (1998) identified a number of effective strategies, such as allowing patrons to save face, asking sober friends to help deal with an intoxicated person, and so on. Wells et al. (1998) also identified ineffective strategies, such as being inconsistent or unfair, not dealing firmly with patrons involved in incidents, using excessive force, and allowing situations to escalate before intervening. In their research on bars, Homel et al. (1997) noted that establishing a cooperative rather than competitive environment has also been found to help prevent aggression.

The *Safer Bars* training protocol (Braun et al., 2000) uses (*i*) empirical data on typical problems encountered in bars, and effective and ineffective behaviors of bar staff noted above; (*ii*) theoretical and empirical research from social psychology on factors that affect aggressive behavior in general (e.g., personal space, body language) (Sears et al., 1991); and (*iii*) techniques developed for police officers and others who work with violent individuals (Albrecht and Morrison, 1992; Breakwell, 1997; Coggans and McKellar, 1995; Garner, 1998; Goldstein, 1983; Nelson, 1994). The training covers the following six broad areas related to preventing aggression and managing problem behavior in barroom contexts: (*i*) understanding how aggression escalates; (*ii*) assessing the situation (including how alcohol intoxication contributes to the situation); (*iii*) keeping cool (i.e., not losing temper); (*iv*) body language (nonverbal techniques); (*v*) responding to problem situations; and (*vi*) legal issues. The training is limited to 25 participants, and the format is primarily group discussion, with overheads and a video to illustrate specific points. Some areas of the training use role play, and the legal section includes a test-yourself quiz. The training also uses a Participant Workbook (that the participants may keep), which reproduces the major points and examples and provides instructions for the role play exercises. The training was revised over several years in consultation with bar owners and staff, a lawyer, police, community health professionals, civic leaders, and liquor licensing officials. During its development, the training was pilot-tested in over 20 licensed premises (Chandler Coutts et al., 2000).

The *Safer Bars* program is currently being evaluated in a randomized control trial (Graham et al., 2000–2003). The evaluation of the impact of *Safer Bars* on actual aggression in bars has not yet been completed; however, the intervention has been completed, and some preliminary results from the training evaluation are presented in the following. In total, 522 staff from 23 bars were trained. The pre- and post-training questionnaires showed a significant improvement on 32 out of 33 of the pre- and post-knowledge–attitude items. All six sections of the training were rated over 8 on average on a scale from 1 (not at all useful) to 10 (extremely useful), with the highest ratings given to the legal section (9.0) and the section on body language (8.9). The different formats of the training (video, role playing, group discussion, and participant workbook) were rated higher than 7 on average, with group discussion rated highest at 8.9. Trainers were rated 4.7–4.8 on a scale of 1 (worst) to 5 (best) on knowledge, encouraging participation, listening, and organization. Finally, 98% of those who completed the feedback questionnaire (89% of all participants) said that they would recommend the training to others.

CONCLUSIONS

Knowledge regarding the relationship between alcohol and aggression is based on the convergence of methods from a variety of research fields and methods. Increasingly, links are being made between the pharmacological effects of alcohol and the actual behavior of persons who are under the influence of alcohol in both experimental and naturalistic contexts. It is also apparent from anthropological, experimental, and observational research that the links between alcohol and aggression are moderated by both the characteristics of the drinker, as well as the social context within which alcohol is consumed.

In this chapter, examples of typical real-life incidents of aggression in the social drinking context of the bar were used to illustrate the potential for various factors to combine in the occurrence of aggression. The examples also demonstrated the various options for prevention. For example, the incident involving groups of young males might have been prevented if the respondent or his friend were less intoxicated and better able to evaluate the risks, if the staff had separated the groups and kept one group inside until the other group dispersed rather than pushing both groups outside, or if the young males involved had stronger taboos regarding physical aggression.

To date, effective prevention strategies are still rare, although successful approaches to preventing alcohol-related aggression have been demonstrated using alcohol policy, alcohol treatment, and interventions directed toward licensed drinking environments. One specific intervention directed toward preventing alcohol-related violence, the *Safer Bars* program, was described to illustrate how prevention programming can be developed from theories and observations regarding alcohol-related aggression and aggression generally. While directed toward the drinking environment, this program also uses knowledge on the effects of alcohol and the characteristics of aggressive drinkers as part of developing an overall comprehensive approach to prevention.

ACKNOWLEDGMENTS

Thanks to Paul Tremblay and Elaine Zibrowski for helpful comments on an earlier draft of this chapter and to Sue Steinback for word processing. Preparation of parts of this chapter was supported by a grant from the National Institute on Alcohol Abuse and Alcoholism (R01AA11505).

REFERENCES

Adalbjarnardottir, S. and Rafnsson, F. D. (2002) Adolescent antisocial behavior and substance use. Longitudinal analyses. *Addict. Behav.* **27**, 227–240.

Albrecht, S. and Morrison, J. (1992) *Contact and Cover: Two-Officer Suspect Control.* Charles Thomas Publisher, Springfield, IL.

Archer, J. (2000) Sex differences in aggression between heterosexual partners: a meta-analytic review. *Psychol. Bull.* **126**, 651–680.

Anderson, A. C. (1982) Environmental factors and aggressive behavior. *J. Clin. Psychol.* **43**, 280–283.

Beck, A., Gilliard, D., Greenfield, L., Harlow, C., Hester, T., Jankowski, L., Snell, T., Stephan, J., and Morton, D. (1993) *Survey of state prison inmates, 1991.* (Bureau of Justice Statistics, Special Report, March 1993, NCJ-136949), U.S. Department of Justice, Washington, DC.

Berkowitz, L. (1993) *Aggression: Its Causes, Consequences, and Control.* McGraw-Hill, New York.

Bettencourt, B. A. and Miller, N. (1996) Gender differences in aggression as a function of provocation: a meta-analysis. *Psychol. Bull.* **119**, 422–447.

Borges, G., Cherpitel, C. J., and Rosovsky, H. (1998) Male drinking and violence-related injury in the emergency room. *Addiction* **93**, 103–112.

Bradford, J. M., Greenberg, D. M., and Motayne, G. G. (1992) Substance abuse and criminal behavior. *Clin. Forensic Psychiatry* **15**, 605–622.

Breakwell, G. M. (1997) *Coping with Aggressive Behaviour.* The British Psychological Society, Leicester.

Brecklin, L. R. and Ullman, S. E. (2002) The roles of victim and offender alcohol use in sexual assaults: results from the national violence against women survey. *J. Studies Alcohol* **63**, 57–63.

Braun, K., Graham, K., Bois, C., Tessier, C., Hughes, S., and Prentice, L. (2000) *Safer Bars Trainer's Guide.* Centre for Addiction and Mental Health, Toronto, Canada.

Burns, T. F. (1980) Getting rowdy with the boys. *J. Drug Issues* **10**, 273–286.

Bushman, B. J. (1997) Effects of alcohol on human aggression: validity of proposed mechanisms, in *Recent Developments in Alcoholism.* (Galanter, M., ed.), Plenum Press, New York, pp. 227–244.

Bushman, B. J. and Cooper, H. M. (1990) Effects of alcohol on human aggression: an integrative research review. *Psychol. Bull.* **107**, 341–354.

Chandler Coutts, M., Graham, K., Braun, K., and Wells, S. (2000) Results of a pilot program for training bar staff in preventing aggression. *J. Drug Educ.* **30**, 171–191.

Chermack, S. T. and Giancola, P. R. (1997) The relation between alcohol and aggression: an integrated biopsychosocial conceptualization. *Clin. Psychol. Rev.* **17**, 621–649.

Chilton, R. and Jarvis, J. (1999) Using the national incident-based reporting system (NIBRS) to test estimates of arrestee and offender characteristics. *J. Quant. Criminol.* **15**, 207–224.

Coggans, N. and McKellar, S. (1995) *The Facts About Alcohol, Aggression, and Adolescence.* Cassell, New York.

Collins, J. J., Kroutil, L. A., Roland, E. J., and Moore-Gurrera, M. (1997) Issues in the linkage of alcohol and domestic violence services, in *Recent Developments in Alcoholism.* (Galanter, M., ed.), Plenum Press, New York, pp. 387–405.

Collins, J. J. and Schlenger, W. E. (1988) Acute and chronic effects of alcohol use on violence. *J. Studies Alcohol* **49**, 516–521.

Cook, P. J. and Moore, M. J. (1993) Violence reduction through restrictions on alcohol availability. *Alcohol Health Res. World* **17**, 151–156.

Daly, M. and Wilson, M. (1998) *Homicide.* Aldine de Gruyter, New York.

Dawson, D. A. (1997) Alcohol, drugs, fighting and suicide attempt/ideation. *Addict. Res.* **5**, 451–472.

De Crespigny, C., Vincent, N., and Ask, A. (1998) *Young Women and Drinking. Vol. 1.* The Flinders University of South Australia School of Nursing, Adelaide, South Australia.

Eagly, A. H. and Steffen, V. J. (1986) Gender and aggressive behavior: a meta-analytic review of the social psychological literature. *Psychol. Bull.* **100,** 309–330.

Fagan, J. (1990) Intoxication and aggression, in *Drugs and Crime.* (Tonry, M. and Wilson, J. Q., eds.), University of Chicago Press, Chicago, pp. 241–320.

Felson, R. B., Baccaglini, W., and Gmelch, G. (1986) Bar-room brawls: aggression and violence in Irish and American bars, in *Violent Transactions. The Limits of Personality.* (Campbell, A. and Gibbs, J. J., eds.), Basil Blackwell, New York, pp. 153–166.

Fromme, K., Katz, E., and D'Amico, E. (1997) Effects of alcohol intoxication on the perceived consequences of risk taking. *Exp. Clin. Psychopharmacol.* **5,** 14–23.

Galanter, M. (Ed.). (1997) *Recent Developments in Alcoholism.* Plenum Press, New York.

Garner, G. W. (1998) *Surviving the Street: Officer Safety and Survival Techniques.* Charles Thomas Publisher, Springfield, IL.

Geen, R. G. (1990) *Human Aggression.* Open University Press, Milton Keynes, England.

Giancola, P. R. (2000) Executive functioning: a conceptual framework for alcohol-related aggression. *Exp. Clin. Psychopharmacol.* **8,** 576–597.

Giancola, P. R., Helton, E. L., Osborne, A. B., Terry, M. K., Fuss, A. M., and Westerfield, J. A. (2002) The effects of alcohol and provocation on aggressive behavior in men and women. *J. Studies Alcohol* **63,** 64–73.

Gibbs, J. J. (1986) Alcohol consumption, cognition and context: examining tavern violence, in *Violent Transactions. The Limits of Personality.* (Campbell, A. and Gibbs, J. J., eds.), Basil Blackwell, New York, pp. 133–151.

Goldstein, A. P. (1983) Behavior modification approaches to aggression prevention and control, in *Prevention and Control of Aggression.* (Goldstein, A. P. and Krasner, L., eds.), Pergamon Press, New York, pp. 156–209.

Graham, K. (1999) *Safer Bars. Assessing and Reducing Risks of Violence.* Centre for Addiction and Mental Health, Toronto, Ontario.

Graham, K. (2000) Preventive interventions for on-premise drinking: a promising but underresearched area of prevention. *Contemp. Drug Problems* **27,** 593–668.

Graham, K., Bois, C., Osgood, D. W., and Gliksman, L. (2000–2003) *Safer Bars: Evaluating an Intervention to Reduce Barroom Aggression.* (NIAAA Project Number: ROI-AA11505).

Graham, K. and Homel, R. (1997) Creating safer bars, in *Alcohol: Minimising the Harm.* (Plant, M., Single, E., and Stockwell, T., eds.), Free Association Press, London, pp. 171–192.

Graham, K., LaRocque, L., Yetman, R., Ross, T. J., and Guistra, E. (1980. Aggression and barroom environments. *J. Studies Alcohol* **41,** 277–292.

Graham, K., Leonard, K. E., Room, R., Wild, T. C., Pihl, R. O., Bois, C., and Single, E. (1998) Current directions in research on understanding and preventing intoxicated aggression. *Addiction* **93,** 659–676.

Graham, K., Schmidt, G., and Gillis, K. (1996) Circumstances when drinking leads to aggression: an overview of research findings. *Contemp. Drug Problems* **23,** 493–557.

Graham, K. and Wells, S. (2001a) "I'm okay. You're drunk!" Self-other differences in the perceived effects of alcohol in real-life incidents of aggression. *Contemp. Drug Problems* **28,** 441–462.

Graham, K. and Wells, S. (2001b) The two worlds of aggression for men and women. *Sex Roles. A Journal of Research* **45,** 595–622.

Graham, K. and Wells, S. (2001c) Aggression among young adults in the social context of the bar. *Addict. Res.* **9,** 193–219.

Graham, K. and Wells, S. "Somebody's Gonna Get Their Head Kicked in Tonight!" Aggression among young males in bars. *Br. J. Criminol.,* in press-a.

Graham, K., Wells, S., and Jelley, J. (2002) The social context of physical aggression among adults. *J. Interpersonal Violence* **17,** 64–83.

Graham, K., Wells, S., and West, P. (1997) A framework for applying explanations of alcohol-related aggression to naturally occurring aggressive behavior. *Contemp. Drug Problems* **24**, 625–666.

Graham, K. and West, P. (2001) Alcohol and crime: examining the link, in *International Handbook of Alcohol Dependence and Problems* (Heather, N., Peters, T. J., and Stockwell, T., eds.), John Wiley and Sons, Sussex, England, pp. 439–470.

Graham, K., West, P., and Wells, S. (2000) Evaluating theories of alcohol-related aggression using observations of young adults in bars. *Addiction* **95**, 847–863.

Graves, T. D., Graves, N. B., Semu, V. N., and Sam, I. A. (1981) The social context of drinking and violence in New Zealand's multi-ethnic pub settings. Research Monograph No. 7, in *Social Drinking Contexts*. (Harford, T. C. and Gaines, L. S., eds.), NIAAA, Rockville, MD, pp. 103–120.

Grupp, L. A. (1980) Biphasic action of ethanol on single units of the dorsal hippocampus and the relationship to the cortical EEG. *Psychophamracology* **70**, 95–103.

Greenfield, L. A. and Henneberg, M. A. (2000) Alcohol, crime and the criminal justice system. Commissioned paper for the Alcohol Policy XII Conference, Alcohol and Crime, Research and Practice for Prevention, Washington, DC.

Gustafson, R. (1993) What do experimental pparadigms tell us about alcohol-related aggressive responding? *J. Studies Alcohol* (**Suppl. 11**), 20–29.

Harford, T. and Muthén, B. O. (2000) Adolescent and young adult antisocial behavior and adult alcohol use disorders: a fourteen-year prospective follow-up in a national survey. *J. Studies Alcohol* **61**, 524–528.

Hauritz, M., Homel, R., McIlwain, G., Burrows, T., and Townsley, M. (1998) Reducing violence in licensed venues through community safety action projects: the Queensland experience. *Contemp. Drug Problems* **25**, 511–551.

Heather, N. (2001) Pleasures and pains of our favourite drug, in *International Handbook of Alcohol Dependence and Problems*. (Heather, N., Peters, T. J., and Stockwell, T., eds.), John Wiley and Sons, Sussex, England, pp. 5–14.

Heinz, A., Mann, K., Weinberger, D. R., and Goldman, D. (2001) Serotonergic dysfunction, negative mood states, and response to alcohol. *Alcohol. Clin. Exp. Res.* **25**, 487–495.

Higley, J. and Linnoila, M. (1997) A nonhuman primate model of excessive alcohol intake: personality and neurobiological parallels of Type 1 and Type II-like alcoholism, in *Recent Developments in Alcoholism*. (Galanter, M., ed.), Plenum Press, New York, pp. 192–226.

Higley, J. D. (2001) Individual differences in alcohol-induced aggression. A nonhuman primate model. *Alcohol Res. Health* **25**, 12–19.

Hoaken, P. N. S., Assaad, J. M., and Pihl, R. O. (1998) Cognitive functioning and the inhibition of alcohol-induced aggression. *J. Studies Alcohol* **September 59**, 599–607.

Hobbs, D., Lister, S., Hadfield, P., Winlow, S., and Hall, S. (2000) Receiving shadows: governance and liminality in the night-time economy. *Br. J. Sociol.* **51**, 701–717.

Homel, R. and Clark, J. (1994) The prediction and prevention of violence in pubs and clubs. *Crime Prevention Studies* **3**, 1–46.

Homel, R., Hauritz, M., Wortley, R., McIlwain, G., and Carvolth, R. (1997) Preventing alcohol-related crime through community action: The Surfers Paradise Safety Action Project. *Crime Prevention Studies* **7**, 35–90.

Homel, R., Tomsen, S., and Thommeny, J. (1992) Public drinking and violence: not just an alcohol problem. *J. Drug Issues* **22**, 679–697.

Hunt, G. P. and Laidler, K. J. (2001) Alcohol and violence in the lives of gang members. *Alcohol Res. Health* **25**, 66–71.

Ireland, C. S. and Thommeny, J. L. (1993) The crime cocktail: licensed premises, alcohol and street offences. *Drug Alcohol Rev.* **12**, 143–150.

Ito, T. A., Miller, N., and Pollock, V. E. (1996) Alcohol and aggression: a meta-analysis on the moderating effects of inhibitory cues, triggering events, and self-focused attention. *Psychol. Bull.* **120**, 60–82.

Jaffe, J. H., Babor, T. F., and Fishbein, D. H. (1988) Alcoholics, aggression and antisocial personality. *J. Studies Alcohol* **49,** 211–218.

Jeavons, C. M. and Taylor, S. P. (1985) The control of alcohol-related aggression: redirecting the inebriate's attention to socially appropriate conduct. *Aggress. Behav.* **11,** 93–101.

Kellermann, A. L. and Mercy, J. A. (1992) Men, women and murder: gender-specific differences in rates of fatal violence and victimization. *J. Trauma* **33,** 1–5.

Kyriacou, D. N., Anglin, D., Taliaferro, E., et al. (1999) Risk factors for injury to women from domestic violence. *N. Engl. J. Med.* **341,** 1892–1898.

Lawrence, C. and Leather, P. (1999) Stereotypical processing: the role of environmental context. *J. Environ. Psychol.* **19,** 383–395.

Lang, E., Stockwell, T., Rydon, P., and Lockwood, A. (1995) Drinking settings and problems of addiction. *Addict. Res.* **3,** 141–149.

Leonard, K. (2000) Domestic violence. What is known and what do we need to know to encourage environmental interventions? Commissioned paper for the Alcohol Policy XII Conference, Alcohol and Crime, Research and Practice for Prevention, Washington, DC.

Leonard, K. E. (1984) Alcohol consumption and escalatory aggression in intoxicated and sober dyads. *J. Studies Alcohol* **45,** 75–80.

Linnoila, V. M. I. and Virkkunen, M. (1992) Aggression, suicidality, and serotonin. *J. Clin. Psychol.* **53,** 46–51.

Lipsey, M. W., Wilson, D. B., Cohen, M. A., and Derzon, J. H. (1997) Is there a causal relationship between alcohol use and violence? in *Recent Developments in Alcoholism.* (Galanter, M., ed.), Plenum Press, New York, pp. 245–282.

MacAndrew, C. and Edgerton, R. B. (1969) *Drunken Comportment. A Social Explanation.* Aldine, Chicago.

MacIntyre, S. and Homel, R. (1997) Danger on the dance floor: a study of interior design, crowding and aggression in nightclubs, in *Policing for Prevention: Reducing Crime, Public Intoxication and Injury.* (Homel, R., ed.), Criminal Justice Press, Monsey, NY, pp. 91–113.

Martin, S. E. (1993) *Alcohol and Interpersonal Violence: Fostering Multidisciplinary Perspectives.* NIH, Rockville, MD.

Martin, S. E. and Bachman, R. (1997) The relationship of alcohol to injury in assault cases, in *Recent Developments in Alcoholism.* (Galanter, M., ed.), Plenum Press, New York, pp. 42–56.

McClelland, D. C., Davis, W. N., Kalin, R., and Wanner, E. (1972) *The Drinking Man. Alcohol and Human Motivation.* Collier-Macmillan Canada, Toronto, Canada.

Miczek, K. A., Weerts, E. M., and DeBold J. F. (1993) Alcohol, benzodiaepine-GABA receptor complex and aggression: ethological analysis of individual differences in rodents and primates." *J. Studies Alcohol* **Suppl. 11,** 170–179.

Miczek, K. A., DeBold, J. F., van Erp, A. M. M., and Tornatsky, W. (1997) Alcohol, GABAa-benzodiazepine receptor complex, and aggression, in *Recent Developments in Alcoholism.* (Galanter, M., ed.), Plenum Press, New York, pp. 139–172.

Miller, M. M. and Potter-Efron, R. T. (1989) Aggression and violence associated with substance abuse. *J. Chem. Depend. Treat.* **3,** 1–36.

Moeller, F. G. and Dougherty, D. M. (2001) Antisocial personality disorder, alcohol and aggression. *Alcohol Res. Health* **25,** 5–11.

Murdoch, D., Pihl, R. O., and Ross, D. (1990) Alcohol and crimes of violence: present issues. *Int. J. Addict.* **25,** 1065–1081.

Neff, J. A., Prihoda, T. J., and Hoppe, S. K. (1991) "Machismo," self-esteem, education and high maximum drinking among Anglo, Black and Mexican-American male drinkers. *J. Studies Alcohol* **52,** 458–463.

Nelson, J. M. (1994) *Teaching Self-Defense: Steps to Success.* Human Kinetics Publishers, Champaign, IL.

Norström, T. (1998) Effects on criminal violence of different beverage types and private and public drinking. *Addiction* **93,** 689–700.

O'Farrell, T. J. and Murphy, C. M. (2002) Behavioral couples therapy for alcoholism and drug abuse: encountering the problem of domestic violence, in *The Violence and Addiction Equation: Theoretical and Clinical Issues in Substance Abuse and Relationship Violence.* (Wekerle, C. and Wall, A.-M., eds.), Brunner-Routledge, New York, pp. 293–303.

O'Farrell, T. J., van Hutton, V., and Murphy, C. M. (1999) Domestic violence before and after alcoholism treatment: a two-year longitudinal study. *J. Studies Alcohol* **60**, 317–321.

Olsson, O. and Wikström, P. H. (1982) Effects of the experimental Saturday closing of liquor retail stores in Sweden. *Contemp. Drug Problems* **11**, 325–353.

Parker, R. (1993) The effects of context on alcohol and violence. *Alcohol Health Res. World* **17**, 117–122.

Parker, R. N. and Rehbun, L.-A. (1995) *Alcohol and Homicide. A Deadly Combination of Two American Traditions.* State University of New York Press, Albany, NY.

Pernanen, K. (1976) Alcohol and crimes of violence, in *The Biology of Alcoholism.* (Kissin, B. and Begleiter, H., eds.), Plenum Press, New York, pp. 351–444.

Pernanen, K. (1991) *Alcohol in Human Violence.* The Guilford Press, New York.

Pernanen, K. (1998) Prevention of alcohol-related violence. *Contemp. Drug Problems* **25**, 477–509.

Peterson, J. B., Rothfleisch, J., Zelazo, P. D., and Pihl, R. O. (1990) Acute alcohol intoxication and cognitive functioning. *J. Studies Alcohol* **51**, 114–122.

Pihl, R. O., Peterson, J. B., and Lau, M. A. (1993) A biosocial model of the alcohol-aggression relationship. *J. Studies Alcohol* **11**, 128–139.

Pihl, R. O. and Ross, D. (1987) Research on alcohol related aggression: a review and implications for understanding aggression. *Drugs Society* **1**, 105–126.

Pihl, R. O. and Peterson, J. (1993) Alcohol and aggression: three potential mechanisms of the drug effect, in *Alcohol and Interpersonal Violence: Fostering Multidisciplinary Perspectives. (Research Monograph No.24).* (Martin, S. E., ed.), NIH, Rockville, MD, pp. 149–159.

Pihl, R. O., Zeichner, A., Niaura, R., Nagy, K., and Zacchia, C. (1981) Attribution and alcohol-mediated aggression. *J. Abnorm. Psychol.* **5**, 468–475.

Pliner, P. and Cappell, H. (1974) Modification of affective consequences of alcohol: a comparison of social and solitary drinking. *J. Abnorm. Psychol.* **83**, 418–425.

Putnam, S. L., Rockett, I. R. H., and Campbell, M. K. (1993) Methodological issues in community-based alcohol-related injury prevention projects: attribution of program effects, in *Experiences with Community Action Projects: New Research in the Prevention of Alcohol and Other Drug Problems.* (Greenfield, T. K. and Zimmerman, R., eds.), Center for Substance Abuse Prevention, Rockville, MD, pp. 31–39.

Radlow, R. (1995) Sociocultural and intrapsychic expectation in the evaluation of drinking comportment. *Exp. Clin. Psychopharmacol.* **3**, 294–297.

Roberts, L. J., Roberts, C. F., and Leonard, K. E. (1999) Alcohol, drugs and interpersonal violence, in *Handbook of Psychological Approaches with Violent Offenders: Contemporary Strategies and Issues.* (VanHasselt, V. B. and Hersen, M., eds.), Kluwer Academic/Plenum Publishers, New York, pp. 493–519.

Room, R. and Rossow, I. (2000) The share of violence attributable to drinking. What do we need to know and what research is needed? Commissioned paper for the Alcohol Policy XII Conference, Alcohol and Crime, Research and Practice for Prevention, Washington, DC.

Rossow, I. (1996) Alcohol related violence: the impact of drinking pattern and drinking context. *Addiction* **91**, 1651–1661.

Rossow, I. (2001) Alcohol and homicide: a cross-cultural comparison of the relationship in 14 European countries. *Addiction* **96(Suppl. 1)**, S77–S92.

Sayette, M. A., Wilson, T., and Elias, M. J. (1993) Alcohol and aggression: a social information processing analysis. *J. Studies Alcohol* **54**, 399–407.

Schmutte, G., Leonard, K., and Taylor, S. (1979) Alcohol and expectations of attack. *Psychol. Rep.* **45**, 163–167.

Sears, D., Peplau, L. A., and Taylor, S. E. (1991) *Social Psychology, 7th ed.* Prentice Hall, New York.

Sharps, P. W., Campbell, J., Campbell, D., Gary, F., and Webster, D. (2001) The role of alcohol use in intimate partner femicide. *Am. J. Addict.* **10,** 122–135.

Sloan, F. A., Reilly, B. A., and Schenzler, C. (1994) Effects of prices, civil and criminal sanctions and law enforcement on alcohol-related mortality. *J. Studies Alcohol* **55,** 454–465.

Steele, C. M. and Josephs, R. A. (1990) Alcohol myopia: its prized and dangerous effects. *Am. Psychologist* **45,** 921–933.

Stockwell, T. (1997) Regulation of the licensed drinking environment: a major opportunity for crime prevention, in *Policing for Prevention: Reducing Crime, Public Intoxication and Injury.* (Homel, R., ed.), Criminal Justice Press. Monsey, New York, pp. 7–33.

Stockwell, T., Lang, E., and Rydon, P. (1993) High risk drinking settings: the association of serving and promotional practices with harmful drinking. *Addiction* **88,** 1519–1526.

Stockwell, T., Somerford, P., and Lang, E. (1991) The measurement of harmful outcomes following drinking on licensed premises. *Drug Alcohol Rev.* **10,** 99–106.

Stritzke, W. G., Patrick, C. J., and Lang, A. R. (1995) Alcohol and human emotion: a multidimensional analysis incorporating startle-probe methodology. *J. Abnorm. Psychol.* **104,** 114–122.

Taylor, S. and Gammon, C. (1976) Aggressive behavior of intoxicated subjects: the effect of third-party intervention. *J. Studies Alcohol* **37,** 917–930.

Tomsen, S. (1997) A top night out—social protest, masculinity and the culture of drinking violence. *Br. J. Criminol.* **37,** 990–1002.

Tuck, M. (1989) *Disorder in the Paired Towns. Drinking and Disorder: A Study of Non-Metropolitan Violence.* Her Majesty's Stationery Office, London, pp. 11–103.

Virkkunen, M. and Linnoila, M. (1993) Brain serotonin, Type II alcoholism and impulsive violence. *J. Studies Alcohol* **11,** 163–169.

Virkkunen, M. and Linnoila, M. (1997) Serotonin in early-onset alcoholism, in *Recent Developments in Alcoholism.* (Galanter, M., ed.), Plenum Press, New York, pp. 173–191.

Washburne, C. (1956) Alcohol, self, and the group. *Q. J. Studies Alcohol* **17,** 108–123.

Wells, S. and Graham, K. (1999) The frequency of third-party involvement in incidents of barroom aggression. *Contemp. Drug Problems* **26,** 457–480.

Wells, S. and Graham, K. (2003) Aggression involving alcohol: relationship to drinking patterns and social context. *Addiction* **98,** 33–42.

Wells, S., Graham, K., and West, P. (1998) "The good, the bad, and the ugly": responses by security staff to aggressive incidents in public drinking settings. *J. Drug Issues* **28,** 817–836.

Wells, S., Graham, K., and West, P. (2000) Alcohol-related aggression in the general population. *J. Studies Alcohol* **61,** 626–632.

White, H. R., Hansell, S., and Brick, J. (1993) Alcohol use and aggression among youth. *Alcohol Health Res. World* **2,** 144–150.

Wiley, J. A. and Weisner, C. (1995) Drinking in violent and nonviolent events leading to arrest: evidence from a survey of arrestees. *J. Criminal Justice* **23,** 461–476.

Wilsnack, R. W., Vogeltanz, N. D., Wilsnack, S. C., and Harris, T. R. (2000) Gender differences in alcohol consumption and adverse drinking consequences: cross-cultural patterns. *Addiction* **95,** 251–265.

Yudko, E., Blanchard, D. C., Henrie, J. A., and Blanchard, R. J. (1997) Emerging themes in preclinical research on alcohol and aggression, in *Recent Developments in Alcoholism.* (Galanter, M., ed.), Plenum Press, New York, pp. 124–138.

Zeichner, A. and Pihl, R. (1979) Effects of alcohol and behavior contingencies on human aggression. *J. Abnorm. Psychol.* **88,** 153–160.

Zeichner, A. and Pihl, R. (1980) Effects of alcohol and instigator intent on human aggression. *J. Studies Alcohol* 41, 265–276.

15
Cognitive-Behavioral Intervention for Childhood Aggressions

Cynthia Hudley

INTRODUCTION

Children who display high levels of overt aggression represent a significant problem in schools, families, and communities. These aggressive children are typically disruptive in school settings, often rejected by their peers, and frequently referred by parents or teachers for mental health services (Kupersmidt and Coie, 1990; Laird et al., 2001). In addition to creating unpleasant social environments for classmates and family members, these children are themselves at substantial risk for a variety of negative social outcomes, including conduct disorder, substance abuse, school failure, and involvement in criminal activity (Loeber, 1990).

The stability of this behavior over time is perhaps the most troubling feature of excessive aggression in childhood (Olweus, 1979). We know that the persistently delinquent adolescent engaged in crimes against persons may well have been the highly aggressive 6-yr-old (Derzon, 2001). Longitudinal data have described a linear pattern of behavioral development that may lead from high rates of problem behaviors in early years (e.g., pushing and hitting in kindergarten) to interpersonal aggression and violence in adolescence and adulthood (e.g., assault with an object) (Patterson, 1992). Overall, a high level of overt childhood aggression that is evident across multiple settings may represent a developmental pathway leading to antisocial and criminal behavior in adolescence and adulthood. Thus, high levels of childhood aggression can result in extremely damaging consequences for perpetrators as well as their victims, families, communities, and the larger society.

Research has uncovered a great deal about the developmental trajectory of aggression, as well as the prevention and reduction of aggressive behavior (Stoff et al., 1997). This chapter will discuss the role of cognitive-behavioral treatment as a strategy to forestall or reduce the development and display of aggressive behavior in children. Preliminary comments on children's social cognitions and a description of social-cognitive theories of childhood aggression will help place this treatment in proper context.

COGNITIVE-BEHAVIORAL PERSPECTIVES ON AGGRESSION

From the cognitive-behavioral perspective, aggression is the behavioral expression of internal processing of external social information. Although contextual influences have been well established as powerful regulators of behavior (Fiske and Taylor, 1991), the likelihood that a given interpersonal event will result in a display of aggressive behavior also depends to a great extent on the child's cognitive processes. Cognitions are one important link between environmental events and behavioral outcomes. The relationship between an antecedent social event and a child's aggressive behavior is mediated through that child's expectations, appraisals, goals, strategy and outcome beliefs, and other cognitive processing mechanisms.

Consider the typical elementary school playground, where children routinely spend time waiting in line to eat, to take a turn in a game, and to engage in any number of other activities. Now imagine that a child is bumped hard from behind by a peer while standing in line. Cognitive processes, including beliefs about the causes of the push and appraisals of response strategies may guide the child to one of several responses that might range from no response at all to throwing a punch at the peer's head. Another child running toward the line will decide whether to push other children aside to claim the place at the front based similarly on cognitive processes, such as beliefs about the outcomes of the behavior and social goals the child wishes to pursue. In short, from the cognitive-behavioral perspective, children's aggression is, in part, produced by the cognitive processing of social cues in the environment.

The cognitive-behavioral perspective has generated a number of theories to describe the development and display of aggressive behavior (Geen and Donnerstein, 1998). These theories may differ in the variables that they posit as central to the process. However, there is general agreement that the personal construction of social events can lead to displays of aggressive behavior, and the interactions between the child and the environment guide the development of aggression. Two key models that are representative of this field provide examples of the focal variables that are generating interesting and important research in childhood aggression.

Social Reasoning

Perhaps the most generative model of children's aggression in the child development literature has been formulated by Dodge and colleagues (Dodge, 1986; Dodge and Crick, 1990). Dodge's model foregrounds the child's information processing skills and abilities and the specific tasks inherent in "on-line" cognition. The model presumes that a child brings a unique set of biological characteristics and memories of past experiences to any given situation; as well, the situation itself provides a set of social cues. The child's response in any situation is the result of processing specific social cues given the constraints of biology and memory.

In its current formulation (Crick and Dodge, 1994), the model posits six discrete tasks that operate in an iterative nonlinear fashion. The first task is encoding. The child is hypothesized to selectively attend to both situational and internal cues according to both the memory store of social experiences and the immediate influence of mood states and biological capacities. Consider teasing on the playground. A child may pay more or less attention to possible onlookers, adults in the area, or their own physical state (e.g., feelings of anger) according to past experiences with teasing, past experiences with the peer, and negative mood generated by a recent reprimand from the teacher.

The second task, interpretation of the social cues, may engage one or several specific processes. Initially, the child may use idiosyncratic social knowledge stored in memory to define or represent the social cues ("teasing means this guy will hit me"). The child may engage in inferences about the causes of events, including attributions of others' intentions. Children also evaluate their own and peers' behavior in prior interactions to help them interpret the current interaction. In our teasing example, the child being teased may attribute hostile intent and expect to succeed in a physical battle. The teaser, based on past experience and personal assessment of verbal skill, may be expecting a lively verbal duel. Thus, the peer's benign but perhaps inappropriate overture is interpreted as a personal threat warranting physical retaliation.

A closely related task is formulating and revising goals, which are defined as orientations toward particular outcomes. Social goals can be the product of multiple influences such as feelings, external sanctions, and peer and cultural norms (Crick and Dodge, 1994). In our teasing example, an appropriate goal for a popular boy might be maintaining social status, but anger produced by an earlier negative interaction might generate a revised goal of physical domination. Goals are key because they influence subsequent responses.

Children will also search for a response, the fourth task in the model. Some responses may be strategies for attaining the specific goal, while others may be more direct reactions to the immediate social stimuli. A clever verbal retort might be the preferred response to teasing in order to maintain social status among friends; a child who feels fear might prefer immediate withdrawal from the situation.

Ultimately, a child will select a response from among possible alternatives based not just on goals, but also on their expectations for the outcomes of the behavior ("what will happen if I run over to the teacher") as well as their evaluation of their efficacy to perform the response ("can I think up a smart comeback"). At this point, the model hypothesizes that the child will enact the selected behavior; the child and the peer will evaluate that response; and the particular interaction will either conclude or the child will begin again to encode social cues.

Social Learning

A model developed by Huesmann and colleagues (Huesmann, 1988; 1998; Huesmann and Eron, 1984) has had significant impact on research in media effects and observational learning of aggression. Huesmann's model foregrounds a child's memory structures and the significance of memory store as a guide to behavior. One key structure, the cognitive script, links together a sequence of expected events and actions. A person's cognitive script provides a map of what will probably happen in a given situation, how to behave appropriately in that situation, and what the outcome will be, given that behavior.

Scripts are built up and maintained in memory through observing models (in both the proximal environment and the media), regular rehearsal, and perceived reinforcement. The child's early learning and reinforcement experiences shape the development and encoding of scripts in memory; more aggressive children are presumed to have unique socialization experiences that result in a larger number of more aggressive scripts stored in memory. Aggressive children rehearse these scripts through fantasy, rumination, and selective attention to aggressive media (e.g., video games), and they perceive the responses of others as reinforcing to the behaviors enacted according to the script.

Normative beliefs, also key cognitive structures guiding aggression in the Huesmann model, are cognitions about what is right for that person. Normative beliefs regulate behavior by specifying what is allowed and what is prohibited in a particular situation.

Thus, they serve as a filter when scripts are being accessed for possible enactment. The mechanisms are self-regulating, in that they are grounded in intrapersonal standards and are most stable across situation and development when enforced by internal sanctions. The presumption is that aggressive children have normative beliefs that condone more aggression in addition to a greater store of aggressive scripts. Thus, on average, when searching the memory store for a script appropriate to a given situation, the aggressive child will favorably evaluate and enact more aggressive scripts than do less aggressive children. In addition, characteristics the child brings to the situation (e.g., angry mood), situational cues (e.g., the presence of a weapon), and the child's interpretation of situational cues (e.g., attributions of hostile intent) can influence the search and enactment of scripts.

COGNITIVE-BEHAVIORAL INTERVENTION TO REDUCE AGGRESSION

Clearly, a variety of cognitive mediators may determine whether a child behaves aggressively. The goal of cognitive-behavioral treatment is to teach children to use cognitive mediators to guide their behavior away from aggressive options and toward nonaggressive responses. Using the tripartite taxonomy of preventive intervention developed by the Institute of Medicine (1994), cognitive-behavioral programs of aggression reduction can generally be categorized as indicated preventive interventions (but *see* Weissberg and Greenberg, 1998, for a discussion of conceptual ambiguity). That is to say, these programs specifically target youth whose current behavior indicates the need for an intervention to prevent subsequent severe behavior disorder. Elevated levels of aggression in middle childhood have long served as a useful screening marker that often prompts a child's inclusion into an appropriate preventive intervention program.

Cognitive-behavioral strategies to reduce children's aggression address specific cognitive deficiencies and distortions in children's social information processing. As such, they are child-focused rather than environmental (home, school community). Robust theory development, as previously described, has generated a rich literature on the social cognitive bases of aggression, which in turn has provided a sound foundation for the creation of preventive interventions. A review of this broad intervention literature is beyond the scope of this chapter (*see* Moeller, 2001; Peters and McMahon, 1996; Robinson et al., 1999, for recent reviews). We now turn to a specific example of a preventive intervention program targeting a theoretically central sequence of disrupted cognitive processing: the interpretation of and response to social cues.

ACCURATELY INTERPRETING THE SOCIAL LANDSCAPE: THE BRAINPOWER PROGRAM

The BrainPower Program, a preventive intervention curriculum for elementary school students, is based on principles of attribution theory. An attribution occurs when an individual assigns a cause to the behavior of others in a social interaction (Weiner, 1986). These causal attributions guide the selection of subsequent courses of action (Kelley, 1973). As such, attributions occupy a prominent position in theories of children's aggression, as described in the prior section. Basic research in developmental psychology has generated a great deal of knowledge about the attributional antecedents of reactive aggression; thus, principles of attribution theory have been especially helpful in designing programs to reduce childhood aggression.

Highly aggressive children often incorrectly attribute deliberately hostile intentions to peers (see Hudley, 1994a for a review), a phenomenon identified in the literature as a hostile attributional bias (Dodge, 1980; Nasby et al., 1980). For example, if asked to imagine being bumped by a peer while walking in the hallway at school, in the absence of any information regarding the cause of the bump excessively aggressive children express certainty that the bump was "on purpose" more than twice as often as a less aggressive peer (Graham et al., 1992; Hudley and Graham, 1993). In ambiguous situations such as this example, overtly aggressive boys in particular make social decisions quickly, ignore available social cues, and endorse retaliatory aggression at higher levels than their less aggressive peers (Crick and Dodge, 1994). As delineated by social cognitive theory, these children attend selectively to social cues, make biased interpretations of the available cues, and act on beliefs that aggressive responding is the appropriate course of action. Thus, excessively aggressive boys may feel quite justified in their choice of behavior based on hastily made biased attributions about the intent of others.

The BrainPower Program (Hudley, 1994b; 2001a; Hudley et al., 1998; Hudley and Friday, 1996; Hudley and Graham, 1993) seeks to modify this dysfunctional pattern of attributional bias. The program model, based on the formulations of Weiner (1986), focuses on volitional control of behavior. If one person in a social interaction believes that the cause of a negative outcome is controllable by the other party (e.g., a bump in the hallway was a volitional act by the peer), anger results, and one possible behavioral response is aggression. In contrast, if the negative outcome is seen as the result of uncontrollable causes (the peer was pushed into you because the hallway was crowded), neither anger nor retaliatory aggression is typically aimed at the peer.

The model generated two fundamental assumptions that guided the curriculum design: aggressive children can learn to recognize accidental causation, and when negative outcomes are attributed to accidental, i.e., nonhostile causes, angry aggressive responses will become unlikely. Using materials and activities appropriate for grades 3–6, the program teaches aggressive children to recognize that sometimes negative social outcomes with peers may be caused by accidental rather than intentionally hostile causes. This attribution retraining is accomplished by demonstrating to student participants that events are not always under the volitional control of themselves or their peers.

Program Characteristics

The 12 lessons in the curriculum cluster into three distinct training components. The first component strengthens aggressive participants' ability to accurately detect others' intentions. A variety of instructional activities train participants to search for, interpret, and properly categorize the verbal, physical, and behavioral cues exhibited by others in social situations. Students are trained to distinguish accidental, helpful, and hostile social cues. For example, one lesson has students act as observers (Intention Detectives) on the playground for 2 days to collect and categorize evidence about the intentions of others. They come back to the group to describe and defend their decisions.

After the participants gain some skills in the interpretation and categorization of social cues, the second component trains participants to first attribute ambiguous negative outcomes to accidental causes. Students are taught to associate inconsistent or uninterpretable social cues with attributions to uncontrollable or accidental causes. For example, students role play ambiguous scenes (e.g., a peer spills milk on your pants in the lunchroom), brainstorm possible causes for the actions, categorizes those causes as deliberate or unintentional, and then decides which causes are more likely given uncertainty about the peer's intent. The group leader guides the discussion to the benefits of attributions to accidental causation and nonaggressive responses (e.g., less chance of getting into trouble with parents or teachers).

The third component links appropriate, nonaggressive behavioral responses to negative social outcomes. Participants make the connection between peer behavior, unbiased thinking, and the need to search for and positively evaluate less verbally and physically aggressive behavioral responses. For example, students generate decision rules about when to enact particular responses (e.g., "When I can't really tell why he did that, I should first act as if it were an accident"). Such decision rules enhance maintenance and generalization of newly acquired attribution patterns by training children to develop patterns of response search and evaluation that are consistent with more benign attributions. As well, this component prompts children to develop decision rules appropriate for potentially dangerous situations (e.g., "If someone threatens me, I should find an adult right away") that minimize the possibility of lethal confrontations.

Participants typically receive BrainPower lessons twice weekly in a small group setting with two trained group leaders. Groups of six students meet in 60-min sessions for a total of 12 sessions in 6 wk. Each group consists of four excessively aggressive and two average nonaggressive students. Average students are included to avoid stigmatizing aggressive participants, provide aggressive students with positive peer models, and allow these average students to reappraise their attitudes and behaviors directed toward the aggressive students as they progress through the program. Such interaction between aggressive students and their peers is considered critical to the generalization of program effects (Bierman, 1986). These opportunities are also useful in counteracting the biased perceptions of peers that are based on a child's reputation for aggression (Hymel, Wagner, and Butler, 1990).

Group leaders are trained to use two primary instructional strategies when presenting the content of the program—coaching and modeling. Coaching involves direct teaching of concepts complemented with discussions, in which each child provides examples as well as opposites of the presented concepts (Asher, 1985). Elements of modeling in the curriculum include role play and video demonstrations. These primary methodologies of coaching and modeling are implemented using multiple lesson formats, including cooperative projects, observation activities, paper–pencil exercises, and outside of class follow-up activities.

Program Effectiveness

The BrainPower Program has been evaluated for effectiveness in a three-step progression. The initial investigation (Hudley, 1994b; Hudley and Graham, 1993) was a rigorously controlled test that reliably linked changes in attributional patterns to changes in children's aggression. Male students, both aggressive and nonaggressive ($n = 108$) in grades 4–6 were randomly assigned to one of three groups: the attributional intervention, a placebo curriculum, or a no-treatment control group. The placebo condition was included to control for possible effects of simple participation in a special pull-out program. Data were collected on participants' attributional patterns, behavior as rated by teachers, and frequency of referral for formal disciplinary activity (e.g., suspension, detention). In addition, aggressive subjects' responses to an actual peer were examined in a cooperative insoluble puzzle task completed within 1 month after the intervention.

Attributions of hostile intention, reported anger, and endorsement of hostile behavior declined significantly only for boys who participated in the preventive intervention program. In the laboratory puzzle task, aggressive boys who had participated in the attributional intervention were significantly less likely to infer that the peer had intentionally caused them to fail than were the other two groups. As well, only aggressive boys in the intervention group were rated by their teachers as significantly less aggressive following the intervention, while changes in teacher ratings of aggression pre- and post-intervention were not significant for either the placebo curriculum or control group. The intervention had no effect on the overall pattern of formal disciplinary action; however, the data aggregated referrals for all types of school infractions (e.g., truancy).

Nonaggressive subjects' judgments of the ambiguous scenarios, teacher ratings of behavior, and patterns of office referrals did not change significantly from pre- to post-intervention as a function of intervention type. It is safe to say that the BrainPower program had no deleterious effects on nonaggressive participants. Taken together, these findings were promising. However, these students were not assessed beyond the period of post-treatment assessment, and this study did not evaluate the efficacy of the intervention in an ecologically valid context.

To address these limitations, the next study implemented the BrainPower Program in schools and examined behavior change over time (Hudley et. al, 1998; Hudley and Friday, 1996). Boys in grades 4–6 at four elementary schools (total $n = 384$) were again randomly assigned to one of three intervention conditions. Procedures were essentially as described in the previous study; however, intervention groups were conducted by two trained instructional aides with experience in small group instruction. As well, students were followed for 12 mo after the end of the intervention.

Again, judgments of intent, self-reported anger, and endorsement of aggressive behavior by students in the attributional intervention declined significantly and remained lower than placebo or control groups at 12-mo follow up. Although statistical analyses of teacher ratings of behavior did not differ significantly by group, an analysis of clinical significance (Jacobsen and Truax, 1991) revealed that significant improvement among individual students in the attributional intervention group occurred at a rate twice that of students in either comparison group from pre- to post-intervention assessment (43%, 21%, and 18% for intervention, placebo, and control conditions, respectively). By the 12-mo follow-up, half of the significantly improved students in the comparison groups had declined to former levels of behavior ratings, compared to only 28% of improved students in the attribution intervention group.

This study also assessed the relationship between attributional change and behavior change by correlating teacher and student measures taken at post-intervention assessment, as well as at 6- and 12-mo follow-ups. For aggressive students in the BrainPower Program, perceptions of intent and teacher ratings of behavior were strongly negatively related at postassessment ($r = -0.51, p > 0.01$), and at 6-mo follow up ($r = -0.28, p > 0.05$). As well, attributions measured at postassessment were related to teacher ratings of behavior at the 6-mo follow-up ($r = -0.24, p > 0.05$). In contrast, relationships between attributions and teacher ratings of behavior were largely nonsignificant for aggressive students in the other two control conditions. Again, no significant changes in attributions or behavior were evident for the nonaggressive students. Overall, these data confirmed that behavior change can be achieved and maintained with an attributional intervention conducted by school staff in school settings.

The final link in this chain of evaluation studies examined the intervention curriculum as one part of a comprehensive youth development program delivered in an after school setting (Hudley, 2001a). Given the overwhelming evidence that multiple interpersonal processes contribute to the display of peer-directed aggression (Greenberg et al., 2001), I reasoned that the most important test would assess the efficacy of this indicated preventive intervention in conjunction with a comprehensive intervention to support the healthy development of children, families, and communities. Initially isolating attributional change provided a direct test of the theoretical relationship between cognition and behavior and established the feasibility of implementing the program with paraprofessionals. Ultimately, however, the program will be most effective as one part of a more comprehensive approach to youth development.

Therefore, the BrainPower Program became one part of the ongoing activities of the 4-H After School Program, a community-based organization that provides services to elementary school-aged children who reside in public housing projects. Youth residing in public housing confront a host of challenges to their academic and social success, as well as to their very safety and survival. Social and economic isolation, high rates of violence, easy availability of drugs, early sexual experience, and minimal academic success all combine to pose a cumulative and highly toxic set of risk factors for youth living in public housing (Dunworth and Saiger, 1994; Hudley, 2001a). The 4-H programs offer a range of youth development activities (e.g., academic tutoring, recreation activities, social skills development, community service projects, and field trips to cultural and recreational events) that are designed to buffer youth against the multiple pervasive risk factors present in their environments.

This research examined whether a universal preventive intervention that incorporated the BrainPower aggression reduction curriculum with boys and girls in an after school setting would forestall the development of high levels of aggression and support perceptions of competence. All 4-H children, both boys and girls, who participated in at least 75% of the BrainPower activities and were present for all data collection activities at the 6-mo and 12-mo follow-ups ($n = 46$), and became the intervention group. Comparison students ($n = 41$), matched on ethnicity, gender, grade, and school cooperation grades, were children living in the same public housing complexes and attending the same elementary schools, but not the afterschool program. The sample was 93% African-American, with mean ages of 9.30 for the 4-H group and 9.28 for the comparison group. Data were collected on participants' attributional patterns, beliefs about aggression, and aggressive behavior as perceived by both teachers and parents. In addition, both students and teachers rated the students' levels of academic competence.

Findings revealed that both boys and girls in 4-H declined steadily in inappropriate perceptions of hostile intent. While girls' scores for the comparison group declined to levels comparable to 4-H boys and girls, comparison boys' scores increased and remained higher than 4-H students. Attitudes endorsing aggression declined for boys in the 4-H program, while comparison boys' attitudes stayed steady at levels higher than the 4-H boys. For girls in the 4-H group, attitudes increased slightly, while comparison girls increased their endorsement of aggression to rates higher than that for all boys. Self-perceptions of academic competence increased overall for 4-H students, but only 4-H girls maintained high levels of self-perceived academic competence by the 12-mo follow-up.

Teacher ratings of behavior for both boys and girls in 4-H improved over time. Comparison girls also improved, while comparison boys' scores declined consistently. Parents' ratings of their children's behavior also improved for 4-H students, but not for the comparison group children. Teacher ratings of academic competence did not differ significantly by group, time, or gender.

Relationships among cognitive and behavioral variables were again strongest for intervention group students. Teacher ratings of behavioral self-control were concurrently related to 4-H students ratings of perceived intent at the 6-mo follow-up ($r = 0.72, p < 0.01$); these measures were unrelated for the comparison group ($r = 0.14$, ns). As well, parent measures of cooperation at the 6-mo follow-up were concurrently related to student attitudes about aggression ($r = 0.64, p < 0.01$) for 4-H students, but not for the comparison students ($r = 0.17$, ns).

The program seemed to counter for all 4-H students a naturally occurring increase in either aggressive behavior (for boys) or attitudes supporting aggression (for girls), and the effect was particularly beneficial for boys. Teachers and parents also perceive the behavior of 4-H students more positively. This program effect was strongest for teacher ratings of boys and parent ratings of girls.

Overall, the BrainPower Program seems to be a promising vehicle for providing specific training in improved social behavior when presented as a universal intervention in combination with the general protective benefits of the youth development program. The improvements in social reasoning and behavior appear to be a function of the special program emphasis embodied by the BrainPower program. These data are consistent with a broad literature (Weissberg and Greenberg, 1998) that demonstrates the effectiveness of comprehensive programming to support children's healthy development, particularly for children who must develop to adulthood in adverse circumstances.

DEVELOPMENTAL, CONTEXTUAL, AND CULTURAL CONSIDERATIONS

The evolution of the Brainpower Program from a precisely focused intervention strategy to an important element in a universal comprehensive youth development program reflects the nature of change in our understanding of how best to forestall the development of children's aggression. We know that children's aggression is influenced by multiple environmental conditions (e.g., extreme poverty), contextual factors (e.g., antisocial peers), and developmental antecedents (e.g., inconsistent harsh parenting); thus, no single treatment strategy can be expected to reliably reduce or prevent childhood aggression. The most effective programs will be those that are not only comprehensive in their scope of activities, but also in the duration of their contact with children and families and the range of contexts in which activities are implemented.

Development and Aggression

Current research and theorizing recognize the significance of developmental and ecological perspectives in understanding and changing behavior. Programs most likely to ensure successful developmental outcomes for children will be those designed to extend over the child's formative years (from early childhood into adolescence). As described at the outset of this chapter, later life antisocial and violent behavior springs from the developmental roots of childhood aggression. The earlier these developmental trajectories are established, the greater the likelihood for behavioral stability and more serious negative outcomes (Moffitt, 1993).

Children are active participants and shapers of their environment from the earliest point in development (Scarr and McCartney, 1983), and later experiences, competencies, and behaviors can be traced, in part, to how individuals selectively create their own social environments. For example, an early bias to perceive hostility in others and respond with inappropriate aggression will lead peers to behave aggressively toward a child, which will in turn increase the child's perceptions of hostility and level of aggressive responding (Hymel et al., 1990). This escalating cycle leads to increased childhood aggression, peer rejection, and ultimately antisocial behavior, delinquency, and other forms of social maladjustment in adolescence and later life (*see* Hudley, 1994a, for a review). Thus, age-appropriate preventive interventions that can support children in developing social competence and successfully managing peer interactions across childhood will have the best chance of interrupting a developmental progression from childhood cognitive bias to adolescent antisocial behavior.

Aggression in Context

Successful comprehensive interventions will also be those that are implemented across an array of the child's socializing environments (home, school, community, peer group). Although the individual actively shapes her own social environment, social psychology has long proven that contextual features also exert powerful control over behavior (Fiske and Taylor, 1991). The family, peer, and school environment can have a substantial impact on the development of aggression; similarly, these multiple environments are critical to the development and maintenance of socially competent behavior. A single intervention targeted solely at a focal child may do little to sustain and reinforce behavioral improvements outside the intervention setting. Concomitant changes in peers' appraisals of the target child, parental discipline and monitoring, and teacher reinforcements and behavior management may be important components of effective cognitive-behavioral interventions to reduce aggression (Lochman and Wells, 1996). For example, the presence of nonaggressive peers in BrainPower intervention groups has helped promote generalization of treatment by changing the way peers view aggressive children. As new strategies for interaction are learned in the intervention groups, nonaggressive peers are able to revise their expectations for excessive aggression from formerly aggressive participants. These revised patterns of interactions should generalize to settings outside the treatment groups and provide a model for other peers when interacting with the formerly aggressive participants.

The current state-of-the-art in preventive interventions to reduce childhood aggression is comprehensive multicomponent programs (Conduct Problems Prevention Research Group, 1999; Greenberg and Kusche, 1993; Lochman and Wells, 1996). These programs strive for a synergistic effect from multiple complementary components that target youth, their families, and the context in which they must function (school and community). However cognitive behavioral intervention strategies provide a unique theoretically grounded active ingredient. These strategies start from the perspective of the individual, and their goal is to strengthen personal capacities and decrease individual limitations. Cognitive-behavioral preventive interventions are designed to enable a child to self-regulate aggressive behavior in the face of environmental stressors (e.g., shoving in the lunch line at school). Embedded in an array of competence-promoting activities that are provided across time and development and that impact multiple social contexts in a child's life, cognitive behavioral strategies can be powerful tools in the service of healthy child development.

Culture and Prevention

Although social cognitive and developmental theory have informed preventive interventions for the reduction of aggression (Weissberg and Greenberg, 1998), a factor largely unexplored in aggression reduction research is the importance of cultural differences in social behavior. However, these differences must surely influence the efficacy of preventive interventions, as social behavior is so completely embedded in cultural context. As preventive interventions move toward comprehensive programs that foster the development of children's social competence the necessity of a culturally informed perspective must not be overlooked (Hudley, 2001b). Although physical violence among children is a universally proscribed behavior, the very essence of social competence is culturally defined.

Relative to prevention research, the active examination of the role of culture in student learning has been occurring in educational research for decades (i.e., research in multicultural education). That work has provided a construct, a culturally relevant pedagogy, that may be illuminating to the field or prevention science. Within this framework, the teacher's task is to incorporate cultural patterns of learning and knowing into instructional practice. In addition, culturally relevant pedagogy, by capitalizing on students' home and community culture, prepares students for collective action for social justice by making visible the lens of culture (Ladson-Billings, 1994).

The equivalent construct for prevention science would be culturally relevant prevention intervention. Such interventions would be designed to empower children within their unique cultural contexts. For example, one appropriate part of a comprehensive preventive intervention might include social skills training. However, rather than imposing a universal (and most likely Eurocentric) standard of social skills, interventions should allow all children to develop those skills appropriate for their own culture and to understand and honor those of other cultures. Cultures differ, for example, in their definitions of appropriate physical proximity across gender and age divides.

Such meta-knowledge would empower children to perceive the construct of social skills itself as flexible and culturally determined and prepare them to effectively fight discrimination based on such superficial characteristics. The converse, culturally irrelevant interventions applied in multicultural contexts, may comprise one of the root causes of inconsistencies in research on prevention efficacy (Botvin et al., 1995). As the Brain-Power Program has developed to serve progressively broader groups of students, I have been mindful of the need to adapt the curricula to best meet the needs of the participants within their unique social and cultural niche.

Stories are now deliberately designed to be very general, with the details of context written to allow students to bring their own lived experiences to the stories. For example, all preadolescents may not hang out in shopping malls; malls are simply not accessible to some communities. A story about two friends meeting may, therefore, not specify exactly where the meeting will take place—participants suggest likely prospects, including the park, the mall, the schoolyard, the gym, and so on. Additionally, one of students' favorite activities involves observing their own social situations (the previously described intention detective activity) and reporting back what kinds of reactions are typical of the people with whom they interact. Students learn simultaneously that not all people interact in the same way, there are many "good" ways to interact with others, and disadvantaging groups because of superficial styles of interaction is a bad idea that inhibits social cooperation. The benefits of this activity are clearly above and beyond the basic goals of the program—to teach children to more accurately infer the intentions of others (i.e., reduce attributional bias). In sum, as prevention efforts move from a focus on disorder (e.g., antisocial behavior) to the enhancement of competence, programs must begin to more explicitly incorporate an understanding of the cultural context in which children's competence will be defined and enacted.

REFERENCES

Asher, S. (1985) An evolving paradigm in social skill training research with children, in *Children's Peer Relations: Issues in Assessment and Intervention.* (Schneider, B., Rubin, K., and Ledingham, J., eds.), Springer-Verlag, New York, pp. 157–171.

Bierman, K. (1986) Process of change during social skills training and its relationship to treatment outcome. *Child Dev.* **57,** 230–240.

Botvin, G., Schienke, S., Epstein, J., Diaz, T., and Botvin, E. (1995) Effectiveness of culturally focused and generic skills training approaches to alcohol and drug abuse prevention among minority adolescents: two-year follow-up results. *Psychol. Addict. Behav.* **9,** 183–194.

Conduct Problems Research Group (1999) Initial impact of the Fast Track prevention trial for conduct problems: I. The high-risk sample. *J. Consult. Clin. Psychol.* **67,** 631–647.

Crick, N. and Dodge, K. (1994) A review and reformulation of social information-processing mechanisms in children's social adjustment. *Psychol. Bull.* **115,** 74–101.

Derzon, J. (2001) Antisocial behavior and the prediction of violence: a meta-analysis. *Psychology in the Schools* **38,** 93–106.

Dodge, K. (1980) Social cognition and children's aggressive behavior. *Child Dev.* **51,** 162–170.

Dodge, K. (1986) A social information processing model of social competence in children, in *The Minnesota Symposium on Child Psychology.* (Perlmutter, M., ed.), Erlbaum, Hillsdale, NJ, pp. 77–125.

Dodge, K. and Crick, N. (1990) Social information-processing bases of aggressive behavior in children. *Pers. Soc. Psychol. Bull.* **16,** 8–22.

Dunworth, T. and Saiger, A. (1994) *Drugs and Crime in Public Housing: A Three-City Analysis.* National Institute of Justice, Washington, DC.

Fiske, S. T. and Taylor, S. E. (1991) *Social Cognition, 2nd ed.* McGraw-Hill, New York.

Geen R. and Donnerstein, E. (1998) *Human Aggression: Theories, Research, and Implications for Social Policy.* Acedemic Press, San Diego, CA.

Graham, S., Hudley, C., and Williams, E. (1992) Attributional and emotional determinants of aggression among African-American and Latino young adolescents. *Dev. Psychol.* **28,** 731–740.

Greenberg, M., Domitrovich, C., and Bumbarger, B. (2001) The prevention of mental disorders in school-aged children: current state of the field. *Prevention Treatment* **4,** Article 1. Available on the World Wide Web: (http://journals.apa.org/prevention/volume4/pre0040001a.html).

Greenberg, M. and Kusche, C. (1993) *Promoting Social and Emotional Development in Deaf Children: The PATHS Project.* University of Washington Press, Seattle.

Hudley, C. (1994a) Perceptions of intentionality, feelings of anger, and reactive aggression, in *Anger, Hostility and Aggression: Assessment, Prevention, and Intervention Strategies for Youth.* (Furlong, M. and Smith, D., eds.), Clinical Psychology Publishing Co., Brandon, VT, pp. 39–56.

Hudley, C. (1994b) The reduction of childhood aggression using the BrainPower Program, in *Anger, Hostility and Aggression: Assessment, Prevention, and Intervention Strategies for Youth* (Furlong, M. and Smith, D., eds.), Wiley & Sons, New York, pp. 313–344.

Hudley, C. (1995) Reducing peer directed aggression in the elementary grades: the effects of an attribution retraining program. Paper presented at the 61st Biennial Meeting of the Society for Research on Child Development, Indianapolis, IN.

Hudley, C. (2001a) Perceived behavioral and academic competence in middle childhood: influences of a community-based youth development program. Retrieved May 20, 2001 from the ERIC/CASS Virtual Library Web site: (http://ericcass.uncg.edu/virtuallib/achievement/1108.html).

Hudley, C. (2001b) The role of culture in prevention research. *Prevention Treatment* **4**, Article 0005c. American Psychological Association. Retrieved May 20, 2001, from (http://journals.apa.org/prevention/volume4/ pre0040005c.html).

Hudley, C., Britsch, B., Wakefield, W., Smith, T., DeMorat, M., and Cho, S. (1998) An attribution retraining program to reduce aggression in elementary school students. *Psychology in the Schools,* **35,** 271–282. Available on-line: (http://www3.interscience.wiley.com/cgi-bin/issuetoc?ID=32112).

Hudley, C. and Friday, J. (1996) Attributional bias and reactive aggression. *Am. J. Prevent. Med.* **12,** 75–81.

Hudley, C. and Graham, S. (1993) An attributional intervention to reduce peer directed aggression among African-American boys. *Child Dev.* **64,** 124–138.

Huesmann, L. R. (1988) An information processing model for the development of aggression. *Aggress. Behav.* **14,** 13–24.

Huesmann, L. R. (1998) The role of social information processing and cognitive schema in the acquisition and maintenance of habitual aggressive behavior, in *Human Aggression: Theories, Research, and Implications for Social Policy.* (Geen, R. and Donnerstein, E., eds.), Academic Press, San Diego, CA, pp. 73–109.

Huesmann, L. R. and Eron, L. (1984) Cognitive processes and the persistence of aggressive behavior. *Aggress. Behav.* **10,** 243–251.

Hymel, S., Wagner, E., and Butler, L. (1990) Reputational bias: view from the peer group, in *Peer Rejection in Childhood.* (Asher, S. and Coie, J., eds.), Cambridge University Press, New York, pp. 156–186.

Institute of Medicine (1994) *Reducing Risks for Mental Disorders: Frontiers for Preventive Intervention Research.* National Academy Press, Washington, DC.

Jacobsen, N. and Truax, P. (1991) Clinical significance: a statistical approach to defining meaningful change in psychotherapy research. *J. Consult. Clin. Psychol.* **59,** 12–19.

Kelley, H. (1973) The process of causal attribution. *Am. Psychologist* **28,** 107–128.

Kendall, P., Ronan, K., and Epps, J. (1991) Aggression in children/adolescents: cognitive-behavioral treatment perspectives, in *The Development and Treatment of Childhood Aggression.* (Pepler, D. and Rubin, K., eds.), Lawrence Erlbaum, Hillsdale, NJ, pp. 341–360.

Kupersmidt, J. and Coie, J. (1990) Preadolescent peer status, aggression, and school adjustment as predictors of externalizing problems in adolescence. *Child Dev.* **61,** 1350–1362.

Ladson-Billings, G. (1994) *The Dreamkeepers.* Jossey-Bass, San Francisco.

Laird, R., Jordan, K., Dodge, K., Pettit, G., and Bates, J. (2001) Peer rejection in childhood, involvement with antisocial peers in early adolescence, and the development of externalizing behavior problems. *Dev. Psychopathol.* **13,** 337–354.

Lochman, J. and Wells, K. (1996) A social-cognitive intervention with aggressive children, in *Preventing Childhood Disorders, Substance Abuse, and Delinquency.* (Peters, R. and McMahon, R., eds.), Sage, Thousand Oaks, CA, pp. 111–143.

Loeber, R. (1990) Development and risk factors of juvenile antisocial behavior and delinquency. *Clin. Psychol. Rev.* **10,** 1–42.

Moeller, T. (2001) *Youth Aggression and Violence: A Psychological Approach.* Erlbaum, Mahwah, NJ.

Moffit, T. (1993) Adolescent-limited and life-course-persistent antisocial behavior: a developmental taxonomy. *Psychol. Rev.* **100,** 674–701.

Nasby, W., Hayden, B., and DePaulo, B. (1980) Attributional bias among aggressive boys to interpret unambiguous social stimuli as displays of hostility. *J. Abnorm. Psychol.* **89,** 459–468.

Olweus, D. (1979) Stability of aggressive reaction patterns in males: a review. *Psychol. Bull.* **86,** 852–875.

Patterson, G. (1992) Developmental changes in antisocial behavior, in *Aggression and Violence Throughout the Life Span.* (Peters, R. D., McMahon, R., and Quinsey, V., eds.), Sage, Newbury Park, CA, pp. 52–82.

Peters, R. and McMahon, R. (Eds.). (1996) *Preventing Childhood Disorders, Substance Abuse, and Delinquency.* Sage, Thousand Oaks, CA.

Robinson, T., Smith, S., Miller, M., and Brownell, M. (1999) Cognitive behavior modification of hyperactivity-impulsivity and aggression: a meta-analysis of school-based studies. *J. Educ. Psychol.* **91,** 195–203.

Scarr, S. and McCartney, K. (1983) How people make their own environments: a theory of genotype-environment effects. *Child Dev.* **54,** 424–435.

Stoff, D., Breiling, J., and Maser, J. (1997) *Handbook of Antisocial Behavior.* Wiley & Sons, New York.

Weiner, B. (1986) *An Attributional Theory of Motivation and Emotion.* Springer-Verlag, New York.

Weissberg, R. and Greenberg, M. (1998) School and community competence-enhancement and prevention programs, in *Handbook of Child Psychology, 5th ed., Vol. 4, Child Psychology in Practice.* (Siegel, I. and Renniger, K. A., eds.), Wiley & Sons, New York, pp. 877–954.

16
Pharmacological Intervention in Aggression

Debra V. McQuade, R. Joffree Barrnett, and Bryan H. King

INTRODUCTION

The recent Surgeon General's Report on Youth Violence, prompted by the Columbine High School tragedy in April 1999, summarizes research relating to youth violence, its causes, and its prevention (2001). In the list of promising therapeutic interventions cited in the report, clinical drug trials are absent. While evidence exists to support the medication treatment of conditions that may predispose to violent behavior, there are simply no studies to date that would support the use of medication specifically to treat youth violence.

Moreover, the appropriate treatment of aggression cannot be reduced to the mere administration of medicine. Not surprisingly, just as aggression is expressed in a variety of ways, and its origins multiply determined, so too, its proper treatment can only flow from a comprehensive approach. The elements of such an approach include a thorough diagnostic assessment, the development of a working therapeutic alliance, and education regarding the illness or condition being treated and the relative risks, potential benefits, and specific goals of treatment and relapse prevention. Typically, therapeutic intervention will be multimodal and include attention to adaptation, coping, and a return to a normal developmental course. Increasingly the importance for treatment to be person or family-driven, culturally competent, and strength-based is being stressed (Henggler and Bourdin, 1990; Lyons et al., 2000), and none of the above is available in pill form.

The following discussion reviews the recent advances in the psychopharmacology of aggression. There are a number of ways such a review could be organized. One strategy is to discuss aggression in the context of specific diagnostic entities and the agents that are helpful for those conditions (Swann, 1999; Weller et al., 1999; Yehuda, 1999). Such an approach can be particularly useful in situations where a drug that might reduce aggression in the context of one diagnosis, for example, depression, could actually make aggression worse in the setting of another condition, like mania. A second less common approach involves adopting a dimensional organization of psychopathology and reviews the utility of various drugs along such a continuum (Hollander, 1999; Kavoussi and Coccaro, 1998b). Such an approach also has merit, particularly where a drug or transmitter system can be linked to dimensional effects. Ethological approaches to aggression are particularly suited to such a schema, where, for example, one would predict that impulsive and predatory aggression might be differentially responsive to drugs. The current review adopts

From: *Neurobiology of Aggression: Understanding and Preventing Violence*
Edited by: M. Mattson © Humana Press Inc., Totowa, NJ

a third common approach, which is simply to organize studies by categories of pharmacologic agents. This strategy is taken here because no single agent, even within a single disease entity, has clear superiority, and the psychopharmacology of aggression does not yet fit particularly well onto a dimensional scaffold. Indeed, the mechanism by which various drugs act, let alone a mechanistic understanding of how they affect aggression, is often unclear. The "dimensional" anti-aggressive effect of a medication, for example, could be as varied as to improve impulse control, reduce irritability, enhance frustration tolerance, or simply to cause sedation. Nevertheless, for each of the classes of medication reviewed below, an attempt is made to highlight diagnostic and dimensional experience when such is possible.

LITHIUM

In his paper reporting the use of lithium in the context of psychotic agitation, Cade (1949) anticipated a potential role for lithium in the treatment of aggression not associated with mania. Cade discriminated its "sedative" or calming effect from a hypnotic one and suggested that lithium might be an alternative to prefrontal leukotomy for the treatment of "restless impulses and ungovernable tempers (in psychopathic mental defectives)." Since Cade's description, numerous case reports and a few open trials have supported his clinical assertion, covering a wide range of ages and diagnoses (including deafness, mental retardation, conduct disorder, schizophrenia, hyperactivity, traumatic brain injury, and antisocial personality disorder) (Miczuk, 1987). Sheard was a pioneer of early studies of aggression and conducted an impressive series of investigations using lithium in nonbipolar "hyperaggressive" prisoners and delinquents (Sheard, 1971, 1975, 1978). His research efforts included one double-blind study in which significant reductions in aggressive behaviors were maintained with lithium relative to placebo, lasting over a 3-mo trial (Sheard et al., 1976), and included some notable instances of remarkable behavioral change (Sheard and Marini, 1978).

In the years since Sheard's seminal work, others have explored the anti-aggressive effects of lithium. Thus, in their review, Fava and Rosenbaum (1993) deem lithium demonstrably effective in the treatment of aggression, not only in populations with bipolar disorder, but with conduct disorder and nonspecific mental retardation as well. Three double-blind placebo-controlled studies (Campbell et al., 1984, 1995; Malone et al., 2000) further support the use of lithium for aggression in children and adolescents diagnosed with conduct disorder. On balance, however, data with children are mixed. Two additional well-controlled studies have yielded negative results (Klein, 1991; Rifkin et al., 1997) although the latter was of only 2 wk duration. In their comprehensive review of 58 published studies describing the use of lithium in children, Alessi and colleagues (1994) characterize aggression as the most widely supported clinical diagnostic entity for which lithium is indicated.

Appropriate concern regarding short- and long-term side effects have likely limited the widespread clinical use of lithium with children, particularly very young children (Hagino et al., 1995), but likely also other special populations, for example, those with co-morbid neurological or medical conditions. This latter group in particular may manifest a heightened potential for neurotoxicity, particularly when lithium is combined with other drugs (Lang and Davis, 2002; Oakley et al., 2001). On the other hand, lithium's well-established record of effectiveness with an ever-growing variety of clinical populations supports consideration of its use. Recently, for example, Prado-Lima and colleagues (2001) have suggested that lithium may reduce maternal child abuse behavior based upon the results of their open trial in eight mothers drawn from a program for abusive parents at the investigators' clinic in Brazil. If replicated and extended, such an intervention might not only reduce aggression in treated parents, but perhaps even lessen the likelihood that their children might be aggressive in adulthood (Clarke et al., 1999).

ADRENERGICS

β-Blockers

Haspel (1995) provided a comprehensive review of available literature on the use of β-blockers for the treatment of aggression, which included 12 uncontrolled studies (involving a total of 87 patients) and four double-blind trials (with 74 patients). In the uncontrolled studies, an impressive 86% of patients were considered successful responders to treatment, despite significant heterogeneity in diagnostic status and a history of treatment-refractory symptoms. The results from controlled studies are consistent with these data. Clinical improvement has been noted for patients diagnosed with dementia (Greendyke and Kanter, 1986; Greendyke et al., 1986), organic brain syndrome (Greendyke et al., 1989), and a population of chronically hospitalized psychiatric patients with a variety of diagnoses (Ratey et al., 1993). Not surprisingly, the documented rates of success in the controlled studies are less robust than those for the open trials.

More recently, Silver and colleagues (1999) demonstrated success with propranolol in treating chronically aggressive hospitalized adults, while Caspi and colleagues (2001) demonstrated success with aggressive patients with schizophrenia, in their randomized controlled trial of pindolol. There are no controlled trials of β-blockers in children, but Connor and associates (1997) have reported a successful open label trial of nadolol with 24 aggressive children and young adults with mental retardation. Additional evidence supportive of β-blockers in ameliorating aggressive symptoms in children includes case reports by Lang and Remington (1994) and Matthews-Ferrari and Karroum (1992), using propranolol and metoprolol, respectively.

The precise mechanism by which β-blockers exert their control on aggressive behaviors is not precisely understood. The affinity that β-blockers have for serotonin (5-HT) receptors is now well recognized (Rabiner et al., 2000), implicating the likely involvement of the central serotonergic system, in addition to—and perhaps even more important than—the adrenergic system (Haspel, 1995). That β-blockers differ in their relative centrally acting vs peripherally acting effects is also considered significant, although a clear relationship between loci of action and clinical effect has yet to be delineated, possibly because less-lipophilic agents do ultimately cross the blood-brain barrier with time and even peripherally mediated changes can result in central nervous system effects (Chugani et al., 1999; Fisher et al., 2002; Uc et al., 2002).

Moreover, it appears that β-blockers frequently must be prescribed in doses higher than needed simply to achieve β-blockade. In their guide on propranolol use in adults, Yudofsky and colleagues (1990) consider doses up to 800 mg daily to be acceptable. Specifically, they recommend a starting dose of 20 mg three times daily, with increases of up to 60 mg/d occurring at intervals of 3 d. Titration is recommended to stop when clinical effect is seen, when side effects (e.g., hypotension or bradycardia) become evident, or when the maximum recommended daily dose is reached.

α-2 Agonists: Clonidine and Guanfacine

The mechanism by which α-2 agonists might attenuate aggression is unclear. Moreover, in many preclinical models of aggression, particularly affective aggression, clonidine actually potentiates aggressive responding. Nevertheless, in recent years, the number of prescriptions of clonidine for children with attention-deficit hyperactivity disorder (ADHD) appears to be growing at an extraordinary rate (Riddle et al., 1999; Zito et al., 2000).

Virtually all of the experience with α-2 agonists and aggression derives from use in children, and available data supporting their use for this indication are meager. Jaselskis and colleagues (1992) reported modest improvement in oppositional behavior in a controlled trial of clonidine involving eight children with autism. Earlier, Hunt and colleagues (1986) had reported some improvement in children with ADHD in a small but randomized clinical trial. There are no controlled studies yet documenting the effectiveness of guanfacine for the treatment of aggression, and the literature supporting the use of α-adrenergic drugs for this purpose generally remains quite limited.

STIMULANTS

Methylphenidate and Dextroamphetamine

Aggressive behavior is common in the context of ADHD, but relatively few studies have specifically aimed to isolate a possible anti-aggressive effect of stimulant medication from potentially broader effects of stimulants on symptom domains in ADHD (Connor et al., 2002; Lyons et al., 2000). That is, are improvements in aggression merely secondary consequences of improvement in impulsivity, hyperactivity, and attention, or is aggression a specific target for drugs like methylphenidate? Two recent reviews address this topic, and the conclusion from each is that the anti-aggressive effect of stimulants appears to be independent of an effect on ADHD symptoms generally (Connor et al., 2002; Lyons et al., 2000; Pine and Cohen, 1999).

Data to support this conclusion include the findings from a controlled trial in 84 children that methylphenidate reduced aggression in those with conduct disorder, independent of whether they also had ADHD (Klein et al., 1997). Moreover, in their recent meta-analysis, Connor and colleagues (2002) calculated an effect size of 0.84 for overt and 0.69 for covert aggression related behaviors in ADHD. These robust effect sizes rival those for stimulants on ADHD core symptoms (Ottenbacher and Cooper, 1983).

ANTICONVULSANTS

Valproate

Since its introduction in 1967 as an antiepileptic agent, valproate (and subsequently divalproex sodium) has come into favor as both an antimanic (for which it carries a Food and Drug Administration (FDA)-approved indication of use) and an anti-aggressive agent. Lindenmayer and Kotsaftis (2000) have recently provided a comprehensive review of its use in the control of violence and aggression. They noted first literature references to this use in 1988, when Kahn et al. (1988) described successful application of valproate to the behavioral control of three patients with organic brain syndrome.

Just 10 yr later, Citrome and colleagues (1998) reported a substantial increase in the use of valproate within the New York State Psychiatric system, for the treatment of chronic nonbipolar hospitalized patients. Their surprising finding was that at the time of their most recent measure, fully 34.1% of inpatients had been prescribed valproate for treatment of aggressive behaviors in the context of a variety of psychiatric diagnoses.

Lindenmayer and Kotsftis (2000) encouraged acknowledgment of this relatively widespread use of valproate and reviewed the available evidence supportive of its efficacy in this regard. They identified 17 reports with a total of 164 patients (10 case reports, three retrospective chart reviews, three open label studies, and one pilot double blind study) and calculated an overall approximate response rate of 77.1%, defining a positive response as a 50% reduction in aggressive symptoms. They stress the essential absence of support from controlled trials, but interpret the available literature to rank-order likely response to valproate as highest in patients with organic brain syndromes, followed by those with dementia or unspecified mental retardation, and finally, mania.

Additional reports have found valproate to be useful for aggression in schizophrenia (Afaq el al., 2002), for aggressive symptom control in autism spectrum disorders (Hollander et al., 2001) and adults with intellectual disability (Ruedrich et al., 1999), in the control of impulsive aggressive behaviors exhibited by patients with personality disorders (Kavoussi and Coccaro 1998a), and in the control of explosive tempers of adolescents (Delito et al., 1998; Donovan et al., 1997, 2000). A number of studies have reported the effectiveness of valproate in ameliorating behavioral disturbances associated with dementia (Buchalter and Lantz, 2001; Haas et al., 1997; Kunik et al., 1998); these reports are particularly impressive in that virtually all of the patients undergoing treatment with valporate experienced improvement in the target symptoms. In his review of this literature, Grossman (1998) concluded that the use of valproate in the elderly population is not only remarkably efficacious, but also advantageous with respect to a balance of symptom control vs drug side effects and/or interactions when compared to alternatives.

In general, consensus among clinical researchers appears to be that valproate's mechanisms of action include an enhancement of central levels of γ-amino butyric acid (GABA), an enhancement of serotonergic neurotransmission, and antikindling effects (Lindenmayer and Kotsaftis, 2001; Stoll and Severus, 1996). Its anticonvulsant effect has prompted some to suggest that it is a particularly good agent to consider for aggression in persons with underlying central nervous system impairments and especially where patients present with electroencephalography (EEG) abnormalities or a seizure disorder (Stoll et al., 1994). Valproate appears to be widely prescribed for aggression in a variety of conditions. Its use for such an indication deserves further study.

Carbamazepine

Carbamazepine has been available since the 1960s, and was used originally as a treatment for trigeminal neuralgia, but has had approval as an anticonvulsant since 1974. Like valproate, it has been used in the treatment of bipolar disorder, although it does not carry a formal indication for same. Also like valproate, investigative extensions of its utility in the control of aggressive behaviors have been ongoing. Recently, it has been suggested that carbamazepine shares with valproate and lithium the effect of causing neuronal inositol depletion (Williams et al., 2002).

Carbamazepine has long been thought to work by control of kindling in the temporal lobe and limbic system (Tunks and Dermer, 1977). Not surprisingly, it has been demonstrated to be successful in the control of aggression associated with abnormalities in these regions (Neppe, 1983; Tunks and Dermer, 1977). Patients with nonspecific EEG abnormalities and concomitant behavioral problems have also been shown on occasion to improve on the drug (Yassa and Dupont, 1983), as have individuals with more significant head injury (Azouvi et al., 1999). A limited number of reports have suggested that aggressive patients with schizophrenia improve with carbamazepine, typically as an adjunct to their antipsychotic therapy (Klein et al., 1984; Neppe 1988; Yassa and Dupont, 1983). Grossman (1998) endorsed the use of carbamazepine as an alternative to valproate with elderly patients with dementia, reporting it to be similarly effective and well-tolerated by this population, albeit with some concern regarding side effects, especially hematological and hepatic. Additional support for the use of carbamazepine for aggression in dementia comes from Tariot and colleagues (1998) Their 6-wk randomized controlled trial comparing carbamazapine to placebo in 51 nursing home residents was terminated after an interim analysis because of the clear superiority of drug over placebo.

The results of studies involving aggressive children have been somewhat variable, so that while Kafantaris and colleagues (1992) were able to report significant reductions of aggression and explosiveness in their open trial of 10 children with conduct disorder, Cueva and colleagues (1996) failed to demonstrated superiority of carbamazepine over placebo for aggression in their double-blind placebo-controlled trial involving 22 children with conduct disorder. Moreover, side effects, including leukopenia, rash, dizziness, and diplopia were common.

Phenytoin

Barratt has contributed significantly to our understanding of aggression and its treatment (Barratt et al., 1997). His studies of impulsive–aggression in male prison inmates have not only suggested a role for phenytoin in the treatment of aggression, but perhaps, more important, underscore the importance of appreciating the behavior in context. Thus, in a controlled trial of phenytoin in 60 inmates who were separated into groups on the basis of whether their aggression was impulsive or premeditated, phenytoin was associated with a significant reduction of impulsive aggression only (Barratt et al., 1997). In a subsequent controlled trial involving 46 men recruited through advertisements describing a potential treatment for problem temper, Stanford and colleagues (2001) similarly found phenytoin significantly more effective than placebo in reducing impulsive aggression. The apparent selectivity for this agent for impulsive aggression is intriguing in light of the recent findings of Keele (2001), demonstrating that phenytoin's inhibition of isolation-induced aggression in the rat is specific to animals rendered hyposerotonemic. Links between impulsivity, aggression, and 5-HT are clearly established (Cherek and Lane, 2001), and it may be that more direct and selective serotonergic drugs will best phenytoin in terms of effectiveness and side-effect burden.

Gabapentin and Lamotrigine

Newer anticonvulsant agents are undergoing investigation with respect to their utility in reducing aggressive behavior. Limited success has been demonstrated with both gabapentin and lamotrigine. Miller (2001) reviewed available data on the use of gabapentin with elderly persons with dementia. Six case reports and two open label trials (representing a total of 24 patients) together provided support for its use. Additionally, Hawkins and colleagues (2000) (retrospective chart review) reported that 23 of 24 nursing home residents with dementia were behaviorally improved with moderate doses of gabapentin. A favorable side effect profile and limited drug interactions may contribute to the further successful use of this drug for this indication.

In the pediatric population, case reports are accumulating that gabapentin can initiate or aggravate aggressive behaviors (Lee et al., 1996; Tallina et al., 1996; Wolf et al., 1995). In these reports, gabapentin was administered for its anticonvulsant effects to a total of 12 children, all with previously diagnosed seizure disorders, but also with co-morbid developmental delay, mental retardation, learning disorders, ADHD, and/or separate and severe neurological conditions. In each case, the children were taking additional anticonvulsant medications. The behavioral effects followed the administration of gabapentin and were reversible upon its withdrawal. When administered for control of epilepsy, otherwise healthy children apparently tolerate the drug well (Appleton et al., 2001; Fisher et al., 2001; Haig et al., 2001). Dose escalation has been proposed as a predisposing factor to behavioral disturbances with children (Besag, 2001). However, caution should be exercised when administering gabapentin for control of aggression in this population, as data are quite limited.

Mixed evidence is also accumulating in support of the use of lamotrigine as an anti-aggressive agent. Devarajan and Dursun (2000) have reported success in its use with an elderly patient with frontal lobe dementia. Davanzo and King (1996) found it effective in the treatment of self-injurious behavior in a neurologically complicated 18 yr old. Uvebrandt and Bausiene (1994) reported that 8 of 13 children with autism experienced a beneficial behavioral effect, when lamotrigine was administered for epilepsy. However, Beran and Gibson (1998) described multiple case reports of adults with intellectual disabilities and seizure disorder who manifested increased aggression when lamotrigine was added to their existing anticonvulsant medication.

SEROTONERGIC AGENTS

Trazodone

Since 1986, three open trials (Pinner and Rich, 1988; Simpson and Foster, 1986; Zubieta and Alessi, 1992) and three case studies (Mashiko et al., 1996; Nguyen and Meyers, 2000; Schneider et al., 1989) have reported the use of trazodone as an anti-aggressive agent. These reports represent a total of 36 patients, with diagnoses including dementia and a variety of other severe neurological problems, but also including one group of healthy children with disruptive behavior disorders. All patients had failed trials of other agents prior to the administration of trazodone. Of this collection of cases, 22 (61%) were reported to have undergone some measurable reduction of aggressive behaviors, while 3 (5%) were reported to have worsened.

Although not considered a first-line treatment, Herrmann (2001) recommends the use of trazodone as an alternative treatment for dementia-associated aggression. As a group, physiatrists who specialize in treatment of traumatic brain injury (TBI) favor the use of trazodone (after carbamazepine and tricyclic antidepressants), prescribing it more frequently than amantadine or β-blockers (Fugatea et al., 1997). These data suggest that neurologically complicated patients and/or patients who have failed to have their aggression controlled with other medications, may respond favorably to this drug.

Clomipramine

Of the tricylclic antidepressants, clomipramine has garnered the most success in modifiying aggression, perhaps because of its serotonergic profile. Recently, Luiselli and colleagues (2000) reported a single case open-label trial comparing clonazepam, propranolol, sertraline, and clomipramine for the treatment of aggression in an adolescent with autism. Clomipramine's effectiveness in reducing the frequency of daily aggressive acts (to 0.9 acts per day) was notably superior to the other medications (reductions to 4.0, 4.2, and 3.2 acts of aggression per day, respectively), and this successful control lasted several months. Two case reports (Luiselli et al., 2001; McDougle et al., 1993) and an additional open label study (Brodkin et al., 1997) further support clomipramine's efficacy in controlling aggression in special populations. In the latter report, 18 out of 35 individuals with pervasive developmental disorders (PDD) showed significant clinical improvement. However, the authors note that 13 of their patients experienced significant adverse effects, including three of whom had seizures during treatment. These authors recommend that a selective serotonin reuptake inhibitor (SSRI), rather than clomipramine, be considered as first-line treatment of individuals with PDD and a seizure history.

Buspirone

Several case studies and open trial reports suggest that buspirone is useful in reducing the aggressive behavior exhibited by adults with developmental disorders (Colella et al., 1992; Gedye 1991; Ratey et al., 1989, 1991; Verhoeven and Tuinier, 1996). Investigators frequently identify an additional benefit to the use of buspirone in this context in that it is unlikely to cause sedation, thus avoiding a frequent and unfortunate side effect of drugs commonly prescribed to this population.

A limited number of case reports indicate that buspirone may be helpful in reducing aggressive symptoms associated with dementia (Holzer et al., 1995), Huntington's disease (Byrn et al., 1994; Findling, 1993), and combined ADHD/conduct disorder (Quiason et al., 1991). A retrospective chart review of 20 adults with TBI (Stanislav et al., 1994) and a single-blind dose escalation study with persons with dementia (Levy et al., 1994) both indicated that buspirone reduced aggression.

Several small studies have examined the use of buspirone in autism, and results are mixed. Realmuto and colleagues (1989) found it useful in the reduction of hyperactivity in two out of the three children in their open trial, again noting the fortuitous absence of side effects. McCormick (1997) performed a double-blind placebo-controlled crossover study with his single patient, and objective evaluations indicated improvement in aggression and self-injurious behavior. Hillbrand (1995) described a remarkable reduction in the extreme violence of an incarcerated adult male with autism and mild mental retardation, with long lasting effects. However, King and Davanzo (1996) noted that the subjects with autism in their study demonstrated worsening in both the frequency and severity measures of their aggressive behaviors when treated with buspirone.

Worsening of aggression resulted in discontinuation of buspirone for four anxious aggressive hospitalized children in an open trial directed by Pfeffer and colleagues (1997); two other children were similarly withdrawn from the drug due to the development of mania. Although an overall reduction in aggression was measured for the group of 25 children enrolled in this open trial, only three children demonstrated improvement in function significant enough to warrant continuation of the drug after study completion.

In general, enthusiasm for buspirone appears to have waned after the flurry of reports in the late 1990s, but it may remain an important consideration in certain circumstances. For example, individuals who appear to become disinhibited and aggressive on even low doses of SSRI may be particularly deserving of study with buspirone given that as a partial agonist (Stanislav et al., 1994), it may exert a moderating (antagonist) influence under such circumstances (Campbell et al., 1992).

Selective Serotonin Reuptake Inhibitors

Within this class of drugs, available choices for treatment now include fluoxetine, sertraline, paroxetine, fluvoxamine, citalopram, and escitalopram. Taken together, this class of medications is receiving increasing support their use in the treatment of aggression, in a variety of contexts.

Since 1995, five double-blind placebo-controlled investigations of fluoxetine, fluvoxamine, citalopram, and paroxetine have been reported. Coccaro and Kavousi (1997) determined that fluoxetine has an anti-aggressive effect superior to placebo when administered to 40 individuals with personality disorders without co-morbid major depression, mania, or schizophrenia. The difference was detected at the end of 4 wk and was sustained throughout measurements taken at the end of months 2 and 3. These results did not appear to be influenced by changes in states of depression, anxiety, or alcohol use.

McDougle and colleagues (1996) reported success with the administration of fluvoxamine to adults with autistic disorder, with improvements in several behavioral measures, including reduction of aggression. Results were maintained at the end of the 12-wk study.

Citalopram was monitored for its effectiveness in controlling aggression in patients with schizophrenia in the double-blind crossover trial reported by Vartiainen and colleagues (1995). Significant reductions in aggressive behaviors were found for each 4-mo block of citalopram administration. Pollack and colleagues (2002) compared citalopram, perphenazine, and placebo for 85 nondepressed patients with dementia. Both drugs demonstrated some superiority to placebo, but only citalopram bested placebo on agitation–aggression and lability–tension measures.

Cherek and colleagues (2002) observed significant reductions in laboratory measures of both impulsivity and aggression in six adult subjects with histories of conduct disorder randomized to treatment with paroxetine for 3 wk. As there were no differences between rates of monetary reinforced responding between the paroxetine and placebo-treated controls, the authors assert that the decreases in aggression and impulsivity associated with active drug are more likely to be specific. That is, it is unlikely that a nonspecific sedative effect of the drug accounts for the observed benefit.

Open trials have recently demonstrated the successful control or reduction of aggression using sertraline in a variety of clinical contexts. These conditions include persons with closed head-injury (Kant et al., 1998; Kim et al., 2001), adults with autistic disorder and PDD (McDougle et al., 1998), and adults with personality disorders and impulsive aggression (Kavoussi et al., 1994). Citalopram has also been reported beneficial for aggressive children and adolescents under outpatient psychiatric care (Armenteros and Lewis, 2002) and for persons with dementia (Pollack et al., 1997).

Kauffman and colleagues (2001) described the successful control of aggression in a 7-yr-old girl with severe PDD with fluvoxamine, while Poyurovsky and colleagues (1995) described similar success with this agent when administered to an 18-yr-old male who was aggressive in the context of his diagnosis of obsessive–compulsive disorder (OCD). Additional case reports have detailed the positive response of four patients with Huntington's disorder to SSRI administration: two responded to fluoxetine (De March and Ragone, 2001) and two responded well to sertraline (Ranen et al., 1996).

The open trials reported by Davanzo et al. (1998) and Constantino et al. (1997) are less enthusiastic. In the first report, 15 institutionalized persons with mental retardation were treated for 4 mo with paroxetine. No reduction in frequency of aggressive behaviors was seen. Severity of aggression was reduced at the end of 1 mo, but this effect was not maintained over the course of the trial. In the Constantino study, the investigators found no significant reduction of aggression toward others associated with fluoxetine, paroxetine, or sertraline at standard doses. Moreover, 13 of their 19 patients were assessed to have statistically significant increases in verbal aggression, in aggression toward self, and in aggression toward objects while taking the SSRI.

There are additional cases in which increased aggression is described for persons taking SSRIs (Wilkinson, 1999). Taken together with the increased aggression noted in the open trials above (and perhaps including the reports of increased aggression with buspirone), this raises suspicion for varying response profiles to SSRIs (or more generally, to serotonergic effects), which may require identification. However, Walsh and Dinan (2001) recently reviewed all published papers linking SSRIs to violence and concluded that evidence does not support a general causal link between SSRI use and increased aggression. They do acknowledge that a small proportion of patients can develop akathisia and/or exhibit increased anxiety under SSRI treatment. Clinical experience with SSRIs in children, particularly those with developmental disabilities, increasingly recognizes a subset of individuals with exquisite sensitivity to this class of drugs.

ANTIPSYCHOTICS

Haloperidol

The success of traditional antipsychotics in the control of aggressive behaviors is well documented (Miczuk, 1987), and their use has been a mainstay of psychiatric clinical practice for many years, particularly in hospital or emergency settings (Buckley, 1999). However, also documented is a growing concern that their success could derive from a far too general sedative effect, as opposed to exerting a moderating influence on physiological mechanisms specific for aggression (Delito et al., 1998). This concern, along with recognition of the potential for worrisome side effects associated with this class of drugs, has led clinicians to search for alternative medications. There is now an increased tendency for clinicians to turn to alternatives in the control of chronic aggression, while acute presentations of aggressive behavior continue to be considered an appropriate setting for the administration of traditional antipsychotics, such as haloperidol and others. Speed of onset, choices for administrative routes (oral, intramuscular, or intravenous) and anticipated likelihood of control continue to support their use in dangerous situations. Acutely, side effect concerns are primarily limited to dystonia and oversedation.

As dystonic reactions are responsive to pharmacological intervention (typically benztropine or diphenhydramine) and the sedative effects may be deep and/or prolonged, but are not commonly life-threatening (in routine emergency setting doses), the risk–benefit ratio regarding the use of these medications for dangerous patients is tipped in its favor.

Despite growing concerns, the use of traditional antipsychotics as prophylactic treatment or for chronic control of aggressive behaviors continues in many settings, especially in nursing homes and other longer-term care facilities. One common application is for the control of aggression associated with dementia. Lonergan and colleagues (2001) recently performed a meta-analysis of available double-blind randomized controlled trials, in which the efficacy of haloperidol was evaluated relative to placebo. Impact on agitation and aggression were independently assessed. Evidence of benefit associated with haloperidol's control of aggression was identifiable, but limited to doses greater than 2 mg/d, and significant side effects (of somnolence and extra-pyramidal symptoms) were common. Additionally, the authors draw the firm conclusion that haloperidol has no effect in successfully controlling agitation for dementia of any severity. This experience raises an interesting practical concern. To the extent that their caretakers may lack skill in distinguishing agitation from aggression, these patients may be subjected to overly zealous and yet ineffective treatment.

Clozapine

Since the introduction of clozapine as the first atypical antipsychotic in 1972, investigators have been interested in the potential for this class of medications to assist in the control of aggressive behaviors. A substantial body of scientific evidence now exists that clozapine has a significant anti-aggressive benefit, especially when administered to aggressive individuals diagnosed with a primary psychotic disorder, such as schizophrenia or schizoaffective disorder (Chengappa et al., 2002; Glazer and Dickson, 1998; Hector, 1998; Lonergan et al., 2001; Rabinowitz et al., 1996; Spivak et al., 1997). However, ample experience has also been reported for aggression in a variety of other conditions including borderline personality disorder (Chengappa et al., 1999), mental retardation (Cohen and Underwood, 1994; Hammock et al., 2001), TBI (Michals et al., 1993), and severe OCD (Steinert et al., 1996), especially when these present with psychotic symptoms. Clozapine has been found to be useful with children and adolescents when they share these diagnoses (Chalasani et al., 2001); and there is limited evidence of its utility for aggressive behavior in autism (Chen et al., 2001; Gobbi and Pulvirenti, 2001).

The anti-aggressive effect of clozapine has been demonstrated to be an independent feature of the drug and not a component of general sedation (Buckley, 1999; Spivak et al., 1997; Volavka et al., 1993). One potential mechanism for this may be its potent antagonism of 5-HT-2 receptors, enhancing serotonergic availability (Kavoussi and Coccaro, 1998b). Combined with its well-deserved reputation for success with the most treatment-resistant of aggressive psychotic patients, this has led some clinical researchers to tout it as a first-line agent in control of these difficult patients (Hector, 1998). However, while it is generally acknowledged to be an effective anti-aggressive agent, its use tends to continue to be limited to aggressive psychotic patients who have failed trials of other antipsychotics, possibly due to its side effect profile.

Risperidone

Risperidone is a second novel antipsychotic whose anti-aggressive benefits have been well investigated. A recent meta-analysis of seven controlled trials document its efficacy in treating aggression associated with primary psychotic illness (Aleman and Kahn, 2001). Several double-blind placebo-controlled studies support its ability to control aggression within various populations, including autism and PDD (McDougle et al., 1998), developmental disabilities (Zarcone et al., 2001), dementia (DeDeyn et al., 1999; Katz et al., 1999), conduct disorder (Findling et al., 2000), and hospitalized adolescents with disruptive behavior disorders and subaverage cognitive abilities (Buitelaar et al., 2001). Other investigators report successful use with hospitalized children with a variety of psychiatric diagnoses, who share aggression as part of their presentation (Buitelaar, 2000; Simeon et al., 2002). A limited number of case reports suggest its effective reduction of aggressive behaviors in patients with borderline personality disorder (Rocca et al., 2002), antisocial personality disorder (Hirose, 2001), and Prader-Willi syndrome (Durst et al., 2000).

Direct comparisons between risperidone and other agents are rare, but one meta-analysis concludes it to be superior to classic antipsychotics (Aleman and Kahn, 2001) in the treatment of aggression in psychotic illness. Sharif and colleagues (2000) reviewed the charts of 24 hospitalized patients with treatment refractory psychotic illness; this group had undergone documented previous trials of both risperidone and clozapine. Aggressive behaviors were responsive to risperidone in 41% of these patients and to clozapine in 71% of them. The authors reaffirm general consensus that clozapine remains the gold standard for treatment-resistant psychotic illness; however, they further assert that risperidone is worthy of a trial with these patients, given its rate of efficacy in combination with its relatively favorable risk–benefit profile.

Olanzapine

Research interest in novel antipsychotics extends to olanzapine, a third alternative from this class. Relative to clozapine and risperidone, there are currently fewer reports regarding its efficacy in ameliorating aggression, and thus, the data are less compelling.

An international collaborative trial has documented olanzapine's superiority to haloperidol in treatment of the positive, negative, and depressive symptoms of schizophrenia (Tollefson et al., 1997). A recent multicenter randomized controlled trial further suggests its tolerability and utility for nursing home residents diagnosed with Alzheimer's dementia (Street et al., 2000). Lower, but not higher doses of olanzapine were associated with improvement in psychotic and behavioral symptoms, although side effects of somnolence and gait disturbance were common.

In their pilot study of olanzapine's use with children diagnosed with PDD, Potenza and colleagues (1999) reported success on many behavioral measures of seven of their eight patients, including measures of aggression. Barnard and colleagues recently reviewed the available research investigations of atypical antipsychotic use in autistic disorders; they reported greater support for risperidone relative to olanzapine within this literature (2002). Decidedly unsupportive of the use of olanzapine for preadolescent children, Krishnamoorthy and King (1998) report that all five children in their open trial discontinued use of the drug because of adverse effects (sedation, weight gain, and/or akathisia) or lack of clinical response. Although it is possible that lower doses would have been

BENZODIAZAPINES

Miczuk (1987) summarizes evidence demonstrating the effective use of benzodiazepines in the short-term control of aggression, which is substantial. Also substantial is the literature documenting paradoxical increases in aggressive behaviors (Miczuk, 1987). Although the incidence of paradoxical reactions is assumed to be low (Gardos, 1980), it can be extreme (Dietch and Jennings, 1988), making reliance on benzodiazepines as a treatment for primary aggression nonoptimal. Chronic administration of benzodiazepines is complicated by tolerance and possible dependence, which is more likely in individuals with a substance history (Pabis and Stanislav, 1996), risks likely not worth taking, given that benzodiazepines have not been demonstrated to be effective in the long-term control of aggression (Salzman, 1988).

Additional side effects to benzodiazepines can include drowsiness, incoordination, diplopia, and decreased mental acuity (Dupont and Saylor, 1992). Fava (1997) cautions against the use of benzodiazepines to treat aggression associated with dementia, noting that the elderly are also vulnerable to disinhibition. In addition, there is some evidence linking benzodiazepine use with hip fractures (Ray, 1989; Wang et al., 2001).

Chronic administration of benzodiazepines to aggressive children and adolescents is generally discouraged. Based on their review of research evidence, Lynn and King (2002) recommend that this class of drug be avoided for both children and individuals with developmental disorders, as relevant case studies have portrayed significant aggressive responses to these medications (Barron and Sandman, 1985; Sheth et al., 1994). Commander and colleagues (1991) suggest that underlying brain insult or damage may predispose children to behavioral disinhibition, as evidenced by their description of four cases. Additionally, there is evidence that adolescents may use the drug for recreational abuse (Pedersen and Lavik, 1991).

Use of short-acting drugs like lorazepam to treat aggression is generally recommended to be limited to acute administration under emergency conditions. In this circumstance, lorazepam continues to be used routinely and current research supports its effectiveness, either given alone or in combination with haloperidol (Salzman et al., 1991). Pabis and Stanislav (1996) extend this recommendation to include any short-term (<2 wk) or occasional as needed use, to control nonpsychotic aggressive episodes.

STEROIDS AND DERIVATIVES

Elevated androgen levels have been associated with male violence, both in (Dabbs et al., 1995) and out (Virkkunen et al., 1994) of prison settings, and normal males experience an increase in aggressive behaviors when administered high doses of testosterone (Kouri et al., 1998). Animal models have reported the successful use of hormonal therapies in the control of aggression (Slavikova et al., 2001). Subsequently, investigators have tried to use a variety of anti-androgens as aggression suppressors. Some success with this approach has been achieved when aggression that is sexual in nature is the focus of control. Medroxyprogesterone acetate and cyproterone acetate (both anti-androgens) have been shown to improve behavioral control in incarcerated men with paraphilias (Kravitz et al., 1995). Ott (1995) reported success treating a male with Kluver-Bucy syndrome with luprolide (an estrogen analog); here control was limited to sexual aggression.

More recently, elderly patients suffering from various dementias have been the focus of study. Orengo and colleagues (2002) have determined that free testosterone levels in elderly men with dementia are positively correlated with nonsexual aggressive behaviors, and several studies have now demonstrated success in reducing nonsexual aggression in this population with anti-androgen treatment. Amadeo (1996) reported the successful control of both verbal and physical aggression in three elderly men with dementia within 4 wk of initiating therapy with medroxyprogesterone acetate and luprolide acetate. Shelton and Brooks (1999) reported two cases of elderly aggressive men with dementia, who improved with administration of conjugated estrogens. Kyomen and colleagues (1991) reported the successful reduction of aggression in two elderly males with dementia with the use of conjugated estrogen and diethylstilbestrol, respectively. Later, these investigators (Kyomen et al., 1999) undertook a double-blind placebo-controlled trial of estrogen therapy with 15 elderly persons with dementia, 13 of whom were female. Within 4 wk of treatment, significant reductions occurred in physical aggression, but there were no changes in verbal or self-directed aggression. In all of these investigations, hormonal therapies were well-tolerated.

SUMMARY AND DISCUSSION

Taken together, a wide variety of drug classes and specific medications appear to be helpful in modifying aggressive behavior. Medications that appear to have the broadest scope of effectiveness across several different entities include, lithium and valproate, SSRIs, antipsychotics, and β-blockers. Other medicines have mixed evidence or have been studied less.

Given the varied choices and the significance of aggression as a clinical problem, it would be tempting to delineate a "best practice" or algorithm of drug treatment for aggression as has been done in the psychopharmacologic management of affective disorders (Ereshevsky, 2001; Hughes et al., 1999; Suppes et al., 2001; Trivedi and Kleiber, 2001). Proposing such an algorithm at this time is probably premature. Many of these studies reviewed used different definitions or ratings of aggression. Thus, as a group, the failure to use a gold standard measure of aggression leads to the inability to assess criterion validity. Furthermore, depending on the study population, aggression might be considered either a trait or a state. Barratt and colleagues' experience (1997), wherein impulsive but not premeditated aggression responded to treatment, is particularly salient in this regard. Currently, an adequate nosology of aggression in humans does not exist because of the lack of consensus, for example, on an ethological or other approach (Barratt and Slaughter, 1998; Eichelman, 1992).

The complexity of studies of the psychopharmacology of aggression is staggering. Even if an adequate conceptualization or organization of aggression in humans existed, and one chose to study a single homogeneous diagnostic entity within this group, the rate or intensity of aggression would likely vary as a function of a wide variety of individual factors. Among such modifying factors would be the individual's sex, neurochemical and hormonal biology, temperament, history of rearing, exposure to abuse or violence, degree of stress, integrity of the central nervous system, living conditions, use of other drugs, and other medical conditions (Borum, 2000; Otto, 2000).

The complicated nature and interaction of risk factors for violence, independent of mental illness, requires a very careful evaluation of any aggressive individual. Often by the time an aggressive patient presents for assessment and treatment, they have accumulated numerous risk factors for violence. Psychopharmacology occurs in the context of minimizing the operation of already accumulated risk factors while preventing the acquisition of others. In some cases, the medication treatment directly targets a risk factor, such as impulsivity or command auditory hallucinations to kill. In other cases, the medication might only serve to make the individual more available to other interventions that then reduce the potential for violence, as in the example of reducing mood instability with lithium to facilitate participation in substance abuse treatment.

The choice of a specific agent to reduce aggression in any individual occurs with a comprehensive understanding of the phenomenology of aggression in that individual. This understanding would include the precursors, illness state, behavioral output, targets of aggression, and sequelae of any aggressive behavior. Thus, for any group of individuals with the same psychopathological condition, the psychopharmacologic target might vary from one individual to another. In a group of individuals with borderline personality disorder, for example, an antidepressant might reduce irritability and hostility associated with subclinical depression, those that exhibit extreme cognitive distortions might respond to an antipsychotic or mood stabilizer, and those with prominent aversive or anxious responses leading to agitation and aggression might respond to an anxiolytic.

Alternatively, a single agent might have utility for the reduction of aggression across a wide variety of conditions in which the different mental illnesses do not share the same fundamental pathophysiology. For example, antipsychotic drugs might reduce aggression in a person with schizophrenia by reducing paranoia, in a person with bipolar illness by reducing mood instability, and in an individual with autism by reducing agitation in response to environmental change leading to aggressive lashing out. These examples illustrate the complexity of the clinical choice of pharmacotherapy given the variability associated with the genesis of aggressive behavior in any individual. A single neurochemical theory of aggression does not account for this clinical complexity.

With respect to the understanding of the usefulness of specific medicines or classes of medicines, such as antipsychotics, investigators need to be able to distinguish between the reduction of aggression due to illness recovery from that exhibited by the unique drug effect on the neurophysiology of aggression. Is there a relationship between different receptor affinities and the success or failure on specific types of aggressive treatment targets? Polypharmacy is increasingly in common use in community settings (Frye et al., 2000; Williams et al., 1999). Advances in our understanding of aggression derived from preclinical studies, particularly insights from animal models, may suggest rational combination drug therapies for aggression in specific contexts. As yet, there are essentially no data with which to determine the differential effectiveness of polypharmacy specifically to treat aggression.

The above review has highlighted the multiplicity of potential pharmacological approaches that may be considered for the treatment of aggression in clinical populations.

We have stressed the complexity of aggressive phenomenology and the complexity of co-occurring variables that need to be considered in the choice of a specific agent. The need and importance for additional research in the pharmacotherapy of aggression cannot be overstated.

REFERENCES

Afaq, I., Riaz, J., Sedky, K., et al. (2002) Divalproex as a calmative adjunct for aggressive schizophrenic patients. *J. Ky. Med. Assoc.* **100**, 17–22.

Aleman, A. and Kahn, R. S. (2001) Effects of the atypical antipsychotic risperidone on hostility and aggression in schizophrenia: a meta-analysis of controlled trials. *Eur. Neuropscyhopharmacol.* **11**, 289–293.

Alessi, N., Naylor, M. W., Ghaziuddin, M., and Zubieta, J. K. (1994) Update on lithium carbonate therapy in children and adolescents. *J. Am. Acad. Child Adolesc. Psychiatry* **33**, 1348–1349.

Amadeo, M. (1996) Antiandrogen treatment of aggressivity in men suffering from dementia. *J. Geriatr. Psychiatry Neurol.* **9**, 142–145.

Appleton, R., Gichtner, K., LaMoreaux, L., et al. (2001) Gapapentin as add-on therapy in children with refractory partial seiures: a 24-week, multicenter, open-label study. *Dev. Med. Child Neurol.* **43**, 269–273.

Armenteros, J. L. and Lewis, J. E. (2002) Citalopram for treatment of impulsive aggression in children and adolescents: an open pilot study. *J. Am. Acad. Child Adolesc. Psychiatry* **41**, 522–529.

Azouvi, P., Jokie, C., Attal, N., et al. (1999) Carbamazapine in agitation and aggressive behavior following severe closed-head injury: results of an open trial. *Brain Inj.* **10**, 797–804.

Barnard, L., Young, A. H., Pearson, J., et al. (2002) A systematic review of the use of atypical antipsychotics in autism. *J. Psychopharmacol.* **16**, 93–101.

Barratt, E. S. and Slaughter, L. (1998) Defining, measuring, and predicting impulsive aggression: a heuristic model. *Behav. Sci. Law* **16**, 285–302.

Barratt, E. S., Stanford, M. S., Felthous, A. R., and Kent, T. A. (1997) The effects of phenytoin on impulsive and premeditated aggression: a controlled study. *J. Clin. Psychopharmacol.* **17**, 341–349.

Barron, J. and Sandman, C. A. (1985) Paradoxical excitement to sedative-hypnotics in mentally retarded clients. *Am. J. Ment. Defic.* **90**, 124–129.

Beran, R. G. and Gibson, R. J. (1998) Aggressive behavior in intellectually challenged patients with epilepsy treated with lamotrigine. *Epilepsia* **39**, 280–282.

Besag, F. M. (2001) Behavioural effects of the new anticonvulsants. *Drug Saf.* **24**, 513–536.

Borum, R. (2000) Assessing violence risk among youth. *J. Clin. Psychol.* **50**, 1263–1288.

Brodkin, E. S., McDougle, C. J., Naylor, S. T., et al. (1997) Clomipramine in adults with pervasive developmental disorders: a prospective open-label investigation. *J. Child Adolesc. Psychopharmacol.* **7**, 109–121.

Buchalter, E. N. and Lantz, M. S. (2001) Treatment of impulsivity and aggression in a patient with vascular dementia. *Geriatrics* **56**, 53–54.

Buckley, P. F. (1999) The role of typical and atypical antipsychotic medications in the management of agitation and aggression. *J. Clin. Psychiatry* **60**, 52–60.

Buitelaar, J. K. (2000) Open-label treatment of risperidone of 26 psychiatrically-hospitalized children and adolescents with mixed diagnoses and aggressive behavior. *J. Child Adolesc. Psychopharmacol.* **10**, 19–26.

Buitelaar, J. K., van der Gaag, R. J., Cohen-Kettenis, P., and Melman, C. T. (2001) A randomized controlled trial of risperidone in the treatment of aggression in hospitalized adolescents with subaverage cognitive abilities. *J. Clin. Psychiatry* **62**, 239–248.

Byrn, A., Martin, W., and Hnatko, G. (1994) Beneficial effects of buspirone therapy in Huntington's Disease. *Am. J. Psychiatry* **151**, 1097.

Cade, J. (1949) Lithium salts in the treatment of psychotic excitement. *Med. J. Australia* **2**, 349–352.

Campbell, A., Baldessarini, R. J.. and Yeghiayan, S. (1992) Antagonism of limbic and extrapyramidal actions of intracerebrally injected dopamine by ergolines with partial D2 agonist activity in the rat. *Brain Res.* **592**, 348–352.

Campbell, M., Kafantaris, V., and Cueva, J. E. (1995) An update on the use of lithium carbonate in aggressive children and adolescents with conduct disorder. *Psychopharmacol. Bull.* **31**, 93–102.

Campbell, M., Small, A. M., Green, W. H., et al. (1984) Behavioral efficacy of haloperidol and lithium carbonate. A comparison in hospitalized aggressive children with conduct disorder. *Arch. Gen. Psychiatry* **41**, 650–656.

Caspi, N., Modai, I., Barak, P., et al. (2001) Pindolol augmentation in aggressive schizophrenic patients: a double blind crossover randomized study. *Int. Clin. Psychopharmacol.* **16**, 111–115.

Chalasani, L., Kant, R., and Chengappa, K. N. (2001) Clozapine impact on clinical outcomes and aggression in severely ill adolescents with childhood-onset schizophrenica. *Can. J. Psychiatry* **46**, 965–968.

Chen, N. C., Bedair, H. X., McKay, B., et al. (2001) Clozapine in the treatment of aggression in an adolescent with autistic disorder. *J. Clin. Psychiatry* **62**, 479–480.

Chengappa, K. N., Ebeling, T., Kang, J. S., Levine, J., and Parepally, H. (1999) Clozapine reduces severe self-mutilation and aggression in psychotic patients with borderline personality disorder. *J. Clin. Psychiatry* **60**, 477–484.

Chengappa, K. N., Vasile, J., Levine, J., et al. (2002) Clozapine: its impact on aggressive behavior among patients in a state psychiatric hospital. *Schizophr. Res.* **53**, 1–6.

Cherek, D. R. and Lane, S. D. (2001) Acute effects of D-fenfluramine on simultaneous measures of aggressive escape and impulsive responses of adult males with and without a history of conduct disorder. *Psychopharmacology* **157**, 221–227.

Cherek, D. R., Lane, S. D., Pietras, C. J., and Steinberg, J. L. (2002) Effects of chronic paroxetine administration on measures of aggressive and impulsive responses of adult males with a history of conduct disorder. *Psychopharmacology* **159**, 266–274.

Chugani, D. C., Niimura, K., Chaturvedi, S., et al. (1999) Increased brain serotonin synthesis in migraine. *Neurology* **53**, 1473–1479.

Citrome, L., Levine, J., and Allingham, B. (1998) Utilizatin of valproate: extent of inpatient use in the New York State Office of Mental Health. *Psychiatr. Q.* **69**, 283–300.

Clarke, J., Stein, M. D., Sobota, M., et al. (1999) Victims as victimizers: physical aggression by persons with a history of childhood abuse. *Arch. Int. Med.* **159**, 1920–1924.

Coccaro, E. F. and Kavoussi, R. J. (1997) Fluoxetine and impulsive aggressive behavior in personality-disordered subjects. *Arch. Gen. Psychiatry* **54**, 1081–1088.

Cohen, S. A. and Underwood, M. T. (1994) The use of clozapine in a mentally retarded and aggressive population. *J. Clin. Psychiatry* **55**, 440–444.

Colella, R. F., Ratey, J. J., and Glaser, A. L. (1992) Paramenstrual aggression in mentally retarded adult ameliorated by buspirone. *Int. J. Psychiatry Med.* **22**, 351–356.

Commander, M., Green, S. H., and Prendergast, M. (1991) Behavioural disturbances in children treated with clonazepam. *Dev. Med. Child Neurol.* **33**, 362–363.

Connor, D. F., Glatt, S. J., Lopez, I. D., et al. (2002) Psychopharmacology and aggression. I: a meta-analysis of stimulant effects on overt/covert aggression-related behaviors in ADHD. *J. Am. Acad. Child Adolesc. Psychiatry* **41**, 253–261.

Connor, D. F., Ozbayrak, K. R., Benjamin, S., et al. (1997) A pilot study of nadolol for overt aggression in developmentally delayed individuals. *J. Am. Acad. Child Adolesc. Psychiatry* **36**, 826–834.

Constantino, J. N., Liberman, M., and Kincaid, M. (1997) Effects of serotonin reuptake inhibitors on aggressive behavior in psychiatrically hospitalized adolescents: results of an open trial. *J. Child Adolesc. Psychopharmacol.* **7**, 31–44.

Cueva, J. E., Overall, J. E., Small, A. M., et al. (1996) Carbamazepine in aggressive dchildren with conduct disorder: a double-blind and placebo-controlled study. *J. Am. Acad. Child Adolesc. Psychiatry* **35**, 480–490.

Czobor, P., Volavka, J., Sheitman, B., et al. (2002) Antipsychotic-induced weight gain and therapeutic response: a differential association. *J. Clin. Psychopharmacol.* **22**, 244–251.

Dabbs, J. M., Carr, T. S., Frady, R. L., and Riad, J. K. (1995) Testosterone, crime and misbehavior among 692 male prison inmates. *Pers. Individual Differences* **18**, 627–633.

Davanzo, P. A., Belin, T. R., Widawski, M. H., and King, B. H. (1998) Paroxetine treatment of aggression and self-injury in persons with mental retardation. *Am. J. Ment. Retard.* **102**, 427–437.

Davanzo, P. A. and King, B. H. (1996) Open trial lamotrigine in the treatment of self-injurious behavior in an adolescent with profound mental retardation. *J. Child Adolesc. Psychopharmacol.* **6**, 273–279.

DeDeyn, P. P., Rabheru, K., Rasmussen, A., et al. (1999) A randomized trail of risperidone, placebo and haloperidol for behavioral symptoms of dementia. *Neurology* **53**, 946–955.

Delito, J. A., Levitan, J., Damore, J., et al. (1998) Naturalistic experience with the use of divalproex sodium on an in-patient unit for adolescent psychiatric patients. *Acta Psychiatr. Scand.* **97**, 236–240.

De March, N. and Ragone, M. A. (2001) Fluoxetine in the treatment of Huntington's disease. *Psychopharmacology* **153**, 264–266.

Devarajan, S. and Dursun, S. M. (2000) Aggression in dementia with lamotrigine treatment. *Am. J. Psychiatry* **157**, 1178.

Dietch, J. T. and Jennings, R. K. (1988) Aggressive dyscontrol in patients treated with benzodiazepines. *J. Clin. Psychiatry* **49**, 184–188.

Donovan, S. J., Stewart, J. W., Nunes, E. V., et al. (2000) Divalproex treatment for youth with explosive temper and mood lability: a double-blind, placebo-controlled crossover design. *Am. J. Psychiatry* **157**, 818–820.

Donovan, S. J., Susser, E. S., Nunes, E. V., et al. (1997) Divalproex treatment of disruptive adolescents: a report of 10 cases. *J. Clin. Psychiatry* **58**, 12–15.

Dupont, R. L. and Saylor, K. E. (1992) Depressant substances in adolescent medicine. *Pediatr. Rev.* **13**, 381–386.

Durst, R., Rubin-Jabotinsky, K., Raskin, S., et al. (2000) Risperidone in treating behavioural disturbances of Prader-Willi syndrome. *Acta Psychiatr. Scand.* **102**, 461–465.

Eichelman, B. (1992) Aggressive behavior: from laboratory to clinic. *Arch. Gen. Psychiatry* **49**, 488–492.

Ereshevsky, L. (2001) The Texas medication algorithm project for major depression. *Managed Care* **10**, 16–17.

Fava, M. (1997) Psychopharmacologic treatment of pathologic aggression. *Psychiatr. Clin. N. Am.* **20**, 427–451.

Fava, M. and Rosenbaum, J. F. (1993) Psychopharmacology of pathologic aggression. *Harv. Rev. Psychiatry* **1**, 244–246.

Findling, R. L. (1993) Treatment of aggression in juvenile-onset Huntington's Disease with buspirone. *Psychosomatics* **34**, 460–461.

Findling, R. L., McNamara, N. K., Branicky, L. A., et al. (2000) A double-blind pilot study of risperidon in the treatment of conduct disorder. *J. Am. Acad. Child Adolesc. Psychiatry* **39**, 509–516.

Fisher, A. A., Davis, M., and Jeffery, I. (2002) Acute delirium induced by metoprolol. *Cardiovascular Drugs Ther.* **16**, 161–165.

Fisher, R. S., Sachdeo, R. C., Pellock, J., et al. (2001) Rapid initiation of gabapentin: a randomized controlled trial. *Neurology* **56**, 743–748.

Frye, M. A., Ketter, T. A., Leverich, G. S., et al. (2000) The increasing use of polypharmacotherapy for refractive mood disorders: 22 years of study. *J. Clin. Psychiatry* **61**, 9–15.

Fugate, L. P., Spacek, L. A., Kresty, L. A., et al. (1997) Measurement and treatment of agitation following traumatic brain injury: II. A survey of the Brain Injury Special Interest Group of the American Academy of Physical Medicine and Rehabilitation. *Arch. Phys. Med. Rehabil.* **78,** 924–928.

Gardos, G. (1980) Disinhibition of behavior by antianxiety drugs. *Psychosomatics* **21,** 1025–1026.

Gedye, A. (1991) Buspirone alone or with serotonergic diet reduced aggression in developmentally disabled adult. *Biol. Psychiatry* **30,** 88–91.

Glazer, W. M. and Dickson, R. A. (1998) Clozapine reduces violence and persistant aggression in schizophrenia. *J. Clin. Psychiatry* **59,** 8–14.

Gobbi, G. and Pulvirenti, L. (2001) Long-term treatment with clozapine in an adult with autistic disorder accompanied by aggressive behavior. *J. Psychiatry Neurosci.* **26,** 340–341.

Greendyke, R. M., Berkner, J. P., Webster, J. C., and Gulya, A. (1989) Treatment of behavioral problems with pindolol. *Psychosomoatics* **30,** 161–165.

Greendyke, R. M. and Kanter, D. R. (1986) Therapeutic effects of pindolol on behavioral disturbances associated with organic brain disease: a double-blind study. *J. Clin. Psychiatry* **47,** 423–426.

Greendyke, R. M., Kanter, D. R., Schuster, D. B., et al. (1986) Propranolol treatment of assaultive patients with organic brain disease. *J. Nerv. Ment. Dis.* **174,** 290–294.

Grossman, F. (1998) A review of anticonvulsants in treating agitated demented elderly patients. *Pharmacotherapy* **3,** 600–606.

Haas, S., Vincent, K., Holt, J., and Lippmann, S. (1997) Divalproex: a possible treatment alternative for demented, elderly aggressive patients. *Ann. Clin. Psychiatry* **3,** 145–147.

Hagino, O. R., Weller, E. B., Weller, R. A., et al. (1995) Untoward effects of lithium treatment in children aged four through six years. *J. Am. Acad. Child Adolesc. Psychiatry* **34,** 1584–1590.

Haig, G. M., Bockbrader, H. N., Wesche, D. L., et al. (2001) Single-dose gabapentin pharmacokinetics and safety in healthy infants and children. *J. Clin. Pharmacol.* **41,** 507–514.

Hammock, R., Levine, W. R., and Schroeder, S. R. (2001) Brief report: effects of clozapine on self-injurious behavior of two risperidone nonresponders with mental retardation. *J. Autism Dev. Disord.* **31,** 109–113.

Haspel, T. (1995) Beta-blockers and the treatment of aggression. *Harvard Rev. Psychiatry* **2,** 274–281.

Hawkins, J. W., Tinklenberg, J. R., Sheikh, J. I., et al. (2000) A retrospective chart review of gapapentin for the treatment of aggressive and agitated behavior in patients with dementias. *Am. J. Geriatr. Psychiatr.* **8,** 221–225.

Hector, R. I. (1998) The use of clozapine in the treatment of aggressive schizophrenia. *Can. J. Psychiatry* **43,** 466–472.

Henggler, S. W. and Bourdin, C. M. (1990) *Family Therapy and Beyond: A Multisystemic Approach to Treating the Behavior Problems of Children and Adolescents.* Brook/Cole Publishers, Pacific Grove, CA.

Herrman, N. (2001) Recommendations for the management of behavioral and psychological symptoms of dementia. *Can. J. Neurol. Sci.* **28,** 96–107.

Hillbrand, M. (1995) The use of buspirone with aggressive behaviors. *J. Autism Dev. Dis.* **25,** 663–664.

Hirose, S. (2001) Effective treatment of aggression and impulsivity in antisocial personality disorder with risperidone. *Psychiatry Clin. Neurosci.* **55,** 161–162.

Hollander, E. (1999) Managing aggressive behavior in patients with obsessive-compulsive disorder and borderline personality disorder. *J. Clin. Psychiatry* **60,** 38–44.

Hollander, E., Dolgoff-Kaspar, R., Cartwright, C., Rawitt, R., and Novotny, S. (2001) An open trial of divalproex sodium in autism spectrum disorders. *J. Clin. Psychiatry* **62,** 530–534.

Holzer, J. C., Gitelman, D. R., and Price, B. H. (1995) Efficacy of buspirone in the treatment of dementia with aggression. *Am. J. Psychiatry* **152,** 812.

Hughes, C. W., Emslie, G. J., Crismon, M. L., et al. (1999) The Texas children's medication algorithm project: report of the Texas consensus conference panel on medication treatment of childhood major depressive disorder. *J. Am. Acad. Child Adolesc. Psychiatry* **38**, 1442–1454.

Hunt, R. D., Minderaa, R. B., and Cohen, D. J. (1986) The therapeutic effect of clonidine in attention deficit disorder with hyperactivity: a comparison with placebo and methylphenidate. *Psychopharmacol. Bull.* **22**, 229–236.

Jaselskis, C. A., Cook, E. H., Jr., Fletcher, K. E., and Leventhal, B. L. (1992) Clonidine treatment of hyperactive and impulsive children with autistic disorder. *J. Clin. Psychopharmacol.* **12**, 322–327.

Kafantaris, V., Cambell, M., Padron-Gayol, M. V., et al. (1992) Carbamazepine in hospitalized aggressive conduct disorder children: an open pilot study. *Psychopharmacol. Bull.* **28**, 193–199.

Kahn, D., Stevenson, E., and Douglas, C. J. (1988) Effect of sodium valproate in three patients with organic brain syndromes. *Am. J. Psychiatry* **145**, 1010–1011.

Kant, R., Smith-Seemiller, L., and Zeiler, D. (1998) Treatment of aggression and irritability after had injury. *Brain Inj.* **12**, 661–666.

Katz, I. R., Jeste, D. V., Mintzer, J. E., et al. (1999) Comparison of risperidone and placebo for psychosis and behavioral disturbances associated with dementia: a randomized double-blind trial—Risperidone Study Group. *J. Clin. Psychiatry* **60**, 107–115.

Kauffmann, C., Vance, H., Pumariega, A. J., and Miller, B. (2001) Fluvoxamine treatment of a child with severe PDD: a single case study. *Psychiatry* **64**, 268–277.

Kavoussi, R. J. and Coccaro, E. F. (1998a) Divalproex sodium for impulsive aggressive behavior in patients with personality disorder. *J. Clin. Psychiatry* **59**, 676–680.

Kavoussi, R. J. and Coccaro, E. F. (1998b) Psychopharmacological treatment of impulsive aggression, in *Neurobiology and Clinical Views on Aggression and Impulsivity*. (Maes, M. and Coccaro, E. F., eds.), John Wiley & Sons, New York, pp. 197–211.

Kavoussi, R. J., Liu, J., and Coccaro, E. F. (1994) An open trial of sertraline in personality disordered patients with impulsive aggression. *J. Clin. Psychiatry* **55**, 137–141.

Keele, N. B. (2001) Phenytoin inhibits isolation-induced aggression specifically in rats with low serotonin. *Neuroreport* **12**, 1107–1112.

Kim, K. Y., Moles, J. K., and Hawley, J. M. (2001) Selective serotonin reuptake inhibitors for aggressive behavior in patients with dementia after head injury. *Pharmacotherapy* **21**, 498–501.

King, B. H. and Davanzo, P. (1996) Buspirone treatment of aggression and self-injury in autistic and nonautistic persons with severe mental retardation. *Dev. Brain Dysfunct.* **9**, 22–31.

Klein, E., Bental, E., Lerer, B., and Belmaker, R. H. (1984) Carbamazepine and haloperidol versus placebo and haloperidol in excited psychosis: a controlled study. *Arch. Gen. Psychiatry* **41**, 165–170.

Klein, R. G. (1991) Preliminary results: lithium effects in conduct disorders, in CME Syllabus and Proceedings Summary, Symposium 2: The 144th Annual Meeting of the American Psychiatric Association, New Orleans LA. American Psychiatric Association, Washington, DC, pp. 119–120.

Klein, R. G., Abikoff, H., Klass, E., et al. (1997) Clinical efficacy of methylphenidate in conduct disorder with and without attention deficit hyperactivity disorder. *Arch. Gen. Psychiatry* **54**, 1073–1080.

Kouri, E. M., Lukas, S. E., Pope, H. G., and Oliva, P. S. (1998) Increased aggressive responding in male volunteers following the administration of gradually increasing doses of testosterone cypionate. *Drug Alcohol Depend.* **40**, 73–79.

Kravitz, H. M., Haywood, T. W. Kelly, J., Wahlstrom, C., et al. (1995) Medroxyprogesterone treatment for paraphilics. *Bull. Am. Acad. Psychiatry Law* **23**, 19–33.

Krishnamoorthy, J. and King, B. H. (1998) Open-label olanzapine treatment in five preadolescent children. *J. Child Adolesc. Psychopharmacol.* **38**, 107–113.

Kunik, M. E., Puryear, L., Orengo, C. A., et al. (1998) The efficacy and tolerability of divalproex sodium in elderly demented patients with behavioral disturbances. *Int. J. Geriatr. Psychiatry* **13**, 29–34.

Kyomen, H. H., Nober, K. W., and Nei, J. W. (1991) The use of estrogen to decrease aggressive physical behavior in elderly men with dementia. *J. Am. Geriatr. Soc.* **39**, 1110–1112.

Kyomen, H. H., Satlin, A., Hennen, J., and Wei, J. Y. (1999) Estrogen therapy and aggressive behavior in elderly patients with moderate-to-severe dementia: results from a short-term, randomized, double-blind trial. *Am. J. Geriatr. Psychiatry* **7**, 339–348.

Lang, C. and Remington, D. (1994) Treatment with propranolol of severe self-injurious behavior in blind, deaf and retarded adolescent. *J. Am. Acad. Child Adolesc. Psychiatry* **33**, 265–269.

Lang, E. J. and Davis, S. M. (2002) Lithium neurotoxicity: the development of irreversible neurological impairment despite standard monitoring of serum lithium levels. *J. Clin. Neurosci.* **9**, 308–309.

Lee, D. O., Steingard, R. J., Cesena, M., et al. (1996) Behavioral side effects of gabapentin in children. *Epilepsia* **37**, 87–90.

Levy, M. A., Burgio, L. D., Sweet, R., et al. (1994) A trial of buspirone for the control of disruptive behaviors in community-dwelling patients with dementia. *Int. J. Geriatr. Psychiatry* **9**, 84–88.

Lindenmayer, J. P. and Kotsaftis, A. (2000) Use of sodium valproate in violent and aggressive behaviors: a critical review. *J. Clin. Psychiatry* **61**, 123–128.

Lonergan, E., Luxenberg, J., and Colford, J. (2001) Haloperidol for agitation in dementia (Cochrane Review), in *The Cochrane Review, Issue 4*. Update Software, Oxford.

Luiselli, J. K., Blew, P., Keane, J., et al. (2000) Pharmacotherapy for severe aggression in a child with autism: "open label" evaluation of multiple medications on response frequency and intensity of behavioral intervention. *J. Behav. Ther. Exp. Psychiatry* **31**, 219–230.

Luiselli, J. K., Blew, P., and Thibadeau, S. (2001) Therapeutic effects and long-term efficacy of antidepressant medication for persons with developmental disabilities. Behavioral assessment in two cases of treatment-resistant aggression and self-injury. *Behav. Modif.* **25**, 62–78.

Lynn, D. and King, B. H. (2002) Pharmacotherapy of aggressive behavior in children and adolescents, in *Practical Child and Adolescent Psychopharmacology*. (Kutcher, S. P., ed.), Cambridge University Press, Cambridge, UK.

Lyons, J. S., Uziel-Miller, N. D., Reyes, F., and Sokol, P. T. (2000) Strengths of children and adolescents in residential settings: prevalence and associations with psychopathology and discharge placement. *J. Am. Acad. Child Adolesc. Psychiatry* **39**, 176–181.

Malone, R. P., Delaney, M. A., Luebbert, J. F., et al. (2000) A double-blind placebo-controlled study of lithium in hospitalized aggression children and adolescents with conduct disorder. *Arch. Gen. Psychiatry* **57**, 649–654.

Mashiko, H., Yokoyama, H., Matsumoto, H., and Niwa, S. (1996) Trazodone for aggression in an adolescent with hydrocephalus. *Psychiatry Clin. Neurosci.* **50**, 133–136.

Matthews-Ferrari, K. and Karroum, N. (1992) Metoprolol for aggression. *J. Am. Acad. Child Adolesc. Psychiatry* **31**, 994.

McCormick, L. H. (1997) Treatment with buspirone in a patient with autism. *Arch. Fam. Med.* **4**, 3688–3670.

McDougle, C. J., Brodkin, E. S., Naylor, S. T., et al. (1998) Sertraline in adults with pervasive developmental disorders: a prospective open-label investigation. *J. Clin. Psychopharmacol.* **18**, 62–66.

McDougle, C. J., Naylor, S. T., Cohen, D. J., et al. (1996) A double-blind placebo-controlled study of fluvoxamine in adults with autistic disorder. *Arch. Gen. Psychiatry* **53**, 1001–1008.

McDougle, C. J., Price, L. H., Volkman, F. R., et al. (1993) Clomipramine in autism: preliminary evidence of efficacy. *J. Am. Acad. Child Adolesc. Psychiatry* **31**, 746–750.

Michals, M. L., Crismon, M. L., Roberts, S., and Childs, A. (1993) Clozapine response and adverse effects in nine brain-injured patients. *J. Clin. Psychopharmacol.* **13**, 198–203.

Miczuk, K. A. (1987) The psychopharmacology of aggression, in *Handbook of Psychopharmacology, Vol. 19.* (Iversen, L. L., Iversen, S.D., and Snyder, S. H., eds.), Plenum Press, New York, pp. 183–328.

Miller, L. J. (2001) Gabapentin for treatment of behavioral and psychological symptoms of dementia. *Ann. Pharmacother.* **35,** 427–431.

Neppe, V. M. (1983) Carbamazepine as adjunctive treatment in nonepileptic chronic inpatients with EEG temporal lobe abnormalities. *J. Clin. Psychiatry* **44,** 326–331.

Neppe, V. M. (1988) Carbamazepine in nonresponsive psychosis. *J. Clin. Psychiatry* **49,** 22–30.

Nguyen, M. and Meyers, W. C. (2000) Trazodone for symptoms of frontal lobe atrophy. *J. Am. Acad. Child Adolesc. Psychiatry* **39,** 1209–1210.

Oakley, P. W., Whyte, I. M., and Carter, G. L. (2001) Lithium toxicity: an iatrogenic problem in susceptible individuals. *Aust. NZ J. Psychiatry* **35,** 833–840.

Orengo, C., Hunik, M. E., Molinari, V., et al. (2002) Do testosterone levels relate to aggression in elderly men with dementia? *J. Neuropsychiatry Clin. Neurosci.* **14,** 161–166.

Ott, B. R. (1995) Leuprolide treatment of sexual aggression in a patient with dementia and the Kluver-Bucy syndrome. *Clin. Neuropharmacol.* **18,** 443–447.

Ottenbacher, K. J. and Cooper, H. M. (1983) Drug treatment of hyperactivity in children. *Dev. Med. Child Neurol.* **25,** 358–366.

Otto, R. K. (2000) Assessing and managing violence risk in outpatient settings. *J. Clin. Psychol.* **50,** 1239–1262.

Pabis, D. J. and Stanislav, S. W. (1996) Pharmacotherapy of aggressive behavior. *Ann. Pharmacother.* **30,** 278–287.

Pedersen, W. and Lavik, N. J. (1991) Adolescents and benzodiazepines: prescribed use, self medication and intoxication. *Acta Psychiatr. Scand.* **84,** 94–98.

Pfeffer, C. R., Jiang, H., and Domeshek, L. J. (1997) Buspirone treatment of psychiatrically hospitalized prepubertal children with symptoms of anxiety and moderately severe aggression. *J. Child Adolesc. Psychopharmacol.* **7,** 145–155.

Pine, D. S. and Cohen, E. (1999) Therapeutics of aggression in children. *Paediatr. Drugs* **1,** 183–196.

Pinner, E. and Rich, C. L. (1988) Effects of trazodone on aggressive behavior in seven patients with organic mental disorders. *Am. J. Psychiatry* **145,** 1295–1296.

Pollack, B. G., Mulsant, B. H., Rosen, J., et al. (2002) Comparison of citalopram, perphenazine, and placebo for the acute treatment of psychosis and behavioral disturbances in hospitalized, demented patients. *Am. J. Psychiatry* **159,** 460–465.

Pollack, B. G., Mulsant, B. H., Sweet, R., et al. (1997) An open pilot study of citalopram for behavioral disturbances of dementia. *Am. J. Geriatr. Psychiatry* **5,** 70–78.

Potenza, M. N., Holmes, J. P., Kanes, S. J., and McDougle, C. J. (1999) Olanzapine treatment of children, adolescents and adults with pervasive developmental disorders: an open-label pilot study. *J. Clin. Psychopharmacol.* **19,** 37–44.

Poyurovsky, M., Halperin, E., Enoch, D., et al. (1995) Fluvoxamine treatment of compulsivity, impulsivity and aggression. *Am. J. Psychiatry* **152,** 1688–1689.

Prado-Lima, P., Knijnik, L., and Padilla, A. (2001) Lithium reduces maternal child abuse behaviour: a preliminary report. *J. Clin. Pharm. Therapeut.* **26,** 279–282.

Quiason, N., Ward, D., and Kitchen, T. (1991) Buspirone for aggression. *J. Am. Acad. Child Adolesc. Psychiatry* **30,** 1026.

Rabiner, E. A., Gunn, R. N., Castro, M. E., et al. (2000) Beta-blocker binding to human 5-HT(1A) receptors in vivo and in vitro: implications for antidepressant therapy. *Neuropsychopharmacology* **23,** 285–293.

Rabinowitz, J., Avnon, M., and Rosenberg, V. (1996) Effect of clozapine on physical and verbal aggression. *Schizophr. Res.* **22,** 249–255.

Ranen, N. G., Lipsey, J. R., Treisman, G., and Ross, C. A. (1996) Sertraline in the treatment of severe aggressiveness in Huntington's disease. *J. Neuropsychiatry Clin. Neurosci.* **8,** 338–340.

Ratey, J. J., Sorgi, P., O'Driscoll, G. A., et al. (1993) Nadolol to treat aggression and psychiatric symptomatology in chronic psychiatric inpatients: a double blind placebo-controlled study. *J. Clin. Psychiatry* **53,** 41–46.

Ratey, J. J., Sovner, R., Mikkelsen, E., and Chmielinski, H. E. (1989) Buspirone therapy for maladaptive behavior and anxiety in developmentally disabled persons. *J. Clin. Psychiatry* **50,** 382–384.

Ratey, J., Sovner, R., Parks, A., and Rogentine, K. (1991) Buspirone treatment of aggression and anxiety in mentally retarded patients: a multiple-baseline, placebo lead-in study. *J. Clin. Psychiatry* **52,** 159–162.

Ratzoni, G., Gothelf, D., Brand-Gothelf, A., et al. (2002) Weight gain associated with olanzapine and risperidone in adolescent patients: a comparative prospective study. *J. Am. Acad. Child Adolesc. Psychiatry* **41,** 337–343.

Ray, W. A. (1989) Benzodiazepines of long and short elimination half life and the risk of hip fracture. *JAMA* **262,** 3303–3307

Realmuto, G. M., August, G. J., and Garfinkel, B. D. (1989) Clinical effect of buspirone in autistic children. *J. Clin. Psychopharmacol.* **9,** 122–125.

Riddle, M. A., Bernstein, G. A., Cook, E. H., et al. (1999) Anxiolytics, adrenergic agents, and naltrexone. *J. Am. Acad. Child Adolesc. Psychiatry* **38,** 546–556.

Rifkin, A., Karajgi, B., Dicker, R., et al. (1997) Lithium treatment of conduct disorders in adolescents. *Am. J. Psychiatry* **154,** 554–555.

Rocca, P., Marchiaro, L., Cocuzza, E., and Bogetto, F. (2002) Treatment of borderline personality disorder with risperidone. *J. Clin. Psychiatry* **63,** 241–244.

Ruedrich, S., Swales, T. P., Fossaceca, C., et al. (1999) Effect of divalproex sodium on aggression and self-injurious behavior in adults with intellectual disability: a retrospective review. *J. Intellect. Disabil. Res.* **43,** 105–111.

Salzman, C. (1988) Use of benzodiazepines to control disruptive behavior in inpatients. *J. Clin. Psychiatry* **49,** 77–87.

Salzman, C., Soloman, D., Miyawaki, E., et al. (1991) Parenteral lorazemapm versus parenteral haloperidol for the control of psychotic disruptive behavior. *J. Clin. Psychiatry* **52,** 177–180.

Schneider, L. S., Gleason, R. P., and Chui, H. C. (1989) Progressive supranuclear palsy with agitation: response to trazodone but not to thiothixine or carbamazepine. *J. Geriatr. Psychiatry Neurol.* **2,** 109–112.

Sharif, Z. A., Raza, A., and Ratakonda, S. S. (2000) Comparative efficacy of risperidone and clozapine in the treatment of patients with refractory schizophrenia or schizoaffective disorder: a retrospective analysis. *J. Clin. Psychiatry* **61,** 498–504.

Sheard, M. H. (1971) The effects of lithium on human aggression. *Nature* **230,** 113–114.

Sheard, M. H. (1975) Lithium in the treatment of aggression. *J. Nerv. Ment. Disord.* **160,** 108–118.

Sheard, M. H. (1978) The effect of lithium and other ions on aggressive behavior, in *Psychopharmacology of Aggression: Modern Problems of Pharmacopsychiatry.* (Valzelli, I., ed.), Karger, Basel, pp. 53–68.

Sheard, M. H. and Marini, J. L. (1978) Treatment of human aggressive behavior: four case studies of the effect of lithium. *Compr. Psychiatry* **19,** 37–45.

Sheard, M. H., Marini, J. L., Bridges, C. I., and Wagner, E. (1976) The effect of lithium on impulsive behavior in man. *Am. J. Psychiatry* **133,** 1409–1413.

Shelton, P. S. and Brooks, V. G. (1999) Estrogen for dementia-related aggression in elderly men. *Ann. Pharmacother.* **33,** 808–812.

Sheth, R. D., Goulden, K. J., and Ronen, G. M. (1994) Aggression in children treated with clobazam for epilepsy. *Clin. Neuropharmacol.* **17,** 332–337.

Silver, J. M., Yudofsky, S. C., Slater, J. A., et al. (1999) Propranolol treatment of chronically hospitalized aggressive patients. *Neuropsychiatry Clin. Neurosci.* **11,** 328–352.

Simeon, J., Milin, R., and Walker, S. (2002) A retrospective chart review of risperidone use in treatment-resistant children and adolescents with psychiatric disorders. *Prog. Neuropsychopharmacol. Biol. Psychiatry* **26**, 267–275.

Simpson, D. M. and Foster, D. (1986) Improvement in organically disturbed behavior with trazodone treatment. *J. Clin. Psychiatry* **47**, 191–193.

Slavikova, B., Kasal, A., Uhlirova, L., et al. (2001) Suppressing aggressive behavior with analogs of allopregnanaolone (epalon). *Steroids* **66**, 99–105.

Spivak, B., Mester, R., Wittenberg, N., et al. (1997) Reduction of aggressiveness and impulsiveness during clozapine treatment in chronic neuroleptic-resistant schizophrenic patients. *Clin. Neuropharmacol.* **20**, 442–446.

Stanford, M. S., Houston, R. J., Mathias, C. W., et al. (2001) A double-blind placebo-controlled crossover study of phenytoin in individuals with impulsive aggression. *Psychiatry Res.* **103**, 193–203.

Stanislav, S. W., Fabre, T., Crismon, M. L., and Childs, A. (1994) Buspirone's efficacy in organic-induced aggression. *J. Clin. Psychopharmacol.* **14**, 126–130.

Steinert, T., Schmidt-Michel, P. O., and Kaschka, W. P. (1996) Considerable improvement in a case of obsessive-compulsive disorder in an emotionally unstable personality disorder, borderline type under treatment with clozapine. *Pharmacopsychiatry* **29**, 111–114.

Stoll, A. L., Banov, M., Kolbrener, M., et al. (1994) Neurologic factors predict a favorable valproate response in bipolar and schizoaffective disorders. *J. Clin. Psychopharmacol.* **14**, 311–313.

Stoll, A. L. and Severus, W. E. (1996) Mood stabilizers: shared mechanisms of action at postsynaptic signal transduction and kindling processes. *Harv. Rev. Psychiatry* **4**, 77–85.

Street, J. S., Clark, W. S., Gannon, K. S., et al. (2000) Olanzapine treatment of psychotic and behavioral symptoms in patients with Alzheimer disease in nursing care facilities: a double-blind randomized placebo-controlled trial-the HGEU Study Group. *Arch. Gen. Psychiatry* **57**, 968–976.

Suppes, T., Swann, A. C., Dennehy, E. B., et al. (2001) Texas medication algorithm project: development and feasibility testing of a treatment algorithm for patients with bipoar disorder. *J. Clin. Psychiatry* **62**, 439–447.

Swann, A. C. (1999) Treatment of aggression in patients with bipolar disorder. *J. Clin. Psychiatry* **60**, 25–28.

Tallina, K. B., Nahuta, M. C., Lo, W., and Tsao, C. Y. (1996) Gabapentin associated with aggressive behavior in pediatric patients with seizures. *Epilepsia* **37**, 501–502.

Tariot, P. N., Erb, R., Podgorski, C. A., et al. (1998) Efficacy and tolerability of carbamazepine for agitation and aggression in dementia. *Am. J. Psychiatry* **155**, 54–61.

Tollefson, G. D., Beasley, C. M., Jr., Tran, P. V., et al. (1997) Olanzapine versus haloperidol in the treatment of schizophrenia and schizoaffective disorders: results of an international collaborative trial. *Am. J. Psychiatry* **154**, 457–465.

Trivedi, M. H. and Kleiber, B. A. (2001) Algorithm for the treatment of chronic depression. *J. Clin. Psychiatry* **62**, 22–29.

Tunks, E. R. and Dermer, S. E. (1977) Carbamazepine in the dyscontrol syndrome associated with limbic system dysfunction. *J. Nerv. Ment. Dis.* **164**, 56–63.

Uc, E. Y., Dienel, G. A., Cruz, N. F., and Harik, S. I. (2002) Beta-adrenergics enhance brain extraction of levodopa. *Mov. Disord.* **17**, 54–59.

U.S. Department of Health and Human Services (2001) *Youth Violence: A Report of the Surgeon General.* U.S. Government Printing Office, Washington, DC.

Uvebrant, P. and Bausiene, R. (1994) Intractable epilepsy in children: the efficacy of lamotrigine treatment, including non-seizure-related benefits. *Neuropediatrics* **25**, 284–289.

Vartiainen, H., Tiihonen, J., Putkonen, A, et al. (1995) Citalopram, a selective serotonin reuptake inhibitor, in the treatment of aggression in schizophrenia. *Acta Psychiatr. Scand.* **91**, 348–351.

Verhoeven, W. M. and Tuinier, S. (1996) The effect of buspirone on challenging behaviour in mentally retarded patients: an open prospective multiple-case study. *J. Intellect. Disabil. Res.* **40,** 502–508.

Virkkunen, M., Kallio, E., Rawlings, R., et al. (1994) Personality profiles and state aggressiveness in Finnish alcoholic violent offenders, fire setters, and healthy volunteers. *Arch. Gen. Psychiatry* **51,** 28–33.

Volovka, J., Zito, J. M., Vitrai, J., and Czobar, D. (1993) Clozapine effects on hostility and aggression in schizophrenia. *J. Clin. Psychopharmacol.* **13,** 287–289.

Walsh, M. T. and Dinan, T. G. (2001) Selective serotonin reuptake inhibitors and violence: a review of the available evidence. *Acta Psychiatr. Scand.* **104,** 84–91.

Wang, P. S., Bohn, R. L., Glynn, R. J., et al. (2001) hazardous benzodiazepine regimens in the elderly; effects of half-life, dosage and duration on risk of hip fracture. *Am. J. Psychiatry* **158,** 892–898.

Weller, E. B., Rowan, A., Elia, J., and Weller, R. A. (1999) Aggressive behavior in patients with attention-deficit/hyperactivity disorder, conduct disorder, and pervasive developmental disorders. *J. Clin. Psychiatry* **60(Suppl. 15),** 5–11.

Wilkinson, D. (1999) Loss of anxiety and increased aggression in a 15-year-old boy taking fluoxetine. *J. Psychopharmacol.* **13,** 420–421.

Williams, C. L., Johnstone, B. M., Kesterson, J. G., et al. (1999) Evaluation of antipsychotic and concomitant medication use patterns in patients with schizophrenia. *Med. Care* **37,** ASS1–ASS6.

Williams, R. S., Cheng, L., Mudge, A. W., and Harwood, A. J. (2002) A common mechanism of action for three mood-stabilizing drugs. *Nature* **417,** 292–295.

Wolf, S. M., Shinnar, S., Kang, H., et al. (1995) Gabapentin toxicity in children manifesting as behavioral changes. *Epilepsia* **36,** 1203–1205.

Yassa, R. and Dupont, D. (1983) Carbamazepine in the treatment of aggressive behavior in schizophrenic patients: a case report. *Can. J. Psychiatry* **28,** 566–568.

Yehuda, R. (1999) Managing anger and aggression in patients with posttraumatic stress disorder. *J. Clin. Psychiatry* **60(Suppl. 15),** 33–37.

Yudofsky, S. C., Silver, J. M., and Hales, R. E. (1990) Pharmacologic management of aggression in the elderly. *J. Clin. Psychiatry* **51,** 29–32.

Zarcone, J. R., Hellings, J. A., Crandal, K., et al. (2001) Effects of risperidone on aberrant behavior of persons with developmental disabilities: I. A double-blind crossover study using multiple measures. *Am. J. Ment. Retard.* **106,** 525–538.

Zito, J. M., Safer, D. J., Reis, S., et al. (2000) Trends in the prescribing of psychotropic medications to preschoolers. *JAMA* **283,** 1025–1030.

Zubieta, J. K. and Alessi, N. E. (1992) Acute and chronic administration of trazodone in the treatment of disruptive behavior disorders in children. *J. Clin. Psychopharmacol.* **12,** 346–351.

Index

A

AD, *see* Alzheimer's disease
ADHD, *see* Attention deficit hyperactivity disorder
Adolescence, *see* Family environment, aggression effects; Media violence; Music videos; Video games
Adrenaline,
    activation and effects,
        attention and vigilance, 97
        circulation, 96, 97
        energy metabolism, 96
        memory, 97
        olfaction, 97
        pain perception, 97
    hypothalamic-pituitary-adrenal axis interactions in aggression, 112, 113
    pathological aggression and alterations in response, 113
    stress response, 93
Adrenoceptors, aggression role,
    α_1-adrenoceptor, 101
    α_2-adrenoceptor, 102, 103, 292
    β-adrenoceptor, 101
    beta-blockers, *see* Beta-blockers
Affective aggression,
    animals, 21
    emotional regulation, 21
    neurobiology,
        amygdala, 22, 23
        frontal cortex, 23, 24
        medial frontal and orbital frontal cortex, 25, 26
        neurotransmitters, 26–28
        periaqueductal gray, 22

Affective defense, animals versus humans, 1
Aggression,
    adolescence, *see* Family environment, aggression effects; Media violence; Music videos; Video games
    alcohol interactions, *see* Alcohol
    behavioral context,
        context and representativeness, 194, 195
        interpretation of contextual cues, 192–194
        problems, 192
    classification, 33, 34
    definitions, 167
    neurobiology, *see specific brain regions and neurotransmitters*
    offensive versus defensive,
        animal studies, *see* Animal aggression
        overview, 73–76
    pharmacotherapy, *see specific drugs*
    psychiatric disorder association, *see also specific disorders*
        epidemiology, 135, 136
        prospects for study, 144
    sex differences, *see* Y chromosome
Aging,
    aggression decline, 154, 155
    neurochemical changes in brain, 154, 155
    stress response, 154
Alcohol,
    aggression risk factors,
        age, 257
        alcoholism, 257, 258

heavy drinking, 258
  males versus females, 124, 257
barroom aggression,
  brawling males, 260, 262
  contributing factors, 261, 262
  sexual overtures, 264
  skid row culture, 263
emotional liability enhancement, 255
environmental modifiers of aggression,
  behavior expectations, 259
  drinking group, 259, 260
  provocation, threat, or triggers, 258, 259
  rewards and punishments, 259
  third parties, 259
$GABA_A$ receptor activation and aggression, 28
group dynamics, 256, 259, 260
increased power concerns, 256
parental alcoholism effects in adolescent aggression, 221, 222
pharmacological effects, 28, 254
public policy and interventions for violence prevention,
  overview, 264, 265
  Safer Bars program,
    risk assessment workbook, 266, 267
    training, 267, 268
relationship between positive and negative effects, 256, 257
risk-taking effects, 255
situation appraisal impairment, 255
violence,
  crime involvement, 253
  escalation risks, 253, 254
Y chromosome and aggression, 124, 257
Alzheimer's disease (AD),
  aggression association,
    management, 155–157
    serotonergic system, 156
    stress role, 156
    transgenic mouse models, 156
  genetics, 155
  pathology, 155
γ-Aminobutyric acid (GABA),
  receptor activation and aggression, 28
  valproic acid effects, 293
Amygdala,
  affective aggression modulation, 22, 23, 34, 35
  anatomy, 8
  central nucleus in aggression inhibition, 9
  functional imaging, 37
  functions, 10
  lesion studies of aggression, 9, 152, 153
  medial nucleus in aggression, 9
  stimulation studies of aggression, 10
Animal aggression,
  affective aggression, 21
  affective defense, 1
  knockout mice, *see specific genes*
  offensive versus defensive aggression differentiation,
    back-attack, 78, 79
    back-defense, 78, 79
    colony model, 80, 81
    competition and frustration tests, 82
    defensive bites, 82
    defensive threat, 83, 84
    neuropharmacological differentiation,
      benzodiazepines, 85, 86
      serotonin, 84, 85
    offensive attack versus defensive behaviors, 79, 80
    overview, 77
    pain-elicited aggression, 83
    resident–intruder model, 81, 82
  primates,
    costs and benefits, 168, 169
    evolutionary considerations, 183
    explanations for aggressive behavior, 168

food and aggression,
  benefits, 173
  coalitionary power, 174
  competition, 174
  costs, 173
  energy for fighting, 174
  function versus dysfunction, 169, 170
killing adults,
  benefits, 177
  coalitions, 178–180
  costs, 178
  dyads, 178
  estrous female numbers as factor, 180
  food availability and distribution, 179, 180
  intraspecific killing, 176, 177
  numerical imbalance assessment, 178, 179
killing infants,
  benefits, 180, 181
  costs, 181
  female dispersal factor effects, 182
  intruders, 182
  male number in group effects, 182
  population density effects, 181
  species distribution and incidence, 180
mating and aggression,
  benefits, 170, 171
  costs, 171
  fertile females and mating opportunities, 171, 172
  male numbers, 173
  operational sex ratio, 172, 173
sex differences, 183
status and aggression,
  benefits, 174, 175
  costs, 175
  kin, 175
  rival number and quality effects, 175, 176
territory and aggression,
  benefits, 176
  costs, 176
  environmental factors, 176
Attachment theory, family environment and aggression effects, 220
Attention deficit hyperactivity disorder (ADHD), aggression and management, 292

B

Bed nucleus of the stria terminalis (BNST), aggression modulation, 10
Benzodiazepines,
  aggression management, 301
  $GABA_A$ receptor activation and aggression, 28
  offensive versus defensive aggression differentiation in animals, 85, 86
  side effects, 301
Beta-blockers,
  aggression management, 291, 292
  dosing, 291, 292
  mechanism of action, 291
  serotonin receptor interactions, 291
Bipolar disorder,
  aggression correlates and risk factors, 139
  prevalence, 139
  types, 139
BNST, see Bed nucleus of the stria terminalis
Borderline personality disorder (BPD),
  aggression correlates and risk factors, 143, 144
  prevalence, 143
BPD, see Borderline personality disorder
BrainPower Program,
  assumptions in curriculum design, 279
  attributions, 278
  characteristics, 279, 280
  effectiveness, 280–283

rationale, 278
Buspirone, aggression management, 296, 297

C

CAH, *see* Congenital adrenal hyperplasia
Calcium-calmodulin kinase II (CamKII), knockout mouse and behavioral effects, 42
CamKII, *see* Calcium-calmodulin kinase II
Carbamazepine,
   aggression management, 293, 294
   mechanism of action, 294
   pediatric use, 294
CBCL, *see* Child Behavior Checklist
CD, *see* Conduct disorder
Cerebrospinal fluid (CSF), 5-hydroxyindoleacetic acid levels, *see* 5-hydroxyindoleacetic acid
Challenge behavior, *see* Animal aggression
Child abuse, effects on adolescent aggression, 218, 219
Child Behavior Checklist (CBCL), 213
Clomipramine, aggression management, 296
Clonidine, aggression management, 292
Clozapine, aggression management, 299
Cognitive-behavioral perspectives, aggression,
   BrainPower Program for childhood intervention,
     assumptions in curriculum design, 279
     attributions, 278
     characteristics, 279, 280
     effectiveness, 280–283
     rationale, 278
   contextual considerations, 284, 285
   culture and prevention, 285, 286
   development and aggression, 283, 284
   overview, 275, 276
   social learning, 277, 278
   social reasoning, 276, 277
Colony model, animal aggression studies, 80, 81
Computer games, *see* Video games
Conduct disorder (CD),
   aggression correlates and risk factors, 141, 142
   prevalence, 141
Congenital adrenal hyperplasia (CAH),
   aggression association and study design, 204, 205
   hormonal dysfunction, 204
CSF, *see* Cerebrospinal fluid
Culture,
   cognitive-behavioral perspectives in aggression prevention, 285, 286
   study design in sex differences of aggression, 201, 202
Cyproterone acetate, anti-androgen therapy, 301, 302

D

Dementia,
   aggression correlates and risk factors, 142, 143
   prevalence, 142
Depression, *see* Bipolar disorder; Major depressive disorder
DES, *see* Diethylstilbesterol
Dextroamphetamine, aggression management in attention deficit hyperactivity disorder, 292
Diethylstilbesterol (DES), aggression studies, 205
Dodge model, aggression formation in childhood, 276, 277

E

Emotion, *see* Affective aggression

F

Family environment, aggression effects,
   attachment theory, 220

Index

biological versus environmental
    effects, 214, 214
biosocial theory, 215–217
broken home, 219
mental disorders of parents, 221, 222
modeling of family–aggression
    interactions, 224, 225
parental violence, 219
perceived parenting styles and
    behavioral problems, 222, 223
physical abuse, 218
protective factors, 224
risk factor studies, overview, 213, 214
sex differences in adolescent
    aggression, 217, 218
sexual abuse, 218, 219
social learning theory, 220
fMRI, *see* Functional magnetic
    resonance imaging
Food, primate aggression,
    benefits, 173
    coalitionary power, 174
    competition, 174
    costs, 173
    energy for fighting, 174
Functional magnetic resonance
    imaging (fMRI),
    orbital frontal cortex role in affective
        aggression, 26
    serotonergic challenge studies, 37

G

GABA, *see* γ-Aminobutyric acid
Gabapentin,
    aggression management, 294, 295
    pediatric use, 295
Glucocorticoids,
    acute effects on aggression, 109–111
    adrenaline-norepinephrine interactions
        in aggression, 112, 113
    aggression management, 153
    brain receptors, 103
    gene regulation, 103
    hyperfunction and inhibited aggres-
        sion, 108
    hypofunction and aggression, 105–107

nongenomic effects, 103, 104
oscillations and rhythmic variations
    in aggression, 108, 109
post-traumatic stress disorder levels,
    107, 111
short- and long-term effects, 104, 105
stress response, 93, 94
Guanfacine, aggression management,
    292

H

Haloperidol,
    aggression management, 298, 299
    side effects, 298, 299
HD, *see* Huntington's disease
5-HT, *see* Serotonin
Huntington's disease (HD),
    behavioral disorders, 159, 160
    neurodegeneration, 158, 159
5-Hydroxyindoleacetic acid,
    cerebrospinal fluid levels,
    aggression studies, 36, 51, 52
    impulsivity studies, 65, 66
    suicide studies, 65, 66
Hypothalamus,
    anterior hypothalamic area and
        aggression, 4
    lateral hypothalamus and aggression
        modulation, 12, 13
    medial hypothalamus and rage
        role, 4, 5
    periaqueductal gray circuit
        neurotransmission, 5
    ventromedial hypothalamus, 5

I

Impulsivity,
    5-hydroxyindoleacetic acid,
        cerebrospinal fluid studies, 65, 66
    suicide and aggression neurochemistry
        similarity, 65–70
Infanticide, primates,
    benefits, 180, 181
    costs, 181
    female dispersal factor effects, 182
    intruders, 182

male number in group effects, 182
population density effects, 181
species distribution and incidence, 180

L

Lamotrigine, aggression management, 295
Lithium,
aggression management, 291, 292
side effects, 291

M

Major depressive disorder (MDD),
aggression correlates and risk factors, 138, 139
Point Subtraction Aggression Paradigm, 137, 138
prevalence, 137
MAOA, *see* Monoamine oxidase A
Mating, primate aggression,
benefits, 170, 171
costs, 171
fertile females and mating opportunities, 171, 172
male numbers, 173
operational sex ratio, 172, 173
MDD, *see* Major depressive disorder
Media violence,
aggression risk factor evidence, 232, 233
erotica effects, 234
exposure extent, 235, 236
games, *see* Video games
history of aggression studies, 231
meta-analysis of aggression effects, 233, 234
music videos, *see* Music videos
prevention of learned violence,
education profession, 243, 244, 248
grass roots efforts, 244, 245
media education groups and Web sites, 245
medical profession, 241–243, 248
public policy and regulations, 245–248
ratings, 246
research prospects, 247, 248
social learning theory, 231, 232
South Africa impact of television, 235
study types and aggression correlation, 236, 237
V-chip, 246
Medial frontal cortex, affective aggression role, 25, 26
Medroxyprogesterone acetate, anti-androgen therapy, 301, 302
Methylphenidate, aggression management in attention deficit hyperactivity disorder, 292
Metrifonate, Alzheimer's disease aggression management, 157
Monoamine oxidase A (MAOA),
gene structure, 44, 45
knockout mouse and behavioral effects, 43, 44, 50, 99
norepinephrine metabolism, 98, 99
polymorphisms, 45, 193
Movies, *see* Media violence
Music videos,
prevention of learned violence,
education profession, 243, 244, 248
grass roots efforts, 244, 245
media education groups and Web sites, 245
medical profession, 241–243, 248
public policy and regulations, 245–248
rap music and sexually-transmitted disease impact, 236–238
violence prevalence and effects, 236

N

NCAM, *see* Neural cell adhesion molecule
NE, *see* Norepinephrine
Neural cell adhesion molecule (NCAM), knockout mouse and behavioral effects, 42
Nitric oxide synthase (NOS), neuronal isoform knockout mouse and behavioral effects, 42

Norepinephrine (NE),
    activation and effects,
        attention and vigilance, 97
        circulation, 96, 97
        energy metabolism, 96
        memory, 97
        olfaction, 97
        pain perception, 97
    acute effects, 99, 100
    aggression control, 98–103
    hypothalamic-pituitary-adrenal
        axis interactions in aggression,
        112, 113
    monoamine oxidase A metabolism,
        98, 99
    oscillations, 99
    receptor subtypes in aggression,
        101–103
    stress response, 93
    trauma and stress reactivity,
        early life, 100
        post-traumatic stress disorder,
        100, 101
NOS, see Nitric oxide synthase
Nucleus accumbens,
    aggression modulation, 11, 12
    limbic-motor integration, 12

O

Obsessive–compulsive disorder (OCD),
    aggression correlates and risk factors,
    141
    prevalence, 141
OCD, see Obsessive–compulsive
    disorder
Olanzapine, aggression management,
    300, 301
Orbital frontal cortex,
    affective aggression role, 25, 26, 35
    functional imaging, 37

P

PAG, see Periaqueductal gray
Parents, see Family environment,
    aggression effects

Parkinson's disease (PD),
    aggression association, 157, 158
    etiology, 157
    neurodegeneration, 157
PD, see Parkinson's disease
Periaqueductal gray (PAG),
    affective aggression role, 22
    anatomy, 1, 2
    defensive rage response role, 1–3
    descending projections, 2, 3
    emotional-expression function
        localization, 4
    hypothalamic circuit
        neurotransmission, 5
    vocalization role, 3
PET, see Positron emission tomography
Phenytoin,
    aggression management, 294
    mechanism of action, 294
Positron emission tomography (PET),
    affective aggression studies
        of frontal cortex, 24
    fenfluramine challenge study
        of suicide victims, 67
    serotonergic challenge studies, 37
Post-traumatic stress disorder (PTSD),
    aggression association,
        correlates and risk factors,
        140, 141
        study design, 140
    aversive stimuli response, 22
    cortisol levels, 107, 111
    norepinephrine stress reactivity,
        100, 101
    prevalence, 139, 140
Prefrontal cortex, aggression modulation,
    11
Primate models, see Animal aggression
PTSD, see Post-traumatic stress disorder

R

Resident–intruder model, animal
    aggression studies, 81, 82
Risperidone, aggression management,
    300

S

Safer Bars program,
  risk assessment workbook, 266, 267
  training, 267, 268
Schizophrenia,
  aggression correlates and risk factors, 136, 137
  prevalence, 136
Selective serotonin reuptake inhibitors (SSRIs),
  aggression management, 297, 298
  offensive versus defensive aggression differentiation in animals, 84, 85
  pediatric use, 297, 298
  types, 297
Septal nuclei,
  anatomy, 6
  fibers of passage, 7
  functions, 6
  lesion studies of aggression, 7, 8
  stimulation studies of aggression, 7
Serotonin (5-HT),
  affective aggression modulation, 26–28, 35, 36
  biosynthesis, 42, 43
  cerebrospinal fluid levels of 5-hydroxyindoleacetic acid, *see* 5-Hydroxyindoleacetic acid
  cortico-cingulo-amygdaloid circuit and emotional regulation, 34–36
  functional brain imaging, 37
  pharmacomodulation of aggression, 37–39
  quantitative trait loci studies of emotion, 39, 40
  receptor subtypes,
    aggression modulation, 37–39
    beta-blocker interactions, 291
    5-HT1A, 41, 42
    5-HT1B, 40–42, 50, 68, 69
    5-HT7, 42
    knockout mouse studies, 38, 40–43, 50
  signal transduction, 42
  suicide, impulsivity, and aggression neurochemistry similarity, 65–70
Serotonin transporter (SERT),
  function, 45, 68
  gene structure and variants, 45, 46, 68
  knockout mouse,
    behavioral effects, 45
    cortical development role, 48–50
  LPR allele and behavioral effects, 46–48, 51–54
SERT, *see* Serotonin transporter
Sex differences, *see* Y chromosome
Single positron emission computed tomography (SPECT), brain activity studies in aggression, 151
Social drinking, *see* Alcohol
Social learning theory,
  aggression formation in childhood, 277, 278
  family environment and aggression effects, 220
  media violence, 231, 232
  neurobiology correlation, 232
Social response reversal (SRR), orbital frontal cortex role, 25, 26
Somatosensory cortex (SSC), serotonin transporter in development, 49, 50
SPECT, *see* Single positron emission computed tomography
SRR, *see* Social response reversal
SSC, *see* Somatosensory cortex
SSRIs, *see* Selective serotonin reuptake inhibitors
Status, primate aggression,
  benefits, 174, 175
  costs, 175
  kin, 175
  rival member and quality effects, 175, 176
Steroid sulfatase (STS), aggression role, 128–130

Stress, *see also* Adrenaline;
    Glucocorticoids; Norepinephrine,
  aggression effects, overview, 95
  aging changes in response, 154
  Alzheimer's disease, 156
  neuroendocrine response,
    93, 94, 111–113
  short- and long-term effects,
    94, 95
Stroke,
  aggression association, 153, 154
  neurodegenerative cascade, 153
STS, *see* Steroid sulfatase
Suicide,
  aggression and impulsivity
    neurochemistry similarity,
    65–70
  5-hydroxyindoleacetic acid,
    cerebrospinal fluid studies,
    65, 66

T

TBI, *see* Traumatic brain injury
Television, *see* Media violence
Territory, primate aggression,
  benefits, 176
  costs, 176
  environmental factors, 176
Testosterone,
  aggression role, 203, 204
  anti-androgen therapy, 301, 302
  embryo effects, 204
TPH, *see* Tryptophan hydroxylase
Traumatic brain injury (TBI),
  aggression changes and management,
    152
  neurodegeneration, 152
Trazodone, aggression management,
  295, 296
Tryptophan, depletion and aggression,
  27, 28
Tryptophan hydroxylase (TPH),
  gene structure, 43
  polymorphisms and aggression,
    43, 67, 68
  serotonin biosynthesis, 42, 43, 67

V

Valproic acid,
  aggression management, 292, 293
  mechanism of action, 293
V-chip, 246
Ventral tegmental area (VTA),
  aggression modulation, 11
Video games,
  aggression effect studies, 239, 240
  content, 238, 239
  historical perspective, 238
  market, 238
  prevention of learned violence,
    education profession,
      243, 244, 248
    grass roots efforts, 244, 245
    media education groups and
      Web sites, 245
    medical profession, 241–243, 248
    public policy and regulations,
      245–248
  rating systems, 239
Violence,
  challenge–aggression situations, 74
  parental violence and adolescent
    aggression, 219
  television and movies, *see* Media
    violence
VTA, *see* Ventral tegmental area

Y

Y chromosome,
  alcohol effects, 257
  contextual problems in aggression
    studies, 195–198, 207
  criminal sex ratio, 119, 183
  family environment and adolescent
    aggression, 217, 218
  history of aggression studies,
    119, 120
  human studies of male phenotypes,
    alcoholism studies, 124
    cultural variation, 201, 202
    evolutionary considerations,
      206, 207

gender bias in child studies, 202, 203
overview, 121
study design issues, 198–200
XYY males, 123, 124
mouse studies of male phenotypes,
   aggression assays, 124, 125
   candidate genes,
      Y^{NPR}, 127, 128
      Y^{PR}, 128–130
   inbred strains, 121, 122
   nonpairing versus pairing region effects, 122, 123, 125–127
   prospects for study, 130, 131
   study design issues, 196–198
   structure, 120, 121
Youth Self Report (YSR), 213
YSR, *see* Youth Self Report